W0107220

Computational Fluid Dynamics

D. Leutloff   R. C. Srivastava (Eds.)

Computational Fluid Dynamics

Selected Topics

With 125 Figures

Springer

Dr. Dieter Leutloff

Technische Hochschule Darmstadt, Fachbereich Mechanik, Hochschulstrasse 1
D-64289 Darmstadt, Germany

Dr. Ramesh C. Srivastava

Gorakhpur University, Department of Mathematics and Statistics
273009 Gorakhpur, India

ISBN-13:978-3-642-79442-1   e-ISBN-13:978-3-642-79440-7
DOI: 10.1007/978-3-642-79440-7

Library of Congress Cataloging-in-Publication Data. Computational fluid dynamics: selected topics / [edited by] D. Leutloff, R. C. Srivastava. p. cm. Papers in honor of Prof. K. G.
Roesner.Includes bibliographical references. ISBN-13:978-3-642-79442-11.Fluid dynamics–Mathematical models. 2. Turbulent boundary layer–Mathematical models. 3. Nonlinear
waves–Mathematical models. 4. Tsunamis–Mathematical models. I. Leutloff, D. (Dieter),
1942- . II. Srivastava, R. C. (Ramesh C.), 1953- . III. Roesner, K. G. TA357.C5878.   1995
532'.05' 015194-dc20   94-48557

This work is subject to copyright. All rights are reserved, whether the whole or part of the
material is concerned, specifically the rights of translation, reprinting, reuse of illustrations,
recitation, broadcasting, reproduction on microfilm or in any other way, and storage in data
banks. Duplication of this publication or parts thereof is permitted only under the provisions
of the German Copyright Law of September 9, 1965, in its current version, and permission for
use must always be obtained from Springer-Verlag. Violations are liable for prosecution
under the German Copyright Law.

© Springer-Verlag Berlin Heidelberg 1995
Softcover reprint of the hardcover 1st edition 1995

The use of general descriptive names, registered names, trademarks, etc. in this publication
does not imply, even in the absence of a specific statement, that such names are exempt from
the relevant protective laws and regulations and therefore free for general use.

Typesetting: Camera ready copy from the author/editor using a Springer TeX macro package
SPIN: 10481402   55/3144 – 5 4 3 2 1 0 – Printed on acid-free paper

Preface

This special volume is dedicated to Professor Karl G. Roesner on the occasion of his 60th birthday on the 16th January 1995. Professor Roesner has a large family of friends all over the world. Some of them have expressed their best wishes and respects which are contained in this volume in the form of scientific contributions. The papers in this volume are mostly in the area of computational fluid dynamics (CFD), which is Professor Roesner's favorite field of research. These papers cover almost all aspects of CFD. They cover diverse topics such as, the tsunami problem, group invariant solution of hydrodynamic equations, non-linear waves, and the modelling of the problem of evaporation-condensation. There is also a survey article on exact solutions for discrete models of the Boltzman equation by one of the foremost experts in this area. Articles are also devoted to turbulent boundary layer problems and quasi-geostrophic drag on a sphere in a rotating fluid. The editors would like to express their gratitude to all the contributors, especially to Soubbaramayer who offered to write the memorial for this volume. We are also thankful to Professor Beiglböck and Mrs. Beiglböck and their colleagues at Springer-Verlag, Heidelberg, for their friendly cooperation in producing this volume.

Darmstadt, July 1994   *R.C. Srivastava*
                                                              D. Leutloff

K.G. Roesner (1994)

Dedication

This volume contains several articles written in honour of Professor Karl G. Roesner, Darmstadt Institute of Technology, on the 60th anniversary of his birthday.

Professor Roesner was born on January 16th, 1935, in Beuthen/Oberschlesien, an industrial town close to the border of Poland, which is now in Poland. There he lived untill January 1945, the end of World War II. One can imagine the situation after the war, where there was no definite tomorrow and no prospective view for the future. This was the time he started his school career. This situation had a great impact on his life.

He started his studies in Physics at the University of Münster, changed to the Georg-August-University of Göttingen and got his diploma in Physics there. Göttingen offered to the students at that time many possibilities to qualify by doing research work. Professor Roesner decided to join the research group of the Max-Planck-Institut für Strömungsforschung, the famous institute where Ludwig Prandtl started to build up the modern Fluid Dynamics. At that time the first digital computers where installed in Germany, and Professor Roesner was one of the first who got used to that type of research tool, especially useful for the numerical analysis of gasdynamics. This early contact with computers had a long-lasting impact on his scientific work. He himself now jokes about that period as the 'stone-age period of computation', where scientists had to perform real physical work to let their programs run on the machine by carrying big packages of punched cards to the card reader.

At the Max-Planck-Institut he finished his dissertation devoted to gasdynamics and obtained his doctorate in Physics in 1967. Some years later he went with his family to Freiburg and became an assistent at the Institute of Applied Mathematics of the Albert-Ludwigs-University, where Henry Görtler encouraged him to qualify and think about an academic career on the basis of a habilitation. In 1976 Professor Roesner finished his habilitation work and moved to Karlsruhe where he was incorporated into the Faculty of Mechanical Engineering as 'Privatdozent'. At the Institute of Fluid Flow and Fluid Machinery he gave lectures in the field of Numerical Fluid Mechanics with the aim to transfer his enthusiasm in numerical simulations of fluid flows to his students and to teach them how to apply this tool carefully and critically. As a physicist he looks at computers as an experimental device with all its preferences and drawbacks which the user has to take into account to get insight into the Physics he wants to investigate.

His early numerical investigation on the method of characteristics was the subject about which he reported on the occasion of the 8th Biennial Symposium on Advanced Problems and Methods in Fluid Dynamics 1967 in Poland. This was at that time the only chance to have contact with colleagues and scientists of the Eastern countries. One of them was the late Professor Yanenko from the Sibirian Branch of the Academy of Sciences of the Soviet Union. This encounter with the Russian mathematicians influenced very much Professor Roesners' further work. In 1970 he joined the Second International Conference on Numerical Methods in Fluid Dynamics at Berkeley, hosted by Maurice Holt. Here he came in contact with Henry Cabannes, Maurice Holt, and many other eminent personalities in Fluid Dynamics. He was included as one of the members of the International Organizing Committee for Computational Fluid Dynamics Conferences. Since that time he has been promoting the idea to make the international family of numerical analysts as large as possible. Professor Roesner is an active member of that committee, representing admirably the European computational fluid dynamicists. He plays an important role in the rigorous selection of papers to be presented at each conference. Having a paper selected by Professor Roesner is a guarantee of scientific high quality.

In 1980 he changed his place of work when he followed an offer to continue teaching and research work at the Darmstadt Institute of Technology in the Department of Mechanics. From this time on he began to design experiments for fluid flows in the laboratory parallel to their numerical simulation. The Workshops on Gases in Strong Rotation, which existed since 1975 and took place every two years, were extended on his instigation to include Seperation Phenomena in Liquid and Gases. Professor Roesner has generously hosted in Darmstadt the first of the new-version workshops. Let me tell you that the 'baby' of Professor Roesner, born in Darmstadt in 1987, has grown and is developing perfectly well. The last of the new series workshops was held in China 1994, and the next one will take place in Brasil in 1996. Hydrodynamic instability phenomena of rotating fluids especially attracted him from the point of view of how to visualize the velocity field without disturbing the flow field. He applied a visual technique on the basis of chemical reaction with photochromic compounds which made it possible to look at the microstructure of the velocity field in totally closed cavities filled with liquid, whilst working on other topics such as boundary layer effects, numerical simulation of shock problems and separation phenomena in two phase flows, which are of great practical interest in the field of environmental problems. He has international scientific collaboration, and scientists from other countries are joining his group from time to time.

He is on the editorial board of the *Computational Fluid Dynamic Journal*, and *Computers and Fluids*. He has translated many important Russian articles into English. He has made available to the West the knowledge of numerics and many branches of fluid dynamics through his translations. An important work by Yu.I. Shokin in Russian, 'The Method of First Differential Approximation' was translated by Professor Roesner and published by Springer-Verlag in 1984.

A few comments about Roesners' human qualities. Very early in his life he was influenced by the destructive effects of war. During his student life in Göttingen he worked as an adviser to the foreign students, helping both those who asked for it and those whom he felt needed it. This idea has developed in him stronger with time. He does not consider lecturing as a mere obligation of an academic teacher but he enjoys teaching, always aiming to transfer some of his own enthusiasm for fluid dynamics to his students. In his every-day relationeships with students and co-workers he shows patience, compassion for personal problems and benevolence for their advancement. He prefers influencing his co-workers by his personal example, rather than by instruction and control. Roesner speaks, in addition to other languages, English, French, and Russian. In this way he has a strong medium of communication and can reach the people of different nations and different cultures.

Finally, I will give a brief glimpse of Professor Roesner's private life. In 1962 Roesner was married to Lotte. A daughter and two sons were born of this very harmonious marriage. Now the children have grown up and are pursuing their professional training. On the many occasions when I was a guest of the Roesners', I admired his great understanding and broad minded view towards his developing children.

During my stay in Darmstadt Mrs. Petra Leutloff invited me one evening to her home together with Professor Roesner and his wife. It was a really nice evening and we freely discussed different problems. Mrs. Petra was serving soft drinks when Mrs. Roesner made a very nice comment about Karl Roesner, my husband is like a 'steam engine'. I am sure this is a most suitable remark about Karl Roesner. Work is steam, work is spirit for him. But Mrs. Roesner is also a strong 'wheel' for this running steam engine.

I express my best wishes to Karl G. Roesner on his 60th birthday, also on behalf of all the authors who contributed to this volume. May he have the same joy in his scientific work in the years ahead. We wish Karl and Lotte many years of health and happiness.

Darmstadt, July 1994   *R.C. Srivastava*

Table of Contents

List of Contributors

N.A. Adams
DLR, Institute for Theoretical
Fluid Mechanics
Bunsenstraße 10
D-37073 Göttingen, Germany

Y. Aihara
Department of Aeronautics and
Astronautics
The University of Tokyo
Tokyo, Japan

A. Averbuch
School of Mathematical Sciences
Tel Aviv University
Tel Aviv 69978, Israel

P. Bar-Yoseph
Computational Mechanics
Laboratory
Faculty of Mechanical
Engineering
Technion, Haifa 32000, Israel

G.K.F. Bärwolff
Technische Universität Berlin
Hermann-Föttinger-Institut
Rudower Chaussee 6
D-12484 Berlin, Germany

M. Ben-Artzi
Institute of Mathematics
Hebrew University
Jerusalem 91904, Israel

A. Birman
Dept. of Physics
Technion-Israel
Haifa 32000, Israel

H. Bischoff
Institut für Wasserbau
Technische Hochschule Darmstadt
D-64287 Darmstadt, Germany

H. Cabannes
Lab. Modélisation en Mécanique,
Associé au CNRS
Université Pierre et Marie Curie
4 Place Jussieu
F-75005 Paris, France

L.B. Chubarov
Institute of Computational
Technologies, Siberian Branch
of the Russian Academy of Science
pr. Lavrentyeva 6
Novosibirsk 630090, Russia

S.V. Coggeshall
Los Alamos National Laboratory,
Los Alamos, NM 87545, USA

A. D'Almeida
Lab. Modélisation en Mécanique,
Associé au CNRS
Université Pierre et Marie Curie
4 Place Jussieu
F-75252 Paris Cédex 05, France

J. Falcovitz
Dept. of Aerospace Engineering
Technion-Israel
Haifa 32000, Israel

C.A.J. Fletcher
CANCES
The University of New South Wales
Sydney 2052, Australia

R. Gatignol
Lab. Modélisation en Mécanique,
Associé au CNRS
Université Pierre et Marie Curie
4 Place Jussieu
F-75252 Paris Cédex 05, France

G. Grötzbach
KfK, Institut für Reaktorsicherheit,
Postfach 3640
D-76021 Karlsruhe, Germany

M. Israeli
Faculty of Computer Science
Technion, Haifa 32000, Israel

E. Kaucher
Institut für Angewandte
Mathematik
Universität Karlsruhe
D-76128 Karlsruhe, Germany

G.S. Khakimsyanov
Institute of Computational
Technologies Siberian Branch
of the Russian Academy of Science
pr. Lavrentyeva 6
Novosibirsk 630090, Russia

L. Kleiser
DLR, Institute for Theoretical
Fluid Mechanics
Bunsenstraße 10
D-37073 Göttingen, Germany

G.A. Maugin
CNRS-Laboratoire de Modélisation
en Mécanique
Université Pierre et Marie Curie
Paris, France

R. Peyret
Laboratoire de Mathématiques,
CNRS URA 168
Université de Nice-Sophia Antipolis
Parc Valrose
F-06108 Nice Cédex 2, France

A. Rizzi
The Royal Institute of Technology
KTH, S-10044 Stockholm, Sweden

Yu.I. Shokin
Institute of Computational
Technologies,Siberian Branch
of the Russian Academy of Science
pr. Lavrentyeva 6
Novosibirsk 630090, Russia

Soubbaramayer
Département des Procédés
d'Enrichissement CEA-Saclay
F-91191 Gif sur Yvette Cédex, France

J.Y. Tu
CANCES
The University of New South Wales
Sydney 2052, Australia

M. Ungarish
Department of Computer Science
Technion, Haifa 32000, Israel

J.M. Vanel
Laboratoire de Mathématiques,
CNRS URA 168
Université de Nice-Sophia Antipolis
Parc Valrose
F-06108 Nice Cédex 2, France

L. Vozovoi
Faculty of Computer Science
Technion, Haifa 32000, Israel

R. Wang
The Computing Center
Academia Sinia
Beijing, China

M. Wörner
KfK, Institut für Reaktorsicherheit,
Postfach 3640
D-76021 Karlsruhe, Germany

J. Xue
Northwestern Normal University
Lanzhou

Continuum Hypothesis in the Computation of Gas-Solid Flows

J.Y. Tu and C.A.J. Fletcher

Centre for Advanced Numerical Computation in Engineering and Science,
The University of New South Wales, Sydney 2052, Australia

1. Introduction

Dilute gas-solid particle flows are encountered in many industrial applications such as coal combustion equipment, cyclone separators and electrostatic precipitators, pneumatic conveying and pulverized coal gasification. There are basically two approaches commonly to predict gas-solid flows: Lagrangian and Eulerian. The fundamental concepts of both approaches and their specific applications have been discussed and reviewed in the literature [1]. In the Lagrangian approach, the motion of single particles are considered and relevant variables are calculated along the particle trajectories. The Eulerian approach treats both gas and particulate flows as continua, and the phases are regarded as two mutually interacting fluids.

When using the Eulerian approach (continuum model) for analysing dilute gas-solid particle flows, one often needs to consider whether the continuum hypothesis is valid in an individual computational control volume, i.e. whether the limiting computational volume is large enough to contain sufficient particles for a stationary average to be determined. The applicability of applying a continuum model for dilute gas-solid particle flows has been discussed in the literature [1,2]. It was suggested [2] that the continuum model is suitable only for low Stokes number and high loading systems. It was estimated [1] that, for instance, for coal particles suspended in air with a loading of unity, the side of a computational control volume which would contain 10^4 particles would be $L/d_p \sim 10^2$ where L is the control volume dimension and d_p is the particle diameter. For particles of $d_p = 100 \mu m$, L would be approximately one centimetre. This dimension of computational volumes seems not to be practical for the numerical calculation of both gas and particulate phases. However, the previous discussions and comments are mainly based on order-of magnitude analyses with a lack of comparison of numerical computations with experiments.

In this paper we use an Eulerian model to predict a turbulent gas-solid particle flow in a 90° bend in comparison with the experimental data obtained by Kliafas and Holt [3]. Since a very dilute suspension was used in the experimental study the question of whether the continuum assumption for the solid phase is satisfactory needs to be addressed. A set of numerical experiments involving comparisons with experimental data are intentionally carried out and arranged in the following way. First the computation is performed on a coarse grid and with a certain particulate loading, where the continuum hypothesis for the particulate phase in individual computational volumes is satisfactory. Then, gradually, the computational grids are refined and the particulate loading is reduced until the continuum condition is not expected to be valid any more. It is interesting that the numerical results, for both the grid refinement and the reduction of particulate loading, do not show any sudden discontinuity or divergence from the case with sufficient particles in individual computational volumes to the case without sufficient particles. All numerical results, as a function of the grid refinement and of the reduction of particulate loading, are almost identical and in good agreement with the experimental data. This observation suggests that we are not required so strictly to follow the continuum hypothesis when using a continuum model. However, the accuracy of numerical predictions using an Eulerian approach will depend on how the boundary conditions, particle-wall collisions, turbulent interactions and other physical phenomena can be appropriately modelled.

2. Governing Equations

In an Eulerian model, both gas and particulate phases are considered as a continuum and a set of Reynolds-averaged conservation equations for the mass and momentum of both phases, gas kinetic energy of turbulence and its dissipation can be written in a generic transport form [4,5]

$$\frac{\partial}{\partial x_i}(A_i \phi) - \frac{\partial}{\partial x_i}(B \frac{\partial \phi}{\partial x_i}) = S \tag{2.1}$$

2.1 Gas Phase

Continuity equation ($\phi = 1$)

$$A_i = \rho_g u_g^i; \qquad B = 0; \qquad S = 0 \tag{2.2}$$

Momentum equation ($\phi = u_g^i$)

$$A_i = \rho_g u_g^i; \qquad B = \mu_{g,eff}; \qquad S = -\frac{\partial P}{\partial x_i} \tag{2.3}$$

Turbulent kinetic energy equation ($\phi = k$)

$$A_i = \rho_g u_g^i; \qquad B = \mu_{g,1} + \frac{\mu_{gt}}{\sigma_k}; \qquad S = P_k - \rho_g \epsilon \tag{2.4}$$

ϵ-equation ($\phi = \epsilon$)

$$A_i = \rho_g u_g^i; \qquad B = \mu_{g,1} + \frac{\mu_{gt}}{\sigma_\epsilon}; \qquad S = \frac{\epsilon}{k}(C_{\epsilon 1} P_k - C_{\epsilon 2} \rho_g \epsilon) \tag{2.5}$$

2

2.2 Particulate Phase

Continuity equation ($\phi = \rho_p$)

$$A_i = \rho_p u_p^i; \qquad B = \rho_p D_p; \qquad S = [\rho_p u_p^i - D_p \frac{\partial \rho_p}{\partial x_i}] \frac{\partial \rho_p}{\partial x_i} \qquad (2.6)$$

Momentum equation ($\phi = u_p^i$)

$$A_i = \rho_p u_p^i \qquad ; \qquad B = \rho_p \nu_p;$$

$$S = \frac{\partial}{\partial x_i}(\rho_p \nu_p \frac{\partial u_p^i}{\partial x_k}) + \frac{\partial}{\partial x_i}[D_p(u_p^i \frac{\partial \rho_p}{\partial x_k} + u_p^k \frac{\partial \rho_p}{\partial x_i})] + F_D + F_G + F_{WM} \qquad (2.7)$$

All the notation in the above equations and details about the gradient models of the particulate phase can be found in [4-6]. In the above governing equations, it is assumed that the particulate phase consists of spherical solid particles of uniform size, and that since the particulate phase is dilute, the interparticle collision is negligible and the particulate behaviour has no influence on the gas flow solution. Moreover, the viscous and pressure terms in the particulate phase momentum equations are neglected for dilute suspensions, but a particle-wall momentum exchange in terms of a momentum source term F_{WM} is included in this study, in order to get a consistent solution with the boundary condition of the particulate phase at the solid wall. The momentum exchange source term is given [5]

$$F_{WM}^N = -C_N[1 + (\bar{e}_p^N)^2]\rho_p \left|W_{p,h}^N\right| W_{p,h}^N (B^N)^2 A_N \qquad (2.8)$$

in the normal direction and

$$F_{WM}^T = -C_T[1 - (\bar{e}_p^T)^2]\rho_p \left|W_{p,h}^T\right| W_{p,h}^T (B^T)^2 A_N \qquad (2.9)$$

in the tangential direction. Here, A_n denotes a face area of the control volume coincident with the wall. $\bar{e}_p^N$ and $\bar{e}_p^T$ are normal and tangential mean restitution coefficients, respectively. $W_{p,h}^N$ and $W_{p,h}^T$ are normal and tangential mean velocities of particulate phase at the distance h away from the wall. B_N and B_T are constants related to the restitution coefficients. C_N and C_T are coefficients which are modelled by

$$C_N = C_m \frac{W_{p,h}^N}{[\sum_{i=1}^{3}(u_{p,h}^i)^2]^{\frac{1}{2}}} \qquad (2.10)$$

where i = 1,2,3 correspond to x, y, and z direction, respectively. C_m is referred to as the particle inertial impacting efficiency which is similar to that obtained by Ilias and Douglas [7] based on the particle inertia (Stokes number St) and the Reynolds number of the gas phase. Referring to their work, for the present case $C_m = 0.9$ is chosen.

3

The coefficient C_T is related to the tangential wall-momentum exchange which is modelled as

$$C_T = \frac{C_N}{y^+} \quad for \quad y^+ \leq 11.63$$

$$C_T = \frac{C_N k}{\ln(E y^+)} \quad for \quad y^+ > 11.63 \tag{2.11}$$

where κ and E are the same as in turbulence model of gas phase [8], and y^+ is a similar definition but for the particulate flow [9]. The effect of both the normal and tangential wall-momentum exchanges on the particulate flow is only considered for control volumes immediately adjacent to the wall.

The gravity force is $F_G = \rho_p g$, where g is the gravitational acceleration. The drag force F_D is defined by

$$F_D^i = \rho_p \frac{f(u_g^i - u_p^i)}{t_p} \tag{2.12}$$

where t_p is the particle relaxation time and f the correction factor selected by

$$\tag{2.13}$$

$$f = \begin{bmatrix} 1 + 0.15 Re_p^{0.687} & 0 < Re_p \leq 200 \\ 0.914 Re_p^{0.282} + 0.0135 Re_p & 200 < Re_p \leq 2500 \\ 0.0167 Re_p & 2500 < Re_p \end{bmatrix}$$

with the particulate Reynolds number defined by

$$Re_p = \frac{|u_g^i - u_p^i| d_p}{\nu_g} \tag{2.14}$$

where ν_g is the kinematic viscosity of the gas phase.

2.3 Boundary Conditions

The boundary conditions at inlet are specified for all dependent variables in both the gas and particulate phases. At outflow the normal gradient of these quantities is set to zero. A 'no-slip' boundary condition is employed for the gas velocity at the wall surface. The boundary conditions at the solid wall for the particulate phase are based on the generalised wall boundary equations which can be written in a generic form [5]

$$a\varphi_w + b[\frac{\partial \varphi}{\partial \eta}]_w = c \quad , \quad \varphi = [W_p^N, W_p^T, \rho_p] \tag{2.15}$$

where η indicates the direction normal to the surface of the wall. The coefficients in the equation can be determined as follows

$$a_N = A^N - B^N; \quad b_N = A^N Kn_h; \quad c_N = 0$$
$$a_T = A^T - B^T; \quad b_T = A^T Kn_h; \quad c_T = 0$$
$$a_\rho = B^N - A^N; \quad b_\rho = A^N Kn_h; \quad c_\rho = (B^N - A^N)\rho_p$$

where

$$A^N = [\frac{1.0 + (-\bar{e}_p^N)^q}{2}]^{\frac{1}{q}}; \qquad A^T = [\frac{1.0 + (\bar{e}_p^T)^q}{2}]^{\frac{1}{q}}$$

$$B^N = [\frac{\bar{e}_p^N + (-\bar{e}_p^N)^q}{1.0 + \bar{e}_p^N}]^{\frac{1}{q}}; \qquad B^T = [\frac{\bar{e}_p^N + (\bar{e}_p^T)^q}{1.0 + \bar{e}_p^N}]^{\frac{1}{q}}$$

where q=2 corresponds to an "energy average". Kn_h is a Knudsen number defined by a gas-particle interaction length [9] divided by the system characteristic length.

3. Numerical Procedure

A versatile non-orthogonal boundary-fitted coordinate grid system is employed. The grids used for the computational domain in Fig. 3.1 are 22x102x22 in the radial, axial and spanwise direction respectively. When a symmetry condition is imposed in the spanwise direction, the actual grids become 22x102x12. The 45 control volumes in the streamwise direction (i.e. 2 degrees each) are used for the 90° bend.

Eq. (2.1) is discretised using a finite volume formulation in generalised coordinate space with the metric information expressed in terms of area vectors. The equations are solved on a non-staggered grid, i.e. all primitive variables are stored at the centroids of the mass control volumes. To approximate the convective terms at faces of the control volumes, a generalised QUICK convective differencing method [10] is used. Second derivatives are evaluated using three-point symmetric formulae. Each governing equation is sequentially relaxed to update one of the primitive variables. A velocity potential correction [11] is introduced to satisfy continuity of gas phase and upgrade the gas pressure using a modified SIMPLEC algorithm [12]. The stored values at the centroids are interpolated and modified to calculate the flow flux at faces of the control volumes using the moment interpolation method [13]. The governing equations for both the gas and particulate phase are solved sequentially at each iteration to obtain all the dependent variables. At each global iteration each equation is iterated, typically 3 to 5 times, using a Strongly Implicit Procedure [6] scheme.

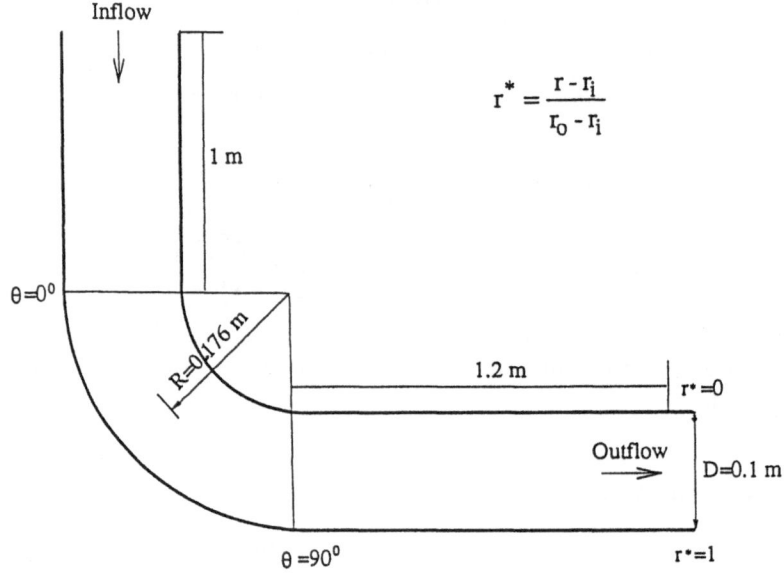

$$r^* = \frac{r - r_i}{r_0 - r_i}$$

Fig. 3.1. Computational domain of a 90° square-sectioned bend

4. Numerical Results

All comparisons presented in this section are referred to the symmetry plane of the bend. The flow conditions which are not repeated in this paper can be found in [3]. A mean normal restitution coefficient $\bar{e}_p^N = 0.9$ and tangential restitution coefficient $\bar{e}_p^T = 0.9$ are chosen for glass beads used in Kliafas and Holt's study.

Fig. 4.1 shows comparisons of computational mean velocity profiles with experimental data for both the gas and particulate phases ($d_p = 50\mu m$) at the bend entrance and the $\theta = 15°$ station. In the same figure, both two and three dimensional computations are presented. It demonstrates that the gas flow is affected (accelerated in the symmetry plane) due to the presence of the walls in the spanwise direction, when comparing the result based on a two-dimensional geometry with those using a three-dimensional geometry. However, only a small influence on the mean particulate flow is observed because of the flat transverse characteristic of the mean velocity profile of the particulate phase. It can also be seen from this figure that there is almost no difference between the two measured data sets of mean particulate velocity with two different bulk velocities since the particle inertia of both cases is very high. It is shown that a satisfactory result for the mean particulate velocity has been obtained by using a two-dimensional model.

In order to address whether the continuum assumption implicit in using an Eulerian model for the particulate phase is satisfactory, a set of numerical

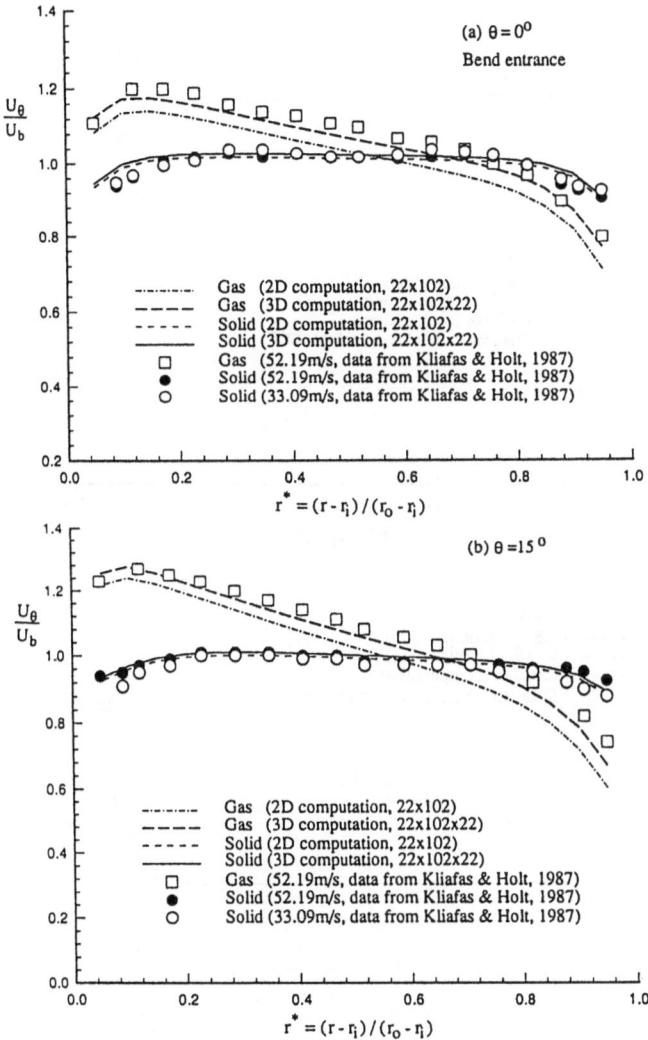

Fig. 4.1. Comparisons of predicted and measured mean streamwise velocity profiles of both gas and particulate phases

computations are intentionally performed in the following way. Initially, we use a particulate loading with a volume fraction of 0.6×10^{-3}, which is 10^4 times higher than the experimental condition, and computational grids of 22×102 in the radial and streamwise directions respectively. The number of particles in a small control volume in this case will be of the order of 10^3. Fig. 4.2 shows a numerical computation compared with the experimental data when using three sets of inlet condition of particulate concentration, where the last one corresponds to the experimental condition. The numerical results

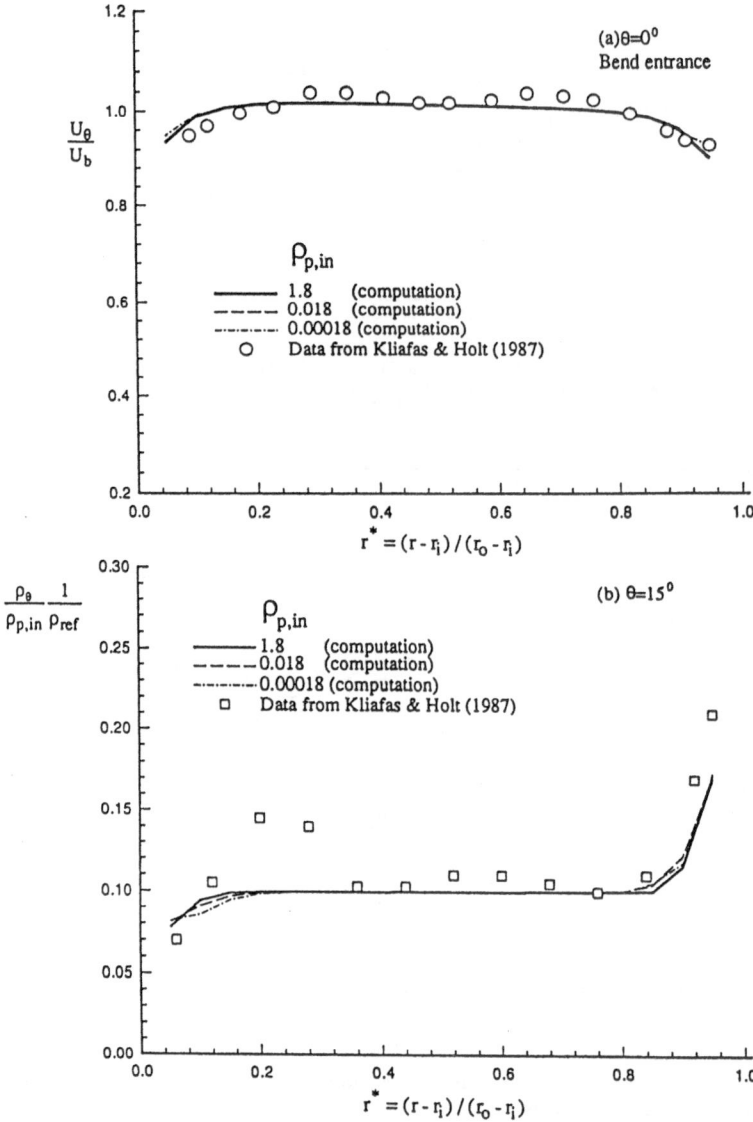

Fig. 4.2. (a) Mean streamwise particulate velocity at bend entrance and (b) particulate concentration distribution at $\theta = 15°$ station, as function of three different inlet conditions of particulate concentration ($U_b = 33.09 m/s$)

are almost identical and in good agreement with the experimental data except for some uncertainty about the experimental particulate concentration near inner wall.

The particulate concentration from the computation is normalised by each inlet value and then by a reference value (ρ_{ref}). This reference value is taken to be 10 for the computation and to be 40 (the number of detected particles)

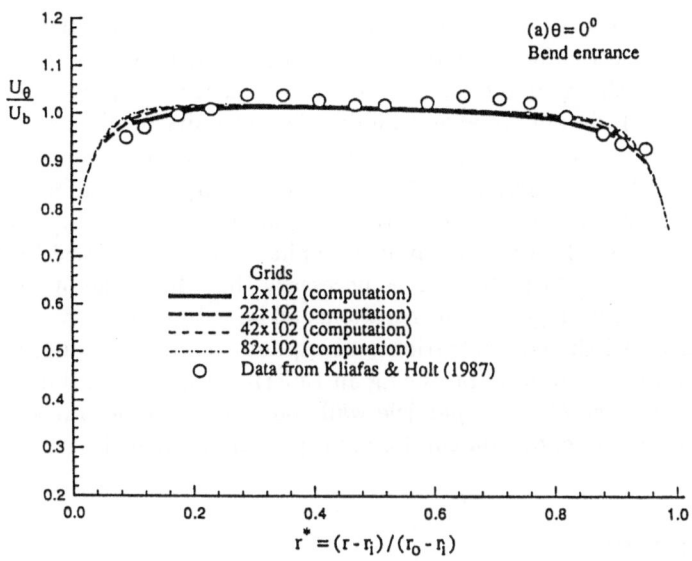

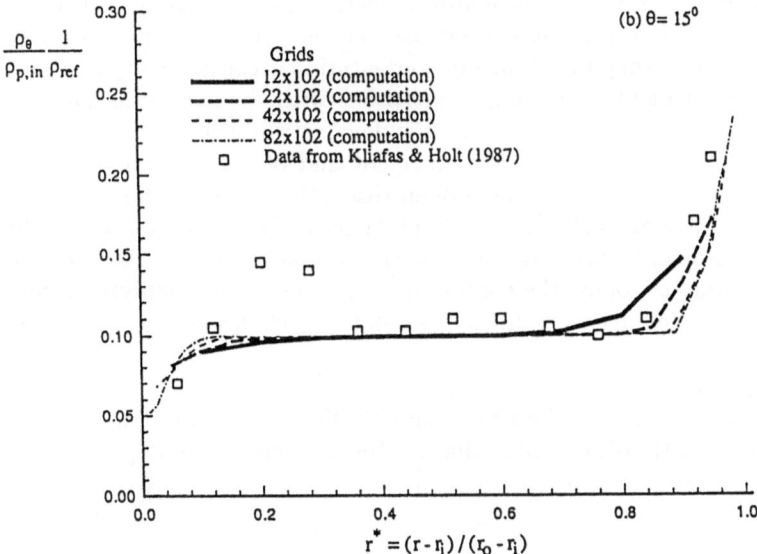

Fig. 4.3. Effect of grid refinements on (a) mean streamwise velocity and (b) concentration

for the measurement. The reference values correspond to the maximum value of the computational results and of the experimental data, respectively. By using this procedure, the particle distribution in a 90° bend from the computation can be compared with measured data.

Fig. 4.3 shows the effect of grid refinement on the computational results of the particulate phase. In this case, a medium particulate loading

$(\rho_{p,in} = 0.0018 kg/m^3)$ is used for which the control volumes in the finest grids cannot contain sufficient particles. It is interesting that the numerical results, for both the grid refinement and the reduction of particulate loading (see Fig. 4.2), do not show any sudden discontinuity or divergence from the case with sufficient particles in individual computational volumes to the case without sufficient particles. This observation suggests that we are not required so strictly to follow the continuum condition when using two-fluid models if the mean values of the particulate phase can be obtained by an experimental technique. In addition it is necessary that the range of solutions, corresponding to the range of concentrations, all lie within one flow regime. In the present case all the concentrations correspond to one-way coupling. The accuracy of numerical predictions using an Eulerian approach will depend on how the boundary conditions, particle-wall collisions, turbulent interactions and other physical phenomena can be appropriately modelled.

5. Conclusions

When an Eulerian approach (continuum model) is used for predicting dilute gas-solid particle flows, one often needs to consider whether the continuum hypothesis is satisfactory in a limiting computational control volume. A set of numerical experiments involving comparisons with the measured data are intentionally carried out in this paper. It is demonstrated that the numerical results, as the function of the grid refinement and of the reduction of particulate loading, do not show any sudden discontinuity or divergence from the case with sufficient particles in individual computational volumes to the case without sufficient particles. This observation suggests that we are not required so strictly to follow the continuum hypothesis when applying a continuum model for dealing with dilute gas-solid particle flows.

Acknowledgment
The authors are grateful to DITARD and Pacific Power (Electricity Commission of New South Wales, Australia) for the financial support provided as part of the DS4PUB project.

References

1. Crowe, C.T., "REVIEW: Numerical Models for Dilute Gas-Particle Flows," Journal of Fluids Engineering, Vol.104, No.3, pp.297-303 (1982).
2. Crowe, C.T., "Two-Fluid versus Trajectory Models; Range of Applicability," Gas-Solid Flows-1986 ASME FED, Vol.35, pp.91-96 (1986).
3. Kliafas, Y and Holt, M., "LDV Measurements of a Turbulent Air-Solid Two-Phase Flow in a 90° bend," Experiments in Fluids, 5, pp. 73-85 (1987).

4. Fletcher, C.A.J, "Gas Particle Industrial Flow Simulation Using RANSTAD," Surveys in Fluid Mechanics (ed. R. Narasimha), Indian Academy of Science, 18, Parts 3 & 4, pp. 657-681 (1993).

5. Tu, J.Y. and Fletcher, C.A.J. "Eulerian Modelling of Dilute Gas-Particle Complex Confined Flows," submitted (1994).

6. Fletcher, C.A.J., "Computational Techniques for Fluid Dynamics," Vol 1: Fundamental and General Techniques; Vol 2: Specific Techniques for Different Flow Categories, 2nd Edition, Springer, Heidelberg (1991).

7. Ilias, S. and Douglas, P.L., "Inertial Impaction of Aerosol Particles on Cylinders at Intermediate and High Reynolds Numbers," Chem. Engng. Sci., 44, pp. 81-99 (1989).

8. Launder, B.E. and Spalding, D.B., "The Numerical Computation of Turbulent Flows," Comp. Methods in Appl. Mech. and Eng., 3, pp. 269-289 (1974).

9. Soo, S. L., "Development of Theories on Liquid Solid Flows," J. Pipeline, 4, pp. 137-145 (1984).

10. Cho, N-H., Fletcher, C.A.J. and Srinivas, K., "Efficient Computation of Wing Body Flows," Lecture Notes in Physics, Springer, 371, pp. 167-191 (1991).

11. Fletcher, C.A.J. and Bain, J.G., "An Approximate Factorisation Explicit Method for CFD," Computers and Fluids, 19, pp. 61-74 (1991).

12. Van Doormal, J.P. and Raithby, G.D., "Enchancements for the SIMPLE method for predicting incompressible fluid flow," Num. Heat Transfer, 7, pp. 147-163 (1984).

13. Rhie, C.M. and Chow, W.L., "Numerical Study of the Turbulent Flow Past an Airfoil with Trailing Edge Separation," AIAA J., 21, pp. 1525 (1983).

Numerical Modelling of Two- and Three-Dimensional External and Internal Unsteady Incompressible Flow Problems

Günter K.F. Bärwolff

Technical University Berlin, Hermann-Föttinger-Institute,
Rudower Chaussee 6, 12484 Berlin, Germany

Summary. In this paper we will discuss numerical methods that solve the full two- and three-dimensional time dependent Navier-Stokes equations and the possibility of coupling the heat conduction equation by the Boussinesq approximation. The solution methods will be applied i) to the investigation of a crystal melt flow caused by free and forced convection as an internal flow problem, and ii) to the full 3d flow around a circular cylinder as a base for LES and as an external flow problem. During the consideration of the two fluid dynamical problem classes the formulation of suitable conservative boundary conditions, especially in the case of 'open' boundaries will be discussed.

1. Introduction

For the Navier-Stokes equations or the Boussinesq equations coupled with the heat conduction equation the complete initial boundary value problems in consideration will be formulated and a spatial finite volume (FV) discretization and a splitting method for the time discretization will be given in sections 2 and 3.

To fulfil the continuity equation at every discrete time level a Poisson equation for the pressure or a pressure correction will be solved. In section 4 some information about the iterative cg/mg solution method will be given. The important question especially for the problem of open boundary conditions of the solvability conditions will also be discussed in section 4. In sections 5 and 6 the two application problems namely:

- the simulation of crystal melt flows and
- the flow around a cylinder

will be examined and some computational results will be shown.

2. The Mathematical Model - The Governing Equations

In investigating of the flow around a cylinder we consider air as an incompressible fluid. In the case of LES we remark that the quantities in consideration are so called 'filtered ones' and instead of the molecular viscosity ν we have to use $\nu_{\text{eff}} = \nu + \nu_{\text{SGS}}$ with the subgrid-scale viscosity ν_{SGS}. For the subgrid-scale viscosity the models of Smagorinski/Lilly, Germano, Lesieur and RNG are in discussion and our further aim is to investigate and to compare the results with the different models. It's important to state that the form of the equations for the LES is the same as the Navier-Stokes equations with a non-constant and nonlinear viscosity.

For the crystal melt flow we consider the melt as a non isothermic, Newtonian incompressible fluid. The viscosity normally depends on the temperature. The coupling of the velocity/pressure field $\mathbf{u}$, p with the temperature field T is realized, first by the viscosity $\nu = \nu(T)$, and second by the outer force term $\mathbf{f}$ of the momentum equation with the density ρ. The density depends on the temperature by:

$$\rho = \rho_0(1 - \hat{\beta}(T - T_0)),$$

where $\hat{\beta}$ is a heat expansion coefficient. ρ_0 and T_0 are suitably chosen reference values of the density and the temperature.

Furthermore the notation p means in this case the pressure over the density but we will call p for simplification the pressure. In the following we will note the governing equations for the problems we want to investigate. For the momentum equation we have:

$$\frac{\partial \mathbf{u}}{\partial t} + \nabla \cdot \mathbf{u}\mathbf{u} = -\nabla p + \nabla(\nu \nabla \mathbf{u}) + \mathbf{f}, \tag{2.1}$$

and the mass continuity equation is of the form:

$$\nabla \cdot \mathbf{u} = 0, \tag{2.2}$$

in our integration region, Ω. For the temperature we have the heat conduction equation including convective terms

$$\frac{\partial T}{\partial t} + \nabla \cdot \mathbf{u}T = \nabla(a\nabla T). \tag{2.3}$$

The coefficient a stands for the temperature conductivity. The equation system (2.1), (2.2), (2.3) will be closed, for example, with the boundary conditions

$$u_i = g_i \quad \text{on} \quad \Gamma_{1u}, \tag{2.4}$$

$$\frac{\partial u_i}{\partial \mathbf{n}} = 0 \quad \text{on} \quad \Gamma_{2u}, \tag{2.5}$$

$$T = h \quad \text{on} \quad \Gamma_{1\theta}, \tag{2.6}$$

$$\frac{\partial T}{\partial \mathbf{n}} = 0 \quad \text{on} \quad \Gamma_{2\theta}, \tag{2.7}$$

and the initial data:

$$u_i = u_{i0}, \qquad\qquad\qquad\qquad\qquad (2.8)$$

$$T = T_0. \qquad\qquad\qquad\qquad\qquad (2.9)$$

Here $\mathbf{n}$ is the normal unit vector on the boundary and $u_i, i = 1, 2, 3$ are the components of the velocity. For the boundary pieces, Γ_{ju}, $\Gamma_{j\theta}$, we require $\Gamma_{1u} \cup \Gamma_{2u} = \Gamma$, $\Gamma_{1u} \cap \Gamma_{2u} = \emptyset$, $\Gamma_{1\theta} \cup \Gamma_{2\theta} = \Gamma$, $\Gamma_{1\theta} \cap \Gamma_{2\theta} = \emptyset$.

Detailed information about the location of Dirichlet or Neumann boundaries and the special initial and boundary data g_i, h, u_{i0}, T_0 is given in the discussion of the specific problems.

3. The Numerical Solution Method

We will give a short overview of the solution method, which was described in detail in the papers [1] and [2].

The spatial discretization of the partial differential equations (2.1), (2.2), (2.3) is done by a finite volume method (FVM). We use a staggered grid formulation, that means four staggered grids. Three grids are used for the velocity components and a fourth one we need for the pressure and the temperature respectively.

As a result of the finite volume discretization we have ordinary differential equations for every finite volume or grid point of the chosen discretization of the spatial computation region Ω closed by the discretization of the initial and boundary conditions.

As an example we present now the discretized equation for a velocity component and the discretized continuity equation. In the case of an (r, φ, z) coordinate system using the Gauss integral theorem we get by integration over a volume element of the φ-grid the following approximation:

$$\frac{\partial v}{\partial t} + \frac{1}{r_j}[v^2_{ji+1k}]_{\overline{\varphi}} + \frac{1}{r_j}[(ruv)_{j+\frac{1}{2}i+\frac{1}{2}k}]_{\overline{r}} + [(wv)_{ji+\frac{1}{2}k+\frac{1}{2}}]_{\overline{z}} + \frac{1}{r_j}(uv)_{ji+\frac{1}{2}k} =$$

$$-\frac{1}{r_j}[p_{ji+1k}]_{\overline{\varphi}} + \frac{1}{r_j}[r_{j+\frac{1}{2}}\nu v_{j+1i+\frac{1}{2}k\overline{r}}]_{\overline{r}} + \frac{1}{r_j^2}[\nu v_{ji+\frac{3}{2}k\overline{\varphi}}]_{\overline{\varphi}} + [\nu v_{ji+\frac{1}{2}k+1\overline{z}}]_{\overline{z}}$$

$$+\frac{\nu}{r_j^2}[u_{ji+1k}]_{\overline{\varphi}} - \frac{\nu}{r_j^2}v_{ji+\frac{1}{2}k} + f_u$$

$$\qquad\qquad\qquad\qquad\qquad (3.1)$$

for the φ-component of the impulse transport equation for the v-velocity component. In the equation (3.1) we have used the notation:

$$[g_{j+\beta i+\alpha k+\gamma}]_{\overline{\varphi}} = \frac{g_{j+\beta i+\alpha k+\gamma} - g_{j+\beta i+\alpha-1k+\gamma}}{\varphi_{i+\alpha} - \varphi_{i+\alpha-1}},$$

and so on. The integration of the continuity equation gives the discretization:

$$\nabla_h \cdot \mathbf{u} := \frac{1}{r_j} v_{ji+\frac{1}{2}k\overline{\varphi}} + \frac{1}{r_j}[(r\,u)_{j+\frac{1}{2}jk}]_{\overline{r}} + w_{jik+\frac{1}{2}\overline{z}} = 0 \,. \tag{3.2}$$

If we understand by $\nabla_h \cdot$ and by ∇_h the FVM representation of the operators $\nabla \cdot$ and ∇ we can write the ordinary differential equations system as

$$\frac{\partial \mathbf{u}}{\partial t} + \nabla_h \cdot \mathbf{u}\mathbf{u} = -\nabla_h p + \nabla_h(\nu \nabla_h \mathbf{u}) + \mathbf{f}, \tag{3.3}$$

$$\frac{\partial T}{\partial t} + \nabla_h \cdot \mathbf{u}T = \nabla_h(a \nabla_h T). \tag{3.4}$$

The discretization of the mass continuity equation:

$$\nabla_h \cdot \mathbf{u} = 0, \tag{3.5}$$

must always be fulfilled and is a restriction on the solution of the ordinary differential equation systems (3.3) and (3.4).

Now we briefly describe the discrete time integration from time level n to time level $n+1$. The time integration of the system (3.3), (3.4), (3.5) is done by the following splitting method:

$$\frac{\tilde{\mathbf{u}} - \mathbf{u}}{\tau} + \sigma_k \nabla_h \cdot \tilde{\mathbf{u}}\tilde{\mathbf{u}} + (1 - \sigma_k)\nabla_h \cdot \mathbf{u}\mathbf{u} = \tag{3.6}$$
$$-\nabla_h p + \sigma_v \nabla_h(\nu \nabla_h \tilde{\mathbf{u}}) + (1 - \sigma_v)\nabla_h(\nu \nabla_h \mathbf{u}) + \mathbf{f},$$

$$\frac{\mathbf{u}^{n+1} - \tilde{\mathbf{u}}}{\tau} = -\nabla_h(p^{n+1} - p), \tag{3.7}$$

with the restriction

$$\nabla_h \cdot \mathbf{u}^{n+1} = 0. \tag{3.8}$$

All quantities without an upper index are taken at the "old" time level n. In the case $\sigma_k = \sigma_v = 0$ $\tilde{\mathbf{u}}$ is explicitly given. If we choose σ_k or σ_v different from zero we will have to solve a linear or nonlinear equation system for $\tilde{\mathbf{u}}$, and this will be done by a Newton method.

If we suppose that $\tilde{\mathbf{u}}$ is given we will have to solve the linear equation system (3.7), (3.8) for $\mathbf{u}^{n+1}$ and $\delta p = p^{n+1} - p$. For this we put equation (3.7) into the equation (3.8) and we get:

$$-\Delta_h \delta p = -\frac{1}{\tau}\nabla_h \cdot \tilde{\mathbf{u}}\,. \tag{3.9}$$

The boundary conditions for δp and $\mathbf{u}^{n+1}$ will be given in section 4.

In the last step we have to solve the equation system

$$\frac{T^{n+1} - T}{\tau} + \nabla_h \cdot \mathbf{u}^{n+1}T = \sigma_T \nabla_h(a \nabla_h T^{n+1}) + (1 - \sigma_T)\nabla_h(a \nabla_h T). \tag{3.10}$$

The choice of τ depends on the spatial discretization parameters, the viscosity (Reynolds number), temperature conductivity number (Prandtl number), the magnitude of the velocity and the weighting parameters $\sigma_{...}$ (see [1], [2]).

4. The Solution of the Equation Systems (3.9) and (3.10) - Solvability Conditions

It is well known that the equation system (3.9) is singular because only the pressure gradient is uniquely determined. If the number of equations (number of finite volumes of the pressure grid of Ω) is equal to n the rank of the coefficient matrix of (3.9) is equal to $n - 1$. For solvability of (3.9) we have to secure the integral or global condition:

$$\int_{\Gamma} \mathbf{n} \cdot \nabla_h \delta p \, d\gamma = \frac{1}{\tau} \int_{\Gamma} \mathbf{n} \cdot \tilde{\mathbf{u}} \, d\gamma , \qquad (4.1)$$

which follows from (3.9) by the application of the Gauss integral theorem. When we use homogeneous Neumann boundary conditions for the pressure correction of the form

$$\mathbf{n} \cdot \nabla_h \delta p = 0 \quad \text{on} \quad \Gamma \qquad (4.2)$$

from the solvability condition (4.1) the necessary condition:

$$\int_{\Gamma} \mathbf{n} \cdot \tilde{\mathbf{u}} \, d\gamma = 0 \qquad (4.3)$$

follows for the predicted velocity $\tilde{\mathbf{u}}$ on the boundary.

In the case of the internal flow problem of modelling of a crystal melt caused by free convection we will not have any problems with the solvability condition (4.1) or (4.3) because $\mathbf{n} \cdot \tilde{\mathbf{u}}$ is equal to zero on the whole boundary and we can work with the homogeneous Neumann boundary condition (4.2) for the pressure correction δp.

We will have more trouble in the case of the existence of open boundary conditions on a part Γ_o of Γ. On the Dirichlet boundary, Γ_d, $\tilde{\mathbf{u}}$ is given by the Dirichlet boundary values of the velocity, for example an inflow profile. After the prediction of $\tilde{\mathbf{u}}$ by solving the system (3.6) and using certain outflow boundary conditions for the normal component of $\tilde{\mathbf{u}}$ on Γ_o, for example:

$$\mathbf{n} \cdot \nabla_h \tilde{\mathbf{u}} = 0,$$

the condition (4.3) is not fulfilled. Normally we will have the situation:

$$\int_{\Gamma} \mathbf{n} \cdot \tilde{\mathbf{u}} \, d\gamma = \epsilon , \qquad (4.4)$$

or,

$$\int_{\Gamma_d} \mathbf{n} \cdot \tilde{\mathbf{u}} \, d\gamma + \int_{\Gamma_o} (\mathbf{n} \cdot \tilde{\mathbf{u}} + \delta) \, d\gamma = 0 , \qquad (4.5)$$

with,

$$\delta = -\frac{1}{G}\epsilon = -\frac{1}{G} \int_{\Gamma} \mathbf{n} \cdot \tilde{\mathbf{u}} \, d\gamma , \qquad (4.6)$$

where $G = \int_\Gamma d\gamma$. That means that the homogeneous Neumann boundary condition (4.2) does not satisfy the solvability condition (4.1). One possibility for overcoming this problem, used in our codes for a few years successfully and also recommended in [8], will now be described.

With the right hand side $-\frac{1}{\tau}\nabla_h \cdot \tilde{\mathbf{u}}$ of equation (3.9) we can control and fulfil the solvability condition by using boundary conditions for δp of the form:

$$\mathbf{n} \cdot \nabla_h \delta p = 0 \quad \text{on} \quad \Gamma_d, \tag{4.7}$$

$$\mathbf{n} \cdot \nabla_h \delta p = -\frac{\delta}{\tau} \quad \text{on} \quad \Gamma_o. \tag{4.8}$$

We can prove the solvability by evaluation of equation (4.1).

It must be said that the presented way of satisfying the condition (4.1) by using the boundary conditions (4.7) and (4.8) is only a sufficient one. The global error of the mass balance with $\tilde{\mathbf{u}}$ was distributed by a local correction of the pressure gradients on Γ_o to control the solvability. It is possible that in some special situations, for example the case of two seperated open boundary sections the described way does not make physical sense because only the global mass conservation and not the local mass conservation was considered. But if we have a single outflow boundary section the method works very well.

The discussed method of boundary condition modification to ensure the solvability of equation (3.9) leads to the boundary conditions for the normal velocity components:

$$\mathbf{n} \cdot \mathbf{u}^{n+1} = \mathbf{n} \cdot \mathbf{w}^{n+1} \quad \text{on} \quad \Gamma_d, \tag{4.9}$$

$$\mathbf{n} \cdot \mathbf{u}^{n+1} = \mathbf{n} \cdot \tilde{\mathbf{u}} + \delta \quad \text{on} \quad \Gamma_o, \tag{4.10}$$

with $\mathbf{w}^{n+1}$ as the given inflow profile for the correction or projection step (3.7), (3.8) of the determination of $\mathbf{u}^{n+1}$.

The solution of equation (3.9) we realize with a cg method preconditioned by an algebraic multigrid method following [7]. When we have δp as a solution of the equation (3.9) we can get $\tilde{\mathbf{u}}$ from (3.7) by an explicit fill in step. The regularization of (3.9) was done by the choice of the pressure at one finite volume of our pressure grid during the multiplication of an element of the main diagonal and the corresponding element of the right hand side vector by a very large number $c_{\text{penalty}} \approx 10^{40}$.

This kind of regularization secures the structure of our finite volume equation system, which is necessary for the application of multigrid methods.

The developed cg/mg method was compared to other solution methods for equation (3.9), especially to solvers for sparse banded symmetrical matrices. The direct solvers for banded matrices consist of two steps. The LU analysis or the inversion is the first step. In the second step we get the solution by a 'fill-in' procedure. This has to be done at every time step. If the matrix does not depend on the time step we have to do the LU analysis only once. If the matrix depends on τ, as in the case of equation (3.10), we have to do the LU analysis when we change τ dynamically. In such cases the cg/mg method is much more efficient than the direct solution. Also in the 3d case, for example, with a grid of ($129 \times 65 \times 129$) finite volumes the cg/mg method solves equation (3.9) faster than the direct method, with sufficient accuracy. In cases of a lower number of finite volumes like in 2d problems, the direct solution methods are faster than the cg/mg methods. Therefore we have implemented the cg/mg method and the direct solver for sparse banded matrices and we use them alternately.

Equation (3.10) is solved for $\sigma_\theta \neq 0$ with the methods developed for the solution of the equation (3.9).

5. The Application of the Solution Methods to Crystal Melt Flow

We will consider two problems. The first problem is a benchmark test for method validation which is intensively discussed in connection with the modelling of crystal melt flows. As a second crystal melt flow we consider the flow of a vertical situated $Bi_{0.5}Sb_{1.5}Te_3$ melt[1].

5.1 Wheeler's Benchmark Problem

Wheeler's benchmark problem is defined in [3] and A.A. Wheeler called up all people dealing with numerical solution of crystal growths problems to solve the benchmark problem. For orientation and comparison of the benchmark results we use the paper [4].

Wheeler considered a cylinder symmetrical problem using the Czochralski crystal growth method. When we use the notations of [4], i.e. (r,z) polar coordinates, t time, u, v, w radial, azimuthal and axial velocity components, p pressure, θ temperature, Gr Grashof's number and Pr Prandtl number, the dimensionless equations are of the form:

$$u_t + (ruu)_r/r + (wu)_z - v^2/r = -p_r + \Delta u - u/r^2 \,, \qquad (5.1)$$

$$v_t + (ruv)_r/r + (wv)_z + uv/r = \Delta v - v/r^2 \,, \qquad (5.2)$$

$$w_t + (ruw)_r/r + (ww)_z = -p_z + \Delta w + Gr\theta \,, \qquad (5.3)$$

$$(ru)_r/r + w_z = 0 \,, \qquad (5.4)$$

$$\theta_t + (ru\theta)_r/r + (w\theta)_z = \Delta\theta/Pr. \qquad (5.5)$$

Δf is for a scalar function f defined by:

$$\Delta f = (r\, f_r)_r/r + (f_\varphi)_\varphi/r^2 + (f_z)_z \,.$$

---

[1] The investigation of crystal growth problems is sponsored by the DARA (registration number 50WM 9207).

19

Since we do not have a charcteristic velocity we use a fictitious velocity $U = \nu_0/L$, the time $t_0 = L^2/\nu_0$ with ν_0 as the viscosity for a reference temperature and a suitable density ρ_0 to obtain a dimensionless problem. The temperatures are normalized by:

$$\theta = \frac{T - T_0}{T_1 - T_0},$$

with the occuring minimal and maximal values, T_0 and T_1 respectively of the temperature.

The boundary conditions are given by

$$
\begin{array}{lll}
u = v = w_r = \theta_r = 0 & \text{for} & r = 0, 0 \leq z \leq \alpha, \\
u = w = 0, v = Re_c, \theta = 1 & \text{for} & r = 1, 0 \leq z \leq \alpha, \\
u = w = \theta_z = 0, v = rRe_c & \text{for} & 0 \leq r \leq 1, z = 0, \qquad (5.6) \\
u_z = v_z = w = 0, \theta = (r - \beta)/(1 - \beta) & \text{for} & \beta \leq r \leq 1, z = \alpha, \\
u = w = \theta = 0, v = rRe_x & \text{for} & 0 \leq r \leq \beta, z = \alpha.
\end{array}
$$

In the above, Re_x and Re_c stand for the rotation of the solid crystal and for the crucible rotation. Fig. 5.1 shows the geometry of the integration region Ω for the benchmark problem.

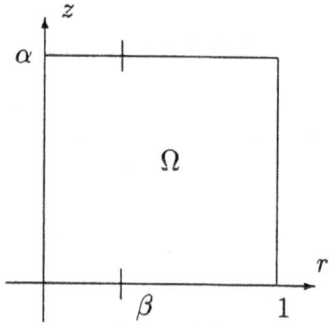

Fig. 5.1. Integration region Ω of the benchmark problem

The benchmark problem sets all derivatives of the quantities in the azimuthal direction to zero, and thus the azimuthal velocity component v is a transport variable, like the temperature, and does not occur in the mass continuity equation. Because of the homogenity in the azimuthal direction, the grid for the azimuthal velocity component is identical with the grid for the pressure and the temperature in the quasi two-dimensional benchmark problem of Wheeler.

The aspect ratio α, which is the ratio of the crucible height to the crucible radius, is fixed to 1. The aspect ratio β, which is the ratio of the crystal radius to the crucible radius, is fixed to 0.4. The Prandtl number Pr is fixed to 0.05. In Table 5.1 the other parameters for the stationary benchmark test

Table 5.1. Benchmark parameter combinations

problem	Gr	Re_x	Re_c
A1	0	10^2	0
A2	0	10^3	0
A3	0	10^4	0
B1	0	10^2	-2.5×10^1
B2	0	10^3	-2.5×10^2
B3	0	10^4	-2.5×10^3
C1	10^5	0	0
C2	10^6	0	0
D1	10^5	10^1	0
D2	10^5	10^2	0
D3	10^5	10^3	0

problems are given. The benchmark problem for all parameter combinations given in Table 5.1 was solved with the above described non-stationary numerical method. For the convective terms of the momentum and the heat transfer equation, central differences are used. As initial velocity data we use $u = v = w = 0$ in the computational region Ω. The initial temperature is set by the solution of the stationary heat conduction problem without convection, using the given boundary conditions for θ. We have used a grid of (64×64) finite volumes.

The computer code was implemented on a RISC6000 IBM 350 workstation. For the presentation and comparison of the results we use plots of streamlines at the levels

$$\Psi = \Psi_{min} + k(\Psi_{max} - \Psi_{min}), \quad k = 1, ..., 9 \text{ and } \Psi = 0 \,,$$

and isotherms at the levels

$$\theta = k/10, \quad k = 1, ..., 9 \,.$$

The stream-function we get by the solution of the equation:

$$- \Psi_{rr} - \Psi_{zz} = (rw)_r - (ru)_z \,, \tag{5.7}$$

with the condition $\Psi = 0$ on the no-slip boundary. Equation (5.7) is solved numerically and the nodes for Ψ are taken at the corners of the pressure grid (see Fig. 5.2). This means that the control volumes of the Ψ grid are replaced by $\delta r/2$ in the r-direction and by $\delta z/2$ in the z-direction relative to the volumes of the pressure grid. This choice allows us to approximate the right hand side of (5.7) in a very natural way.

The computation of the steady state in the case of the problems B1, B2, B3, C1, C2, D1, D2 and D3 was realized with the nonstationary code described above after a few hundred time steps. To reach the steady state of the problems A1, A2 and A3 we need many time steps.

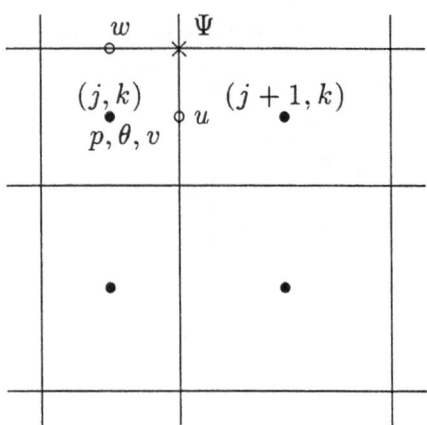

Fig. 5.2. Location of the quantities in the $\varphi = const$ cut of the grid

The Figs. 6.2, 6.3 ... 6.7 show the results of the benchmark problem solution for some parameter combinations. Comparison with the results of [4] gave no essential differences in the minimal and maximal values of the stream function and in the graphs of isolines.

Only in the case A1 we find a difference. We get the value -2.3×10^{-1} and in [4] the result -2.3×10^{1} was given. This must be a mistake or a misprint in [4] because all other comparisons showed only very small differences.

5.2 The Numerical Simulation of Vertically Situated ($Bi_{0.5}Sb_{1.5}Te_3$) Melt Zones

The governing equations of this problem are similar to the equations of Wheeler's benchmark problem. Only in the momentum equations we have to take into consideration temperature dependent viscosity, $\nu = \nu(\theta)$. We will write the equations in the 3d form:

$$u_t + (ruu)_r/r + (uv)_\varphi/r + (wu)_z - v^2/r = -p_r + \nabla \cdot (2\mu \mathbf{S_r}) - 2\mu S_{\varphi\varphi}/r \quad (5.8)$$

$$v_t + (ruv)_r/r + (vv)_\varphi/r + (wv)_z + uv/r = -p_\varphi/r + \nabla \cdot (2\mu \mathbf{S_\varphi}) + 2\mu S_{r\varphi}/r \quad (5.9)$$

$$w_t + (ruw)_r/r + (vw)_\varphi/r + (ww)_z = -p_z + \quad (5.10)$$

$$(ru)_r/r + v_\varphi/r + w_z = 0 \quad (5.11)$$

$$\theta_t + (ru\theta)_r/r + (v\theta)_\varphi + (w\theta)_z = (r\varsigma\theta_r)_r/r + (\varsigma\theta_\varphi)_\varphi/r^2 + (\varsigma\theta_z)_z \quad (5.12)$$

for $(r, \varphi, z) \in \Omega$. For ς we have $\varsigma = \nu(\theta)/(\nu_0 Pr)$ and μ stands for $\nu(\theta)/\nu_0$. The components of the vectors $\mathbf{S_r} = (S_{rr}, S_{r\varphi}, S_{rz})$, $\mathbf{S_\varphi} = (S_{r\varphi}, S_{\varphi\varphi}, S_{\varphi z})$ and $\mathbf{S_z} = (S_{rz}, S_{\varphi z}, S_{zz})$ are defined as:

$$S_{rr} = u_r, \qquad S_{\varphi\varphi} = (v_\varphi + u)/r, \qquad S_{zz} = w_z,$$
$$S_{r\varphi} = \tfrac{1}{2}(v_r + u_\varphi/r - v/r), \quad S_{\varphi z} = \tfrac{1}{2}(w_\varphi/r + v_z), \quad S_{rz} = \tfrac{1}{2}(u_z + w_r).$$
$$(5.13)$$

$\nabla \cdot (\mathbf{S})$ is in the (r, φ, z) coordinate system defined for a vector $\mathbf{S} = (S_1, S_2, S_3)$ as:

$$\nabla \cdot (\mathbf{S}) = (rS_1)_r/r + (S_2)_\varphi/r + (S_3)_z \ .$$

The boundary conditions are given by

$$
\begin{array}{lll}
u = w_r = \theta_r = 0 & \text{for} & r = 0, 0 \le z \le \alpha, \varphi \in (0, 2\pi), \\
u = w = 0, \theta = 4z(\alpha - z)/\alpha^2 & \text{for} & r = 1, 0 \le z \le \alpha, \varphi \in (0, 2\pi), \\
u = w = \theta = 0, & \text{for} & 0 \le r \le 1, z = 0, \varphi \in (0, 2\pi), \quad (5.14) \\
u = w = 0, \theta = 0 & \text{for} & 0 \le r \le 1, z = \alpha, \varphi \in (0, 2\pi), \\
v = 0 & \text{on} & \text{the whole boundary.}
\end{array}
$$

α is the ratio of the height H to the radius R of the cylindrical ampulla. The integration region Ω is a cylinder $\{(r, \varphi, z)|r \in (0,1), \ \varphi \in (0, 2\pi), \ z \in (0, \alpha)\}$ (q.v. Fig. 5.1 - a cut at $\varphi = const.$). The Grashof and the Prandtl numbers are definded as:

$$Gr = g\hat{\beta}\delta T R^3/\nu_0^2 \ , \qquad (5.15)$$

$$Pr = \nu/\kappa \ , \qquad (5.16)$$

with the relations

$$\nu_0 = \nu(T_0) \quad , T_0 = 613°C \ , \qquad (5.17)$$

$$\nu = (c - dT) * 10^{-7} \, m^2/s \quad , c = 11.6, \, d = 1.3 * 10^{-2}/1°C \ , \qquad (5.18)$$

$$\kappa = 4.4 * 10^{-6} m^2/s \ , \qquad (5.19)$$

$$\hat{\beta} = 9.6 * 10^{-5} K^{-1} \ , \qquad (5.20)$$

$$\delta T = (5 + H/1mm) \, K \ . \qquad (5.21)$$

θ is defined as $\theta = (T - T_0)/\delta T$. In Table 5.2 we summarize the investigated parameter combinations for the simulation of the $Bi_{0.5}Sb_{1.5}Te_3$ melt zones. The height of the ampulla, H, and the acceleration, g, are the only variable parameters of our problem. The radius R is fixed at $4mm$. α and the temperature difference δT are functions of H. In the table we also list the Rayleigh number Ra_w defined by Müller et al. [5] as:

Table 5.2. Investigated parameter constellations

problem	$H \, [mm]$	$g \, [m/s^2]$	α	$\delta T \, [K]$	Gr	Ra_w
P1	8	9.81	2	13	0.5943×10^4	0.0078×10^6
P2	12	9.81	3	17	0.7772×10^4	0.0519×10^6
P3	16	9.81	4	21	0.9600×10^4	0.2028×10^6
P4	20	9.81	5	25	1.1429×10^4	0.5895×10^6
P5	24	9.81	6	29	1.3258×10^4	1.4180×10^6
P6	8	9.81×10^{-4}	2	13	0.5943×10^0	0.7847×10^0
P7	12	9.81×10^{-4}	3	17	0.7772×10^0	5.1949×10^0
P8	16	9.81×10^{-4}	4	21	0.9600×10^0	2.0282×10^1
P9	20	9.81×10^{-4}	5	25	1.1429×10^0	5.8947×10^1
P10	24	9.81×10^{-4}	6	29	1.3258×10^0	14.1790×10^1

$$Ra_w = \alpha \, g \hat{\beta} H^3 \delta T / (\nu_0 \kappa). \qquad (5.22)$$

However, Müller made his investigations only for the constant aspect ratio $\alpha = 2$.

We have used a grid of (64×64) finite volumes, as in the case of Wheeler's benchmark solution. For all cases P1 ... P10 stationary solutions exist. The solutions are strictly two-dimensional. 3d effects did not occur ($v = 0$ and $\frac{\partial \xi}{\partial \varphi} = 0$ for all variables ξ). This means it is not necessary for the investigated problems to use the genuine 3d formulation. It's only necessary to consider the equations (5.8), (5.10), (5.11), (5.12) without the terms with derivatives in the φ-direction and terms, which contain the velocity component v. The Figs. 6.8, 6.9,..., 6.13 show isolines of the stationary temperature and streamfunction field (streamlines). In the case of microgravitation (P6, ..., P10) very small or no influence of the flow field due to the temperature field occur and the convection was dominated by the heat conduction ($Pr = 0.082$). The temperature field is symmetrically relative to the symmetry line $z = \alpha/2$.

Further investigations were done with "artificial" parameter combinations of the problem of $Bi_{0.5}Sb_{1.5}Te_3$ melt zone. The reference [5] details some parameter studies for typical flows in crystal growth. The above defined aspect ratio α is a constant ($\alpha = 2$). For our problem class with the given boundary conditions it is only possible to study the influence of Ra_w.

Table 5.3. Investigated "artificial" parameter combinations

problem	$H (= D) \, [mm]$	$\delta T \, [K]$	Gr	Ra_w
K1	8	13	0.0059×10^6	0.0078×10^6
K2	16	21	0.0768×10^6	0.1014×10^6
K3	24	29	0.3579×10^6	0.4726×10^6
K4	32	37	1.0826×10^6	1.4294×10^6
K5	40	45	2.5715×10^6	3.3953×10^6

We realize the variation of Ra_w by the variation of H and R. All other parameters we take from the above defined problem with $g = 9.81 \, m/s^2$ and we summarize the investigated parameter combinations in Table 5.3. We have solved the problems K1, K2, ... on a grid of (64×64) finite volumes. For the problems K1, K2, K3 and K4 stationary solutions exist. For problem K5 we find periodic oscillations and only a time-periodic solution exists. In Figs. 6.14 ... 6.19 some results are illustrated.

For all the following figures of θ-isolines and streamlines we use the reference form and orientation of Fig. 5.3.

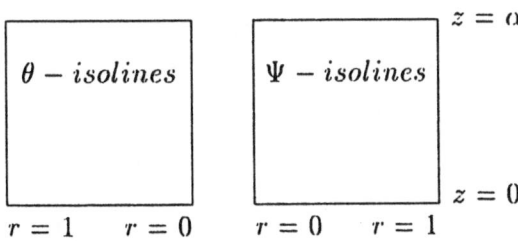

Fig. 5.3. Notation for the graphical output

5.3 Further Investigations

The problems solved and discussed in section 4 require some further studies. For the problems P1, P2, ... we investigate the 3d problem caused by the nonhomogeneous distribution of the temperature parabolic profiles in the azimuthal direction. A boundary condition of the form:

$$\theta(\varphi, z) = \theta_0(\varphi)4z(\alpha - z)/\alpha^2 \qquad \text{for} \qquad r = 1, 0 \le z \le \alpha, \qquad (5.23)$$

with

$$\theta_0(\varphi) = 1 + abs(\pi - \varphi)/\pi * \Delta\theta_0 \qquad \text{for} \qquad 0 \le \varphi \le \pi, \qquad (5.24)$$

with $\Delta\theta_0 = 0.1$ will be considered. The supposition $\frac{\partial \xi}{\partial \varphi} = 0$ is not valid for the transport quantities ξ and the corresponding 3d terms do not vanish.

For the problem P1 with $\Delta\theta_0 = 0.15$ in the Figs. 6.20 ... 6.25 the 3d results of the computation of the stationary solution for the cuts $\varphi = 0$, $\pi/2$, π, $3\pi/2$ are given. For this computation we have used a grid of $(27 \times 19 \times 27)$ finite volumes in the (r, φ, z) coordinate system. To reach the steady state we need 2 hours on a RISC6000 workstation (IBM 350).

The grid we used is coarse but the investigations are continuing and we use now grids of $(33 \times 25 \times 33)$ or more finite volumes.

Furthermore, it is neccesary for the problems K1, K2, ... to investigate the question of occuring 3d effects with increasing Rayleigh numbers. For Rayleigh numbers greater than those of the problems K4 and K5 we can expect 3d effects also in the case of cylinder symmetric geometrical and technological conditions.

This is currently under investigation.

6. The External Flow Around a Cylinder

The governing equations for the flow around a circular cylinder were written down in the (r, φ, z)-coordinate system. The cylinder with the radius R_c and the length L_z is situated with the axis in the spanwise direction (z-direction).

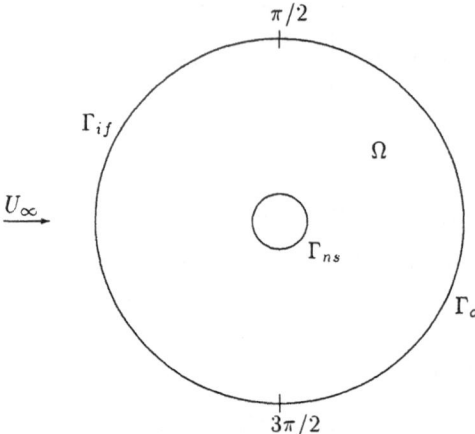

$\pi/2$

Γ_{if}

Ω

U_∞

Γ_{ns}

Γ_o

$3\pi/2$

Fig. 6.1. Cut across the (fictitious) integration region Ω at $z = const$

With the characteristic velocity U_∞, a length R_c, and molecular viscosity ν, we can write the dimensionless governing equation system without body forces as:

$$u_t + (ruu)_r/r + (uv)_\varphi/r + (wu)_z - v^2/r = -p_r + \nabla \cdot (2\mu\mathbf{S_r}) - 2\mu S_{\varphi\varphi}/r \quad (6.1)$$

$$v_t + (ruv)_r/r + (vv)_\varphi/r + (wv)_z + uv/r = -p_\varphi/r + \nabla \cdot (2\mu\mathbf{S_\varphi}) + 2\mu S_{r\varphi}/r \quad (6.2)$$

$$w_t + (ruw)_r/r + (vw)_\varphi/r + (ww)_z = -p_z + \nabla \cdot (2\mu\mathbf{S_z}) \quad (6.3)$$

$$(ru)_r/r + v_\varphi/r + w_z = 0, \quad (6.4)$$

with $\mu = \nu_{\text{eff}}/(\nu Re)$ and the Reynolds number $Re = U_\infty R_c/\nu$. We consider a fictitious cylindrical integration region Ω with the radius $\zeta = R_o/R_c$, namely,

$$\Omega = \{(r, \varphi, z) | r \in (1, \zeta), \varphi \in (0, 2\pi), z \in (0, \lambda)\},$$

with $\zeta = R_o/R_c \geq 15$, $\lambda = L_z/R_c$ (q.v. Fig. 6.1). For the formulation of the boundary conditions we divide the boundary into:

$$\Gamma_{ns} = \{(r, \varphi, z) | r = 1, \varphi \in (0, 2\pi), z \in (0, \lambda)\},$$

$$\Gamma_{if} = \{(r, \varphi, z) | r = \zeta, \varphi \in (\pi/2, 3\pi/2), z \in (0, \lambda)\},$$

$$\Gamma_o = \{(r, \varphi, z) | r = \zeta, \varphi \in (3\pi/2, \pi/2), z \in (0, \lambda)\},$$

$$\Gamma_p = \{(r, \varphi) | r \in (1, \zeta), \varphi \in (0, 2\pi)\}.$$

For the case of a cylinder without endplates the boundary conditions:

$$
\begin{array}{lll}
\mathbf{u} = 0 & \text{on} & \Gamma_{ns} \\
u = cos(\varphi) + \chi, v = -sin(\varphi) + \chi, w = 0 + \chi & \text{on} & \Gamma_{if}, \\
\frac{\partial^2 \mathbf{u}}{\partial x_\infty^2} = 0 & \text{on} & \Gamma_o, \\
\mathbf{u}(r, \varphi, 0) = \mathbf{u}(r, \varphi, \lambda) & \text{on} & \Gamma_p,
\end{array} \quad (6.5)
$$

are used to realize periodic boundary conditions in the spanwise direction. x_∞ is the direction of undisturbed "inflow"-velocity, or the direction of the main stream. χ is, in the case of LES, a disturbance, for example, a white noise with an amplitude of 10^{-4}. In the case of laminar flow simulation, χ is equal to zero.

The Figs. 6.26, 6.27, ..., 6.31 show some results of the laminar flow simulation for a Reynolds number of 1000 with a grid of $(93 \times 93 \times 27)$ finite volumes in the (r, φ, z) coordinate system. The figures show only the near-cylinder region. In Fig. 6.31, isolines of the vorticity are drawn.

For the used parameters, 3d effects did not occur and the field of velocity component w, in the spanwise direction, is equal to zero.

With the developed solution methods for the Navier-Stokes equation and with the experience of the investigation of 3d laminar flows we have just started investigating LES for turbulent flows around bluff bodies in complex geometries.

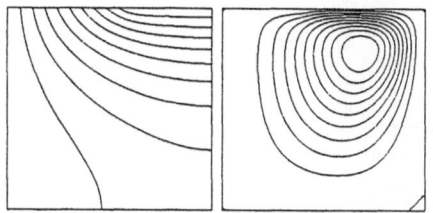

Fig. 6.2. A2 - stat. isolines for θ and Ψ

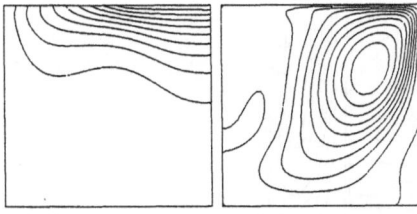

Fig. 6.3. A3 - stat. isolines for θ and Ψ

Fig. 6.4. B2 - stat. isolines for θ and Ψ

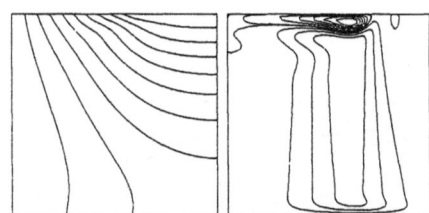

Fig. 6.5. B3 - stat. isolines for θ and Ψ

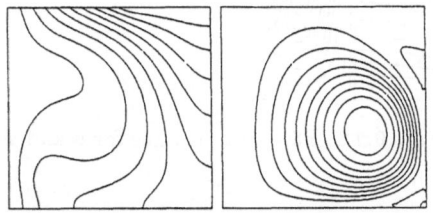

Fig. 6.6. C2 - stat. isolines for θ and Ψ

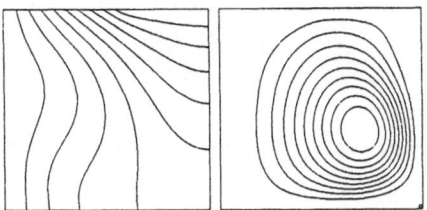

Fig. 6.7. D2 - stat. isolines for θ and Ψ

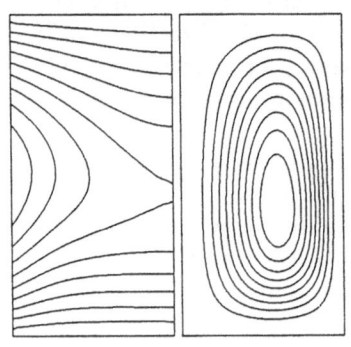 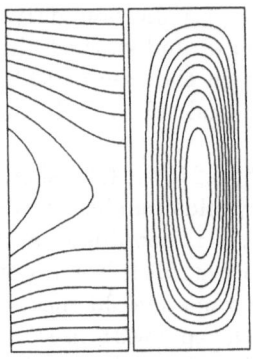

Fig. 6.8. P1 - stat. isolines for θ and Ψ **Fig. 6.9.** P2 - stat. isolines for θ and Ψ

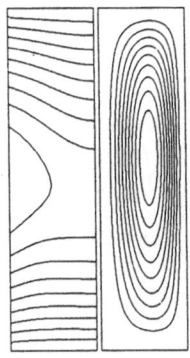 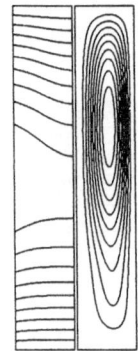

Fig. 6.10. P3 - stat. isolines for θ and Ψ **Fig. 6.11.** P5 - stat. isolines for θ and Ψ

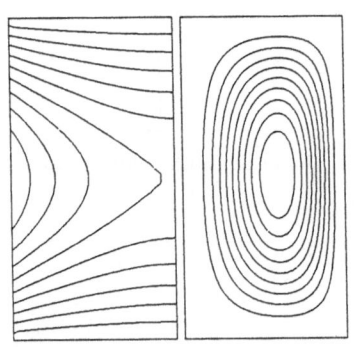 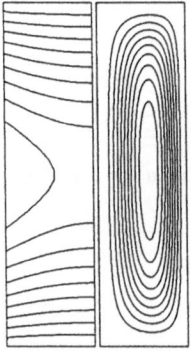

Fig. 6.12. P6 - stat. isolines for θ and Ψ **Fig. 6.13.** P8 - stat. isolines for θ and Ψ

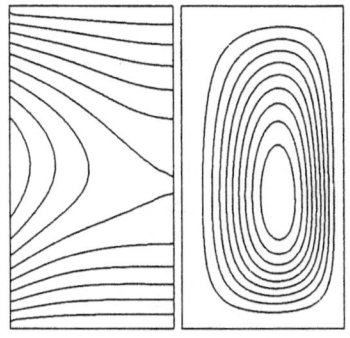

Fig. 6.14. K1 - stat. isolines for θ and Ψ

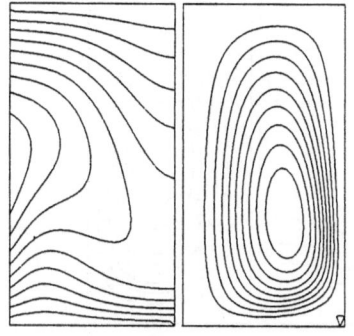

Fig. 6.15. K2 - stat. isolines for θ and Ψ

Fig. 6.16. K4 - stat. isolines for θ and Ψ

Fig. 6.17. K5 - stat. isolines for θ and Ψ

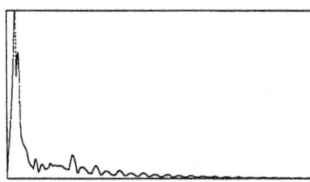

Fig. 6.18. K4 - time history of $\|\frac{\partial \theta}{\partial t}\|$

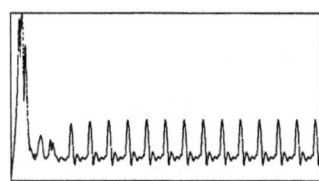

Fig. 6.19. K5 - time history of $\|\frac{\partial \theta}{\partial t}\|$

29

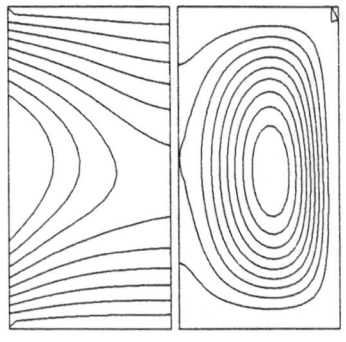

Fig. 6.20. P1/3d - $\varphi = 0$

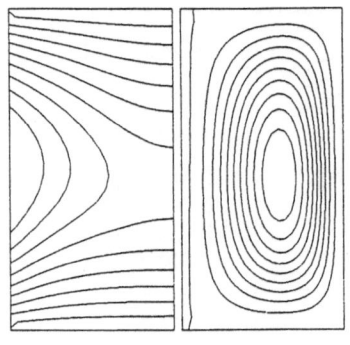

Fig. 6.21. P1/3d - $\varphi = \pi/2$

Fig. 6.22. P1/3d - $\varphi = \pi$

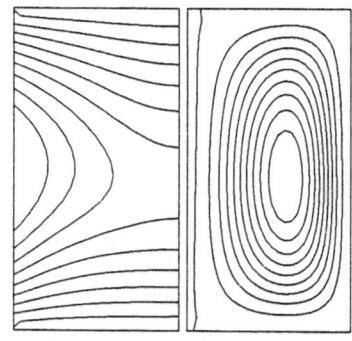

Fig. 6.23. P1/3d - $\varphi = 3\pi/2$

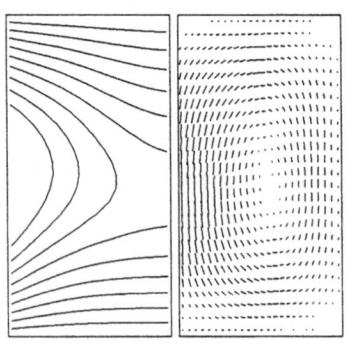

Fig. 6.24. P1/3d $(u_{r,z}) - \varphi = 0$

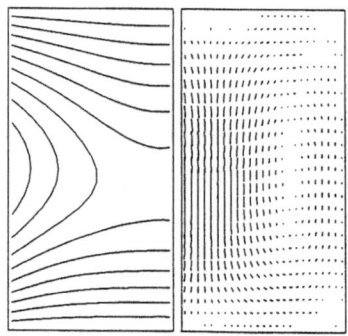

Fig. 6.25. P1/3d $(u_{r,z}) - \varphi = \pi$

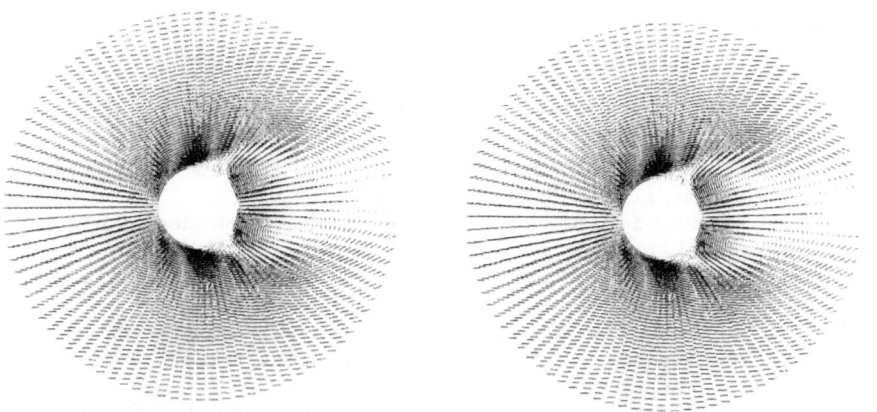

Fig. 6.26. Cylinder flow after 4000 steps **Fig. 6.27.** Cylinder flow after 14000 steps

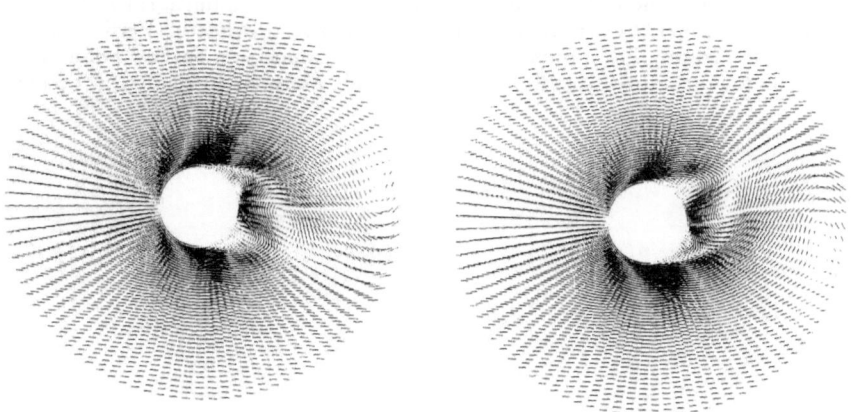

Fig. 6.28. Cylinder flow after 20000 steps **Fig. 6.29.** Cylinder flow after 21000 steps

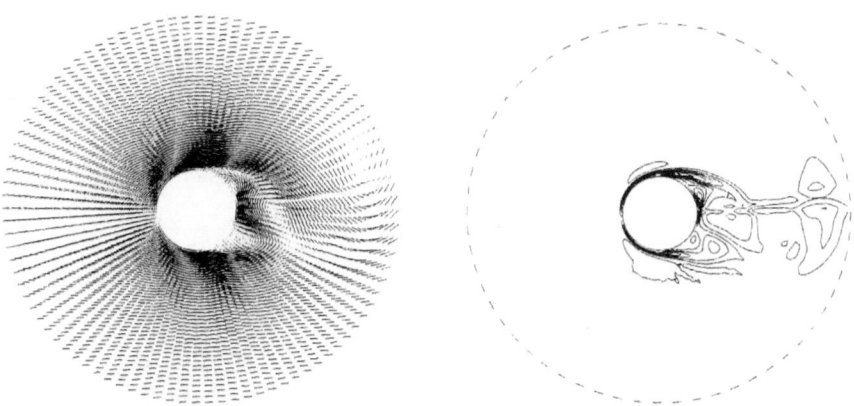

Fig. 6.30. Cylinder flow after 22000 steps **Fig. 6.31.** Cylinder flow after 22000 steps

References

1. Bärwolff, G. and Seifert, G., Efficient 2D and 3D Navier-Stokes solver, paper of the 5. ISCFD Sendai/Japan, 1993
2. Naue, G. and Bärwolff, G., Transportprozesse in Fluiden, Deutscher Verlag für Grundstoffindustrie, Leipzig 1992
3. Wheeler, A.A., Four test problems for the numerical simulation of flow in Czochralski crystal growth, Journal of Crystal Growth 102 (1990) 691
4. Bückle, U. and Schäfer, M., Benchmark results for numerical simulation of flow in CZ growth, Journal of Crystal Growth 126 (1293) 682-694
5. Müller, G., Neumann, G. and Matz.H., A two-Rayleigh-number model of bouyancy-driven convection in vertical melt growth configurations, Journal of Crystal Growth 84 (1987) 36-49
6. Baumgartl, J., Budweiser, W., Müller, G. and Neumann, G., Studies of buoyancy driven convection in a vertical cylinder with parabolic temperature profile, Journal of Crystal Growth 97 (1989) 9-17
7. Fuhrmann, J., On the convergence of algebraically defined multigrid methods, Preprint No. 3, Institut für Angewandte Analysis und Stochastic, Berlin, 1992.
8. Sani, R. and Gresho, P.M., Summary of two minisymposia on outflow boundary conditions for incompressible flow, Int. J. Numer. Methods in Fluids (in press)

Numerical Experiments in Double–Diffusive Convection

R. Peyret and J.M. Vanel

Laboratoire de Mathématiques, Université de Nice-Sophia Antipolis
Parc Valrose, 06108 Nice Cédex 2, France

Summary. We present and discuss some numerical experiments in double–diffusive convection where a layer of fluid, stably stratified by concentration, is subject to heating. By considering first the case where the fluid is heated from below we justify the need of highly accurate numerical approximations, like spectral methods. The spectral Chebyshev method for solving the equations governing the motion of a double–diffusive fluid is then briefly described. Finally, the case where the fluid is laterally heated is discussed in more details. In particular, the formation of convective layers followed by the merging of some of them is discussed from the numerical results.

1. Introduction

Double–diffusive convection deals with the motion of a fluid resulting from the combined effect of gravitational forces and diffusion of two components with different molecular diffusivities. These two components are very often the temperature and the salinity because of the various possible applications to oceanography. More generally, double–diffusive convection concerns temperature and concentration of any solute, or even concentrations of two components. The number of components may be also larger than two and the associated physical phenomenon is then called "multi–diffusive convection". We refer to articles by Turner [1], [2] for a general discussion of such problems.

The physical phenomena associated with double–diffusive convection are much richer and, at the same time, much more complicated than those encountered in usual one–component convection. As it may be expected, this physical complexity leads naturally to numerical difficulties. The main difficulty is connected with the disparity of the molecular diffusivities, which is often more than one order of magnitude: for example, the ratio of saline to thermal diffusivities is about 10^{-2}. To this disparity corresponds a disparity in diffusive times, that is in diffusive velocities, leading to stiff problems whose numerical solution is always delicate.

For an illustration of some peculiar behaviour associated with double–diffusive convection we consider the classical problem of the instability of a layer of fluid heated and concentrated from below (Fig. 1.1). In this situation the effect of the concentration $\bar{S}$ is stabilizing but it is not the case for the temperature $\bar{T}$. Even if the resulting distribution of the density $\bar{\varrho}$ defined by

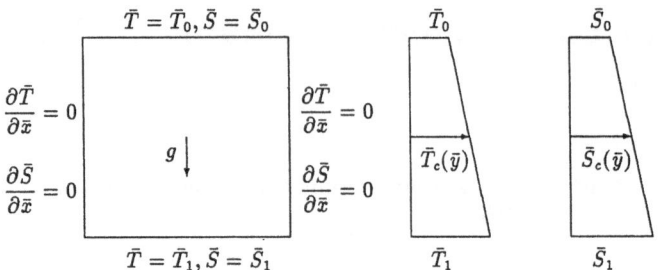

Fig. 1.1. Stratified fluid heated from below

the equation of state [1]

$$\bar{\varrho} = \bar{\varrho}_r[1 - \alpha(\bar{T} - \bar{T}_r) + \beta(\bar{S} - \bar{S}_r)] \qquad (1.1)$$

is "stable", that is the fluid is denser in its lower part, the situation may be unstable under infinitesimal perturbations because of the release of the potential energy contained in the destabilizing component $\bar{T}$. This problem has been the object of a large number of studies; among them we quote the laboratory experiments described in [3], [4], [5] the theoretical analysis developed in [3], [6], [7], [8] and the numerical studies reported in [9], [10], [11].

Let us consider, as in [11], the case where the fluid is enclosed in a square cavity with noslip walls. For given values of the Prandtl number Pr and of the diffusivity ratio τ, various flow patterns, steady or not, can be obtained according to the values of the thermal Rayleigh number R_T and the solutal Rayleigh number R_S but also to the nature of the initial conditions.

Some of them are discussed in [11]; for now we only want to describe one of these numerical experiments and to justify the interest of using an highly accurate method for solving the associated mathematical problem. The dimensionless Navier–Stokes equations in vorticity ω – streamfunction ψ formulation associated with transport–diffusion equations for the temperature T and the concentration S are solved with a Chebyshev–collocation method developed in [11] and which will be briefly described in the next Section. The calculations were done with 21 × 21 collocation points and the time–step was $\Delta t = 10^{-3}$, the time-accuracy of the scheme being $O(\Delta t^2)$. Figs. 1.2a and 1.3 show the time evolution (the reference time is the thermal diffusive time based on the side of the square container) of some significative flow quantities: ω_{max} and ψ_{max} are respectively the values of the vorticity and the streamfunction at the collocation points where their absolute values are maximum; N_T are N_S are respectively the thermal and solutal Nusselt numbers averaged along the lower horizontal wall. These results correspond to $Pr = 1$, $\tau = 0.1$, $R_T = 5000$ and $R_S = 2000$. The initial condition is

---

[1] See the list of symbols. The bars refer to dimensional quantities.

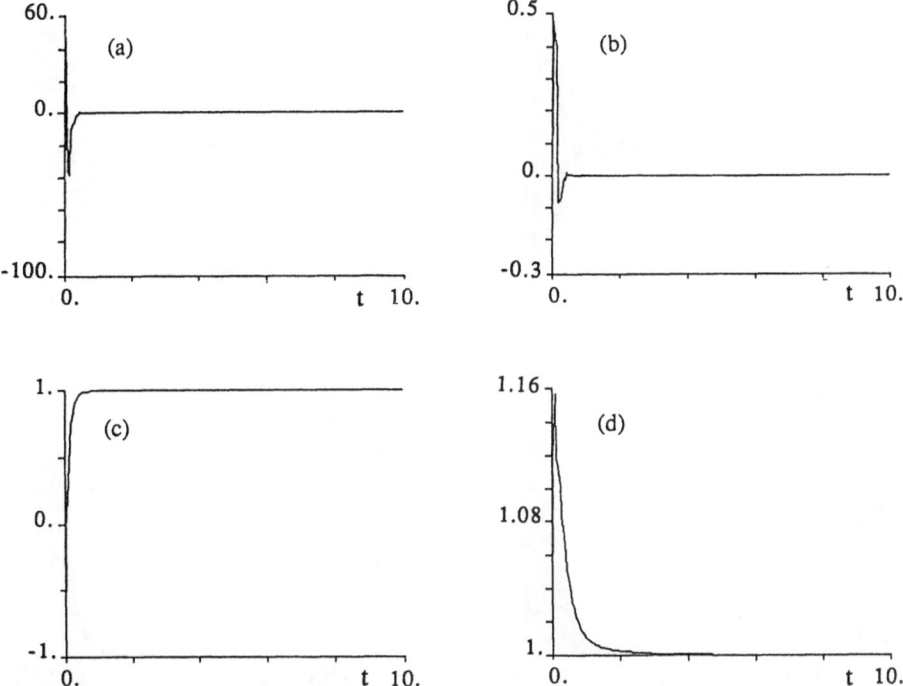

Fig. 1.2. Evolution in time ($0 \leq t \leq 10$) of (a) Maximum vorticity ω_{max}, (b) Maximum streamfunction $-\psi_{max}$, (c) Thermal Nusselt number N_T, (d) Solutal Nusselt number N_S

the conductive solution [null velocity, $T = T_c(y)$, $S = S_c(y)$] with a strong perturbation in temperature defined by

$$T = T_c + \sin \pi x . \sin \pi y. \tag{1.2}$$

Fig. 1.2a shows the results for $0 \leq t \leq 10$. It may be seen that, after a short transient phase, the flow seems to have reach a steady state which is the rest. More precisely, at $t = 10$, the maximum values of the time–derivatives are:

$$|\partial \omega / \partial t| \simeq 5 \times 10^{-10}, |\partial \psi / \partial t| \simeq 10^{-11}, |\partial T / \partial t| \simeq 4 \times 10^{-13}, |\partial S / \partial t| \simeq 3 \times 10^{-7}.$$

These residuals are usually accepted as characterizing a "numerical" steady-state solution. However, in the present case, this is not the ultimate solution. The time–integration has been pursued in order to get a temporal convergence towards the asymptotic state with a higher precision, in better agreement with the "infinite" spectral accuracy. As a matter of fact a careful examination of the time signals reveals an oscillatory behaviour with a very small amplitude. This amplitude becomes to increase near $t = 40$ although remaining relatively small: in Fig. 1.3 showing the time signals for

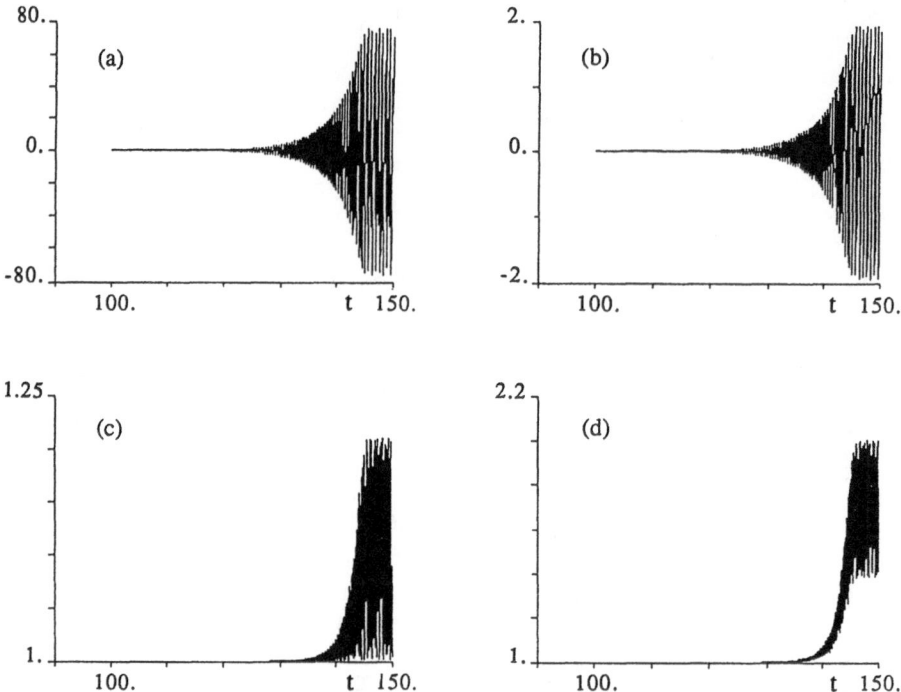

Fig. 1.3. Evolution in time ($100 \leq t \leq 150$) of (*a*) Maximum vorticity ω_{max}, (*b*) Maximum streamfunction $-\psi_{max}$, (*c*) Thermal Nusselt number N_T, (*d*) Solutal Nusselt number N_S

$100 \leq t \leq 150$, the maximum values of the derivatives are $|\partial\omega/\partial t| \simeq 5 \times 10^{-3}$ and $|\partial S/\partial t| \simeq 5 \times 10^{-5}$ (and smaller values for ψ and T) at $t = 100$. Then, the amplitude of the oscillations increases more rapidly as far as a time–periodical flow is obtained. The calculations have been continued up to $t = 250$ confirming the persistency of the periodical solution, at least up to this time! This example, which is not unusual in convective instability problems (see the GAMM workshop on oscillatory convection at low Prandtl number [12]), shows how careful and prudent we must be when calculating flow fields in which bifurcating states are apt to occur. The association of highly accurate numerical methods like spectral methods and very powerful computers permitting long time integration with a high resolution has made possible the study of complex unsteady flows which were not attainable in the past. The double–diffusive convection is an example of such a situation. In the following of this paper we first briefly describe the mathematical model and its numerical solution using a Chebyshev spectral method. Then a numerical experiment concerning the formation and evolution of convective layers in a stratified fluid laterally heated will be presented and discussed in details.

2. The Motion Equations and Their Numerical Solution

The equations governing the two–dimensional unsteady double–diffusive convection, within Boussinesq's approximation, can be written in dimensionless variables as:

$$\frac{\partial T}{\partial t} + \mathbf{V}.\nabla T - \nabla^2 T = 0, \tag{2.1}$$

$$\frac{\partial S}{\partial t} + \mathbf{V}.\nabla S - \tau \nabla^2 S = 0, \tag{2.2}$$

$$\frac{\partial \omega}{\partial t} + \mathbf{V}.\nabla \omega - Pr\nabla^2 \omega = Pr(R_T \frac{\partial T}{\partial x} - R_S \frac{\partial S}{\partial x}), \tag{2.3}$$

$$\nabla^2 \psi + \omega = 0, \tag{2.4}$$

where the velocity $\mathbf{V} = (u, v)$ with $u = \partial\psi/\partial y$, $v = -\partial\psi/\partial x$. In the above equations T is the temperature, S is the concentration, $\omega = \partial v/\partial x - \partial u/\partial y$ is the vorticity and ψ is the streamfunction. The space variables x and y (the gravity is directed in the negative y–direction) are made dimensionless with the characteristic length d (defined according to the considered problem), the time with d^2/κ_T, the velocity with κ_T/d, the vorticity and the streamfunction with κ_T/d^2 and κ_T, respectively. The characteristic quantities are $\Delta\bar{T}$ for temperature and $\Delta\bar{S}$ for concentration, these depending on the considered problem. The diffusivity ratio τ, the Prandtl number Pr, the thermal Rayleigh number R_T and the solutal Rayleigh number R_S are defined respectively by:

$$\tau = \frac{\kappa_S}{\kappa_T} \quad , \quad Pr = \frac{\nu}{\kappa_T} \quad , \quad R_T = \frac{g\alpha\Delta\bar{T}d^3}{\kappa_T\nu} \quad , \quad R_S = \frac{g\beta\Delta\bar{S}d^3}{\kappa_T\nu}. \tag{2.5}$$

We refer to the list of symbols for the definition of the quantities entering in these parameters.

The equations (2.1)-(2.4) are solved in a rectangular domain with the boundary conditions

$$\psi = 0, \tag{2.6a}$$

$$\frac{\partial\psi}{\partial n} = 0, \tag{2.6b}$$

for noslip walls (n=unit vector normal to the boundary) and

$$a_T T + b_T \frac{\partial T}{\partial n} = q_T \quad , \quad a_S S + b_S \frac{\partial S}{\partial n} = q_S, \tag{2.7}$$

where the constants a_T, b_T, a_S, b_S, q_T and q_S depend on the considered problem. The initial conditions at $t = 0$ are:

$$\mathbf{V} = \mathbf{V}^0 = (u^0, v^0) \quad , \quad \omega = \omega^0 = \frac{\partial v^0}{\partial x} - \frac{\partial u^0}{\partial y}, \tag{2.8}$$

$$T = T^0 \quad , \quad S = S^0. \tag{2.9}$$

The problem (2.1)-(2.9) is solved by means of the spectral Chebyshev method developed in [16] and [11]. The discretization with respect to time makes use of a three-level finite-difference approximation of the time derivatives:

$$\left(\frac{\partial \phi}{\partial t}\right)^{n+1} \cong \frac{3\phi^{n+1} - 4\phi^n + \phi^{n-1}}{2\Delta t}, \qquad (2.10)$$

where ϕ^n is the approximation of any variable at time $n\Delta t$. The diffusive and the buoyant terms are evaluated at time level $n+1$, while the nonlinear terms of type $\mathbf{V}.\nabla \phi$ are treated in an explicit manner using an Adams–Bashforth extrapolation:

$$(\mathbf{V}.\nabla \phi)^{n+1} \cong \{\mathbf{V}.\nabla \phi\}^{n,n-1} = 2(\mathbf{V}.\nabla \phi)^n - (\mathbf{V}.\nabla \phi)^{n-1}. \qquad (2.11)$$

According to these approximations, the time–discretization of the equations (2.1)-(2.4) writes as:

$$(\frac{3}{2\Delta t} - \nabla^2)T^{n+1} = -\{\mathbf{V}.\nabla T\}^{n,n-1} + \frac{4T^n - T^{n-1}}{2\Delta t}, \qquad (2.12)$$

$$(\frac{3}{2\Delta t} - \tau\nabla^2)S^{n+1} = -\{\mathbf{V}.\nabla S\}^{n,n-1} + \frac{4S^n - S^{n-1}}{2\Delta t}, \qquad (2.13)$$

$$(\frac{3}{2\Delta t} - Pr\nabla^2)\omega^{n+1} = -\{\mathbf{V}.\nabla \omega\}^{n,n-1} + \frac{4\omega^n - \omega^{n-1}}{2\Delta t} + Pr(R_T\frac{\partial T}{\partial x} - R_S\frac{\partial S}{\partial x})^{n+1},$$
$$(2.14)$$

$$\nabla^2\psi^{n+1} + \omega^{n+1} = 0. \qquad (2.15)$$

This scheme is second–order accurate in time. The accuracy could be easily increased by considering more time–levels in (2.10) and (2.11). For example, a third–order scheme involving the time–levels $n+1, n, n-1$ and $n-2$ is used in [13]. The explicit treatment of the nonlinear terms leads to a constraint on the size of the time–step [14]. Such a constraint can be avoided by considering a fully implicit scheme but an iterative procedure is then needed for solving the resulting nonlinear problem [15].

Because of the explicit treatment of all the convective terms, T^{n+1} and S^{n+1} can be calculated independently of the dynamical variables ω^{n+1} and ψ^{n+1}. Therefore, at each time–cycle the problem reduces to the solution of uncoupled Helmholtz problems for the temperature T^{n+1} and the concentration S^{n+1}, then to the solution of a Stokes–type problem for the vorticity ω^{n+1} and the streamfunction ψ^{n+1}.

The spatial approximation makes use of Chebyshev polynomial in both directions. In the tau method [16] the unknowns are the coefficients of the truncated series expansion in Chebyshev polynomials, while in the colloca-tion method [11], the unknowns are the values of the dependent variables at the collocation points defined as the extreme of the Chebyshev polynomial of highest degree. After introducing the boundary conditions deduced to (2.6)–(2.7), the resulting algebraic systems are solved by a matrix diagonalization procedure ([17] for the tau method, [11] for the collocation method). The evaluation of the nonlinear terms arising in the right–hand–sides of (2.12)-(2.14) is made through the usual pseudospectral technique consisting in calculating derivatives in the spectral space and products in the physical space, both spaces being connected with FFT.

The prescription of the boundary conditions (2.7) for the temperature and the concentration does not present difficulty. On the other hand, the main difficulty is the treatment of the noslip conditions (2.6) at the walls involved in the Stokes–type problem. This classical problem is completely removed thanks to the influence matrix technique [11], [16]. The technique consists first in splitting the solution $(\omega^{n+1}, \psi^{n+1})$ in two parts:

$$\omega^{n+1} = \tilde{\omega} + \omega' \quad , \quad \psi^{n+1} = \tilde{\psi} + \psi'.$$

The first part $(\tilde{\omega}, \tilde{\psi})$ is the solution of the complete equations (2.14)-(2.15) with the Dirichlet condition (2.6a) for $\tilde{\psi}$ and the zero Dirichlet condition for $\tilde{\omega}$. Then, the second part (ω', ψ') is solution of the homogeneous equations deduced from (2.14)-(2.15); the boundary conditions are $\psi' = 0$ and $\omega' = \lambda$ where λ is unknown and must be determined so that the normal derivative boundary condition $\partial \psi^{n+1}/\partial n = 0$ (Eq. (2.6b)) is satisfied. The technique is described in [11], [16], it leads to an algebraic system of the type:

$$M \, \Lambda = S$$

for determining the vector $\Lambda = (\lambda_1, \ldots, \lambda_N)^T$ where the λ_j's are the values of λ at the collocation points on the boundary. The matrix M is the influence matrix. The rank of this matrix is examined in [11], [16], [18] and [19]. It was found that the matrix M has four null eigenvalues when the noslip condition is prescribed on the four sides of the computational domain like here. In the tau method considered in [16] the matrix is made regular by approximating the vorticity with Chebyshev polynomials whose the degree is smaller of two units than the degree of those approximating the streamfunction. In the case of the collocation technique [11], the same result is obtained by omitting four collocation points on the boundary when constructing the influence matrix. Interesting results concerning the choice of these points are proven in [19], in which the case of mixed stress–free and no-slip walls is also studied.

The use of the influence matrix technique leads to the solution of a large number of Dirichlet problems for Helmholtz equations. Most of them serve for the construction of the influence matrix, they are time–independent and are solved in a preprocessing stage performed before to start the time–integration. Therefore, the influence matrix is inverted in this preprocessing stage and its inverse is stored. Then, at each time step, the solution of the Stokes–type problem requires only the solution of 4 Helmholtz equations with Dirichlet conditions, which reduces to matrix products thanks to the diagonalization procedure done once and for all in the preprocessing stage. So, the resulting algorithm is highly efficient and is well suited to vectorization.

3. Lateral Heating of a Stratified Fluid

In this section we are interested in the formation and evolution of convective cells when a stratified layer of fluid is subject to a horizontal temperature gradient. This problem has been the object of several laboratory experiments in various situations: wide tank or narrow slot, wall suddenly or progressively heated [20]–[25],. ... The stability has been studied analytically in the linear and nonlinear cases [20], [26]–[29] and numerically in [30], [31]. Finally, we ascertain the small number of numerical solutions of the full nonlinear problem using the Navier–Stokes equations. The difficulty and the cost of such computations are the cause of this apparent lack of interest. These drawbacks can be reduced by assuming a periodicity in the vertical direction like in the finite–difference solutions presented in [21], [32], [33]. The case of a finite rectangular tank has been the object of a pioneering work using finite–differences in [21]. The first highly accurate solution obtained by a Chebyshev spectral method has been presented in [34] and some of the results obtained in [34] have been reproduced in [35] using again a spectral approximation. In the present configuration the fluid is initially at rest in a rectangular tank of height H and width L (Fig. 3.1); it is stably statified by means of a constant negative concentration gradient

$$\bar{\phi}_0 = \frac{d\bar{S}}{d\bar{y}} = \frac{\bar{S}_0 - \bar{S}_1}{H}.$$

The initial temperature of the fluid is constant and equal to $\bar{T}_0$. We are interested in the motion of the fluid when the left vertical wall ($\bar{x} = 0$) is suddenly heated and then maintained at the constant temperature $\bar{T}_1 > \bar{T}_0$. The right side of the tank is kept at $\bar{T}_0$ and the horizontal walls are adiabatic. In the subsequent motion, the concentration at the horizontal walls is kept fixed at its initial value ($\bar{S}_1$ at the bottom and $\bar{S}_0$ at the top with $\bar{S}_0 < \bar{S}_1$). The flux of concentration is maintained equal to zero at both vertical walls. Concerning the fluid velocity, the noslip conditions are prescribed on the whole boundary.

For the definition of the dimensionless variables, we choose $d = L$, $\Delta\bar{T} = \bar{T}_1 - \bar{T}_0$ and $\Delta\bar{S} = \bar{S}_1 - \bar{S}_0$. The dimensionless temperature T and concentration S are respectively defined by $T = (\bar{T} - \bar{T}_0)/\Delta\bar{T}$ and $S = (\bar{S} - \bar{S}_0)/\Delta\bar{S}$. One important physical parameter is the Rayleigh number Ra based on the length scale (see [21], [30])

$$\eta = \frac{\alpha\Delta\bar{T}}{\beta|\bar{\phi}_0|} = \frac{\alpha\Delta\bar{T}}{\beta\Delta\bar{S}}H, \tag{3.1}$$

which characterizes the height to which a heated fluid particle would rise in the initial solute gradient $\bar{\phi}_0$, therefore

$$Ra = \frac{g\alpha\Delta\bar{T}\eta^3}{\kappa_T\nu} = \frac{R_T^4}{A^3 R_S^3}, \tag{3.2}$$

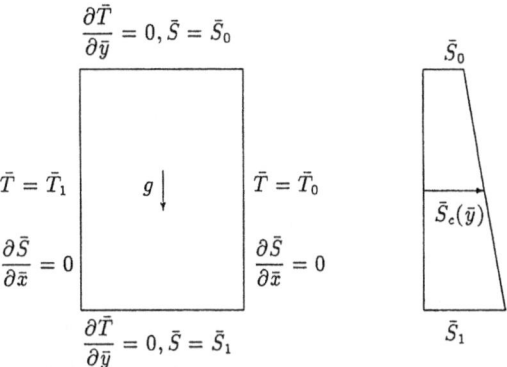

Fig. 3.1. Stratified fluid laterally heated ($\bar{T}_1 > \bar{T}_0$)

where $A = L/H$ is the aspect ratio of the tank and R_T and R_S are respectively the thermal and the solutal Rayleigh number defined in (2.5). Therefore the dimensionless parameters defining the problem are $Ra, r = \eta/L = R_T/(AR_S), A = L/H, \tau = \kappa_S/\kappa_T$ and $Pr = \nu/\kappa_T$. The problem has been solved using the tau–Chebyshev method briefly described in the section 2. Some numerical results with $\tau = 10^{-2}, Pr = 7$ corresponding to salted water have been presented in [34]. We discuss here results obtained for the same Prandtl number ($Pr = 7$) but for a larger value of τ namely $\tau = 0.05$, the other parameters being $A = 1/4$, $Ra = 2.5 \times 10^4$ and $r = 2/3$. To these values correspond $R_T = 84,375$ and $R_S = 506,250$. The calculations were performed with 41 Chebyshev polynomials in the horizontal direction and 121 in the vertical one, the time–step being $\Delta t = 5 \times 10^{-5}$.

From the observations of the results we can distinguish two phases in the flow motion. The first phase is the successive appearance of convective layers which, after some time, fill up completely the tank. The second phase is characterized by the merging of some layers leading ultimately to a quasi–steady state with a much smaller number of cells. Only the first phase was calculated in [34] because the possible merging would be too long to obtain within a reasonable time due to the small value of τ considered ($\tau = 10^{-2}$). However, the behaviour of the flow in this first phase in found to be similar whatever the value of τ used ($\tau = 0.01$ or $\tau = 0.05$). Fig. 3.2 shows the instantaneous streamlines and the iso–concentration lines during the first phase of the motion (the heated wall is at left). At the beginning of the sudden heating, the fluid near the heated wall presents a strong upward motion, leading to the appearance of an horizontal component of the concentration gradient which combines with the contribution of the horizontal temperature gradient so that the resulting density is almost statically stable. At the bottom of the container the up–going fluid possesses an excess of kinetic energy because its concentration is nearly constant; on the contrary in the mid–height of the region the fluid remains stratified while it goes up and has to exert a larger

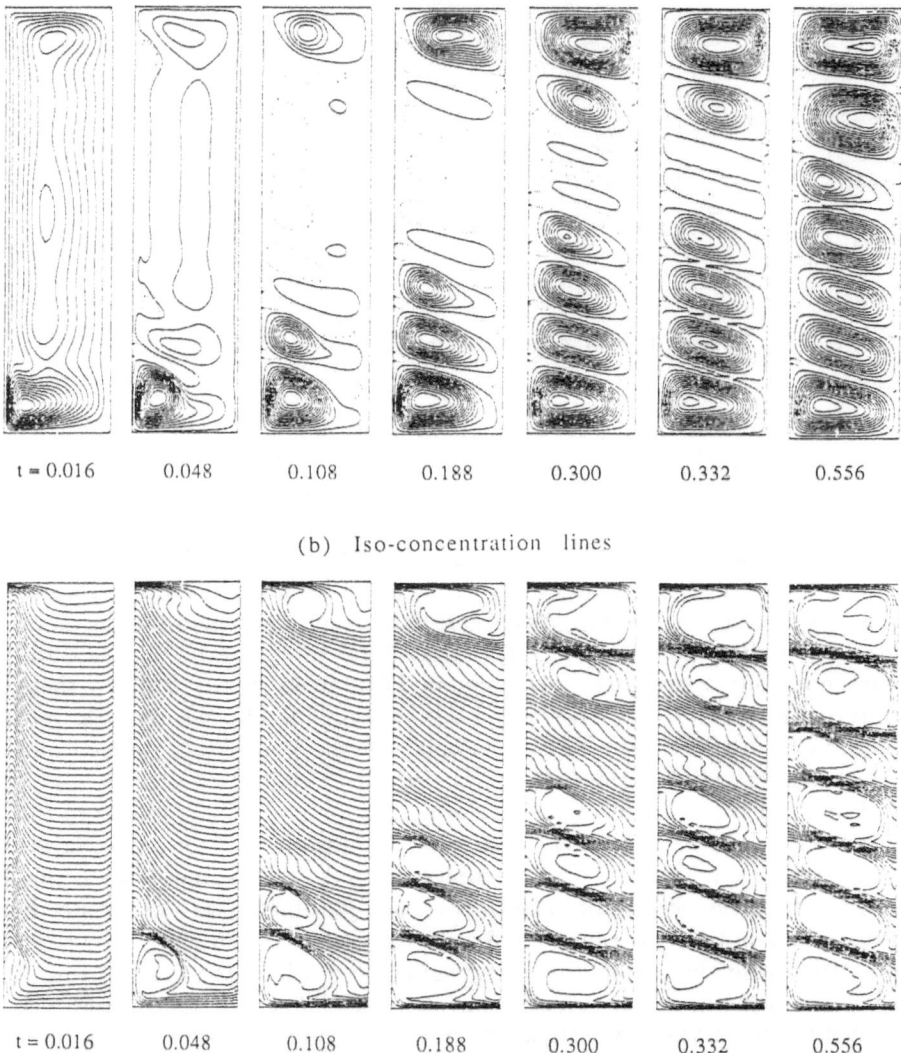

(a)   Streamlines

t = 0.016   0.048   0.108   0.188   0.300   0.332   0.556

(b)   Iso-concentration   lines

t = 0.016   0.048   0.108   0.188   0.300   0.332   0.556

Fig. 3.2. Streamlines and iso-concentration lines: first phase of the formation of layers

work against gravitation. This explains why the first convective cell appears at the bottom of the tank near the heated wall. Such a behaviour has been experimentally observed for exemple in [21], [22]. The physical process above described was first observed in [36] and applied to the present problem in [20]. After the first cell has been created, it acts like a boundary on the fluid located above and a second layer forms by a similar mechanism, and so on.

42

One observes concurrently a layer formation in cascade starting from the top but the process is less efficient: the formation and growth of the successive layers is less fast. These phenomena recur until the whole tank is filled with co–rotating convective layers. However when the Rayleigh number Ra is lowered, the layers form more and more slowly, and for small enough values of Ra, the middle part of the container remains free of cells, even for quite large times.

(a)   Streamlines

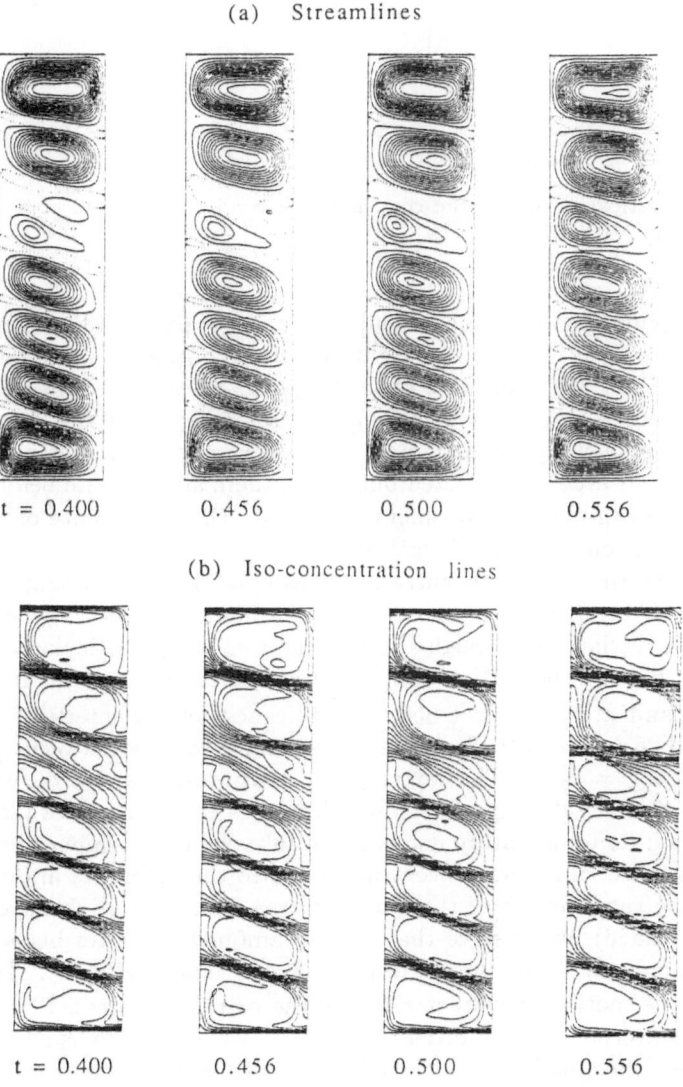

t = 0.400   0.456   0.500   0.556

(b)   Iso-concentration lines

t = 0.400   0.456   0.500   0.556

Fig. 3.3. Streamlines and iso-concentration lines: final phase of the formation of layers

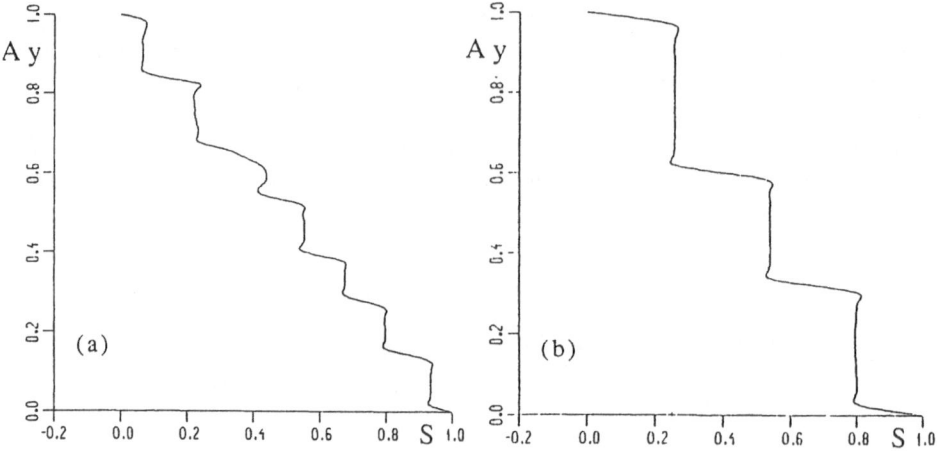

Fig. 3.4. Vertical profile of the concentration S at $x = 0.5$: (a) $t = 0.556$, (b) $t = 4.015$

The calculations reported in [34] for the same parameters except that τ was equal to 10^{-2} shown that eight cells were obtained at the end of the first stage. Here, likely because of the larger thickness of the interfaces when $\tau = 0.05$, eight cells cannot enter in the total height of the container. In the Fig. 3.3a showing the final stage of the layers formation, it may be seen that eight cells are really created but two of them amalgamate before to have completely been formed. We shall discuss later the dependence of the height h_c of the cells on the rising length scale η.

The patterns of iso–concentration lines (Fig. 3.3b) show that concentration in each layer becomes more and more homogeneous as time increases with sharp variations through the interfaces. This is particularly visible in Fig. 3.4a showing the vertical profile of the concentration at the mid–section $x = 1/2$ and at time $t = 0.556$. The thickness of the interface is varying between $0.25\ h_c$ and $0.45\ h_c$ while it was found to vary between $0.1\ h_c$ and $0.15\ h_c$ for $\tau = 10^{-2}$. The layers slope downward towards the right. The reason is that the density of the fluid decreases as it progresses far from the heated wall: its temperature decreases to take the local temperature while it conserves more longer its concentration due to the difference in diffusivities. Concerning the intensity of the motion at the final stage of the layers formation ($t = 0.556$), we observe that the streamfunction varies between -0.31 and 7.08, and the vorticity between -4.77×10^2 and 1.37×10^3. The second phase of the motion is characterized by the merging process. Fig. 3.5 shows the flow patterns from $t = 0.704$ to $t = 4.015$ when the flow is nearly steady. First of all, the smaller cells located in the mid–part of the container disappear, then the bottom cell and finally the top cell disappear also. The merging phenomenon is usually observed in laboratory experiments associated with

(a)   Streamlines

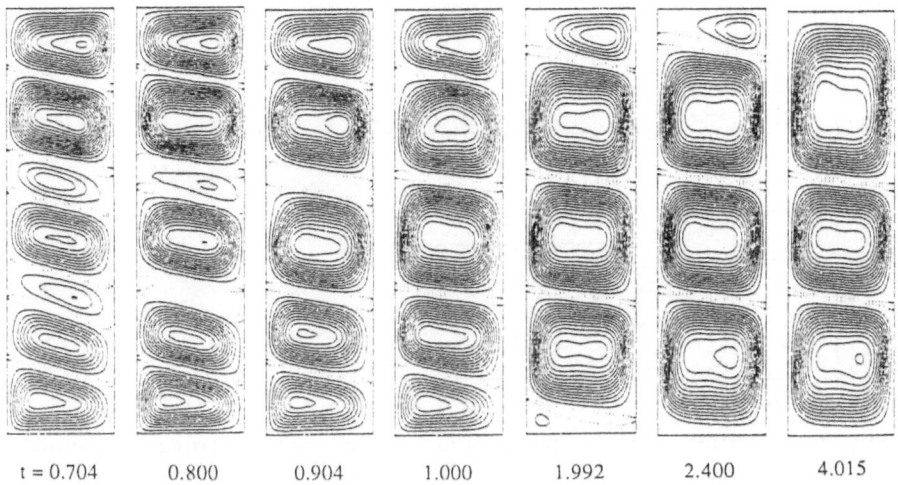

t = 0.704   0.800   0.904   1.000   1.992   2.400   4.015

(b)   Iso-concentration   lines

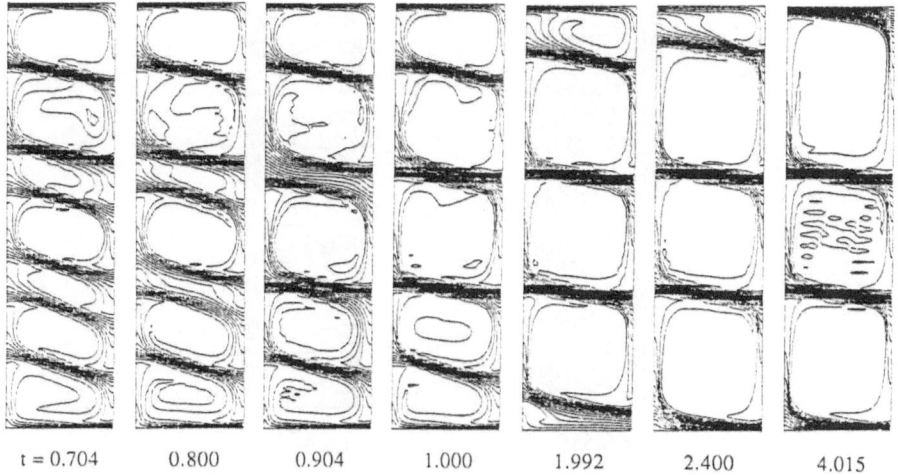

t = 0.704   0.800   0.904   1.000   1.992   2.400   4.015

Fig. 3.5. Streamlines and iso-concentration lines: merging and final state

two mechanisms depending on the considered physical situation: interface breakdown and interface migration. The present mechanism seems to be an interface migration as it can be seen in Fig. 3.6 which shows the time history of the location of the interfaces. This location y_c is characterized by the maximum values of the derivative $|\partial S/\partial y|$ at the vertical section $x = 1/2$. It is clear that merging is obtained through a vertical (upward and downward) motion of the interfaces. The number of the layers decreases from eight

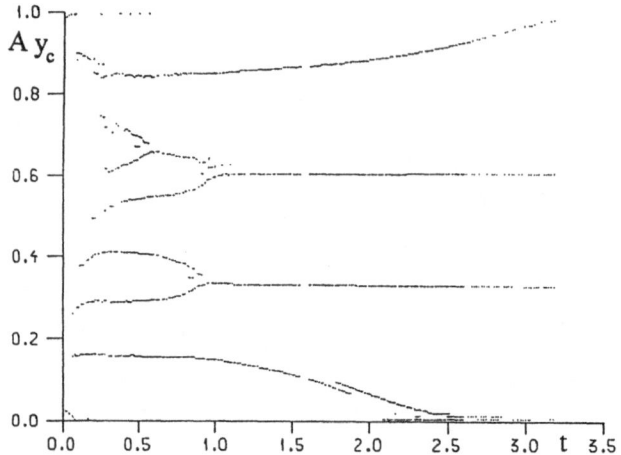

Fig. 3.6. Location y_c of maximums of $|\partial S/\partial y|$ at $x = 0.5$ showing the motion of interfaces

t = 0.556   t = 4.015

Fig. 3.7. Isotherms at $t = 0.556$ and $t = 4.015$

to three. Note that the first merging (at $t \simeq 0.52$) appearing in this figure was in fact explained above as the impossibility for the eighth layers to find room within the height of the tank. It is interesting to observe that the two ultimate interfaces were already at the same height as early as $t = 1$. At the final computed state ($t = 4.015$), the concentration S (Fig. 3.4b) is constant in each layer with a large gradient through the interfaces and near the horizontal walls (the concentration is kept fixed at these walls). The temperature field (Figs. 3.7 and 3.8) is stratified within each layer with a positive (and

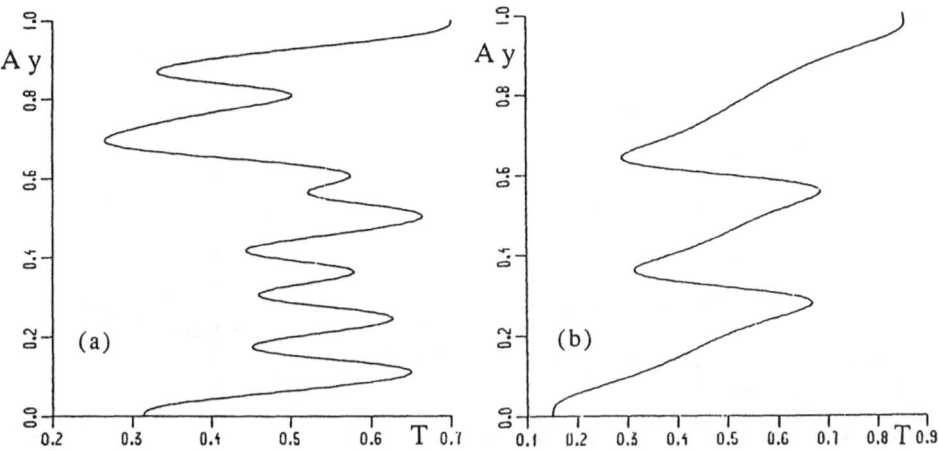

Fig. 3.8. Vertical profile of the temperature T at $x = 0.5$: (a) $t = 0.556$, (b) $t = 4.015$

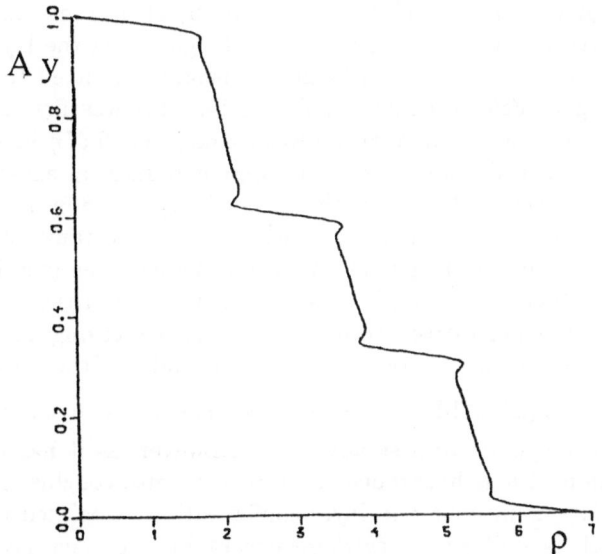

Fig. 3.9. Vertical profile of the density ϱ at $x = 0.5$ and $t = 4.015$

constant) gradient and exhibits a boundary layer at both vertical walls (the temperature is kept fixed at these walls). The vertical profile of the resulting density is shown in Fig. 3.9. Finally, the motion of the fluid in each layer is illustrated by the vertical profile (at $x = 1/2$) of the horizontal velocity u (Fig. 3.10). At time $t = 4.015$, the streamfunction varies between -0.17 and 13.85, and the vorticity between 5.41×10^2 and 1.87×10^3.

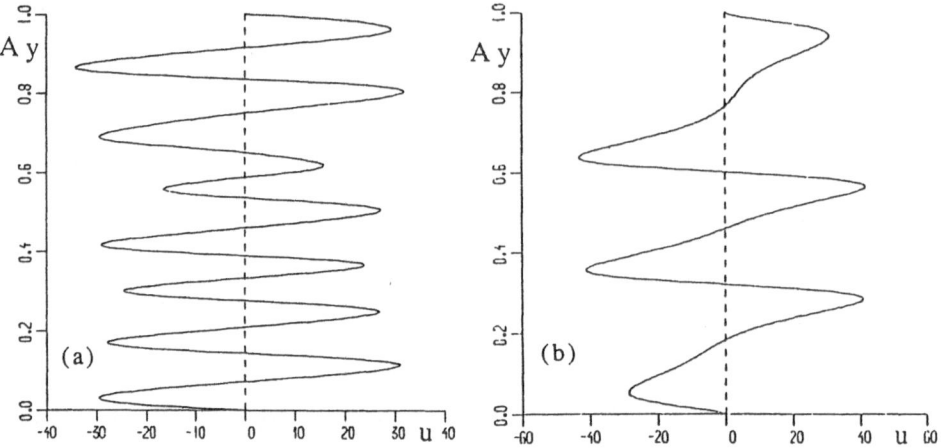

Fig. 3.10. Vertical profile of the horizontal velocity u at $x = 0.5$: (a) $t = 0.556$, (b) $t = 4.015$

An important question which has been the object of numerous discussions in the literature is the dependence of the height h_c of the layers on the physical parameters. However, this height is usually admitted to be of the order of the length scale η defined in Eq. (3.1). This was first established in [21] where h_c was found to vary between 0.68η and 0.97η in laboratory experiments using a wide tank. From the present numerical experiments, it is found that the average height of the layers is $h_c = 0.86\eta$ at the end of the first phase of the motion (Fig. 3.3) when the whole tank has been filled with layers. Then, the height of the layers increases as merging occurs until the three–cell configuration is attained. At the end of the computation ($t = 4.015$) the time–derivatives were continuously decreasing and relatively small. For example, the derivative of the maximum value of the vorticity satisfies $(\dfrac{\partial max(\omega)}{\partial t}/max(\omega) \simeq 10^{-3}$. Therefore, we may think that the three–cell configuration corresponds to a steady state. However, as it has been pointed out in Section 2, it is hazardous to draw a definite conclusion. Another possible final state could be a one–layer configuration associated with a concentration which is uniform everywhere except in boundary layers at the horizontal walls. In laboratory experiments (e.g. [27]) conducted in a tank with impermeable walls, the number of cells continuously decreases as time is increasing until a one–cell state is obtained with a completely uniform concentration. The same behaviour would be numerically obtained by prescribing the concentration flux equal to zero on the whole boundary.

Our numerical experiments, even those conducted with a smaller aspect ratio ($A = 1/6$), did not exhibit the simultaneous formation of the layers along the heated wall as found in the laboratory experiments [20], [21], [22]. As a matter of fact, there exists two concurrent mechanisms responsible for the creation of cells: the first one is the sequential phenomenon due to the presence of horizontal walls above described and the other is an instability mechanism leading to the simultaneous occurence of cells. This instability needs some time to develop so that, if the ratio H/η is too small, the whole tank will be filled by cells created by the first mechanism before the instability could manifest. This is likely the situation of the present numerical calculations. A careful examination of the streamlines during the first phase of the motion (Fig. 3.2a) shows, in the region still free of cells at $t = 0.188$, some dotted lines ($\psi = 0$) delimiting the location of the future cells visible at $t = 0.3$. However, it is hard to decide if this pattern is the result of an instability or the influence of the neighbouring layers. Further numerical experiments performed in a much higher tank would be necessary. It is interesting to note that some numerical studies assuming periodicity in the vertical direction [21], [37] are able to simulate the instability above mentioned.

4. Conclusion

Numerical experiments in double–diffusive convection have been performed by means of spectral Chebychev methods. These experiments illustrate the interest of using highly accurate numerical methods for calculating complex flow patterns arising in double diffusion. We think that such kind of flows taking place inside simple geometries but rather sensitive to the errors associated with the numerical approximation constitutes the ideal field of application of the spectral methods. However, the presence of inner thin regions where some quantities, like the concentration, strongly vary requires the consideration of a large number of collocation points to accurately represent these steep gradients. This problem can be removed by using a domain decomposition method [40], [38], possibly supplemented by a local self–adaptive coordinate transform [39]. However, the implementation of such a combination remains quite delicate. On the other hand, the domain decomposition method for handling the difficulty associated with the representation of thin inner layers in double-diffusive convection has been used with success in various physical situations, namely effect of a hot spot [40], [41] or heat source [42] in a stratified layer of fluid and effect of diffusion in a two–layer configuration [43].

Acknowledgment

The authors wish to thank A. Mofid for his help in the preparation of the manuscript. The calculations have been done on the CRAY computer of CCVR.

List of Symbols

A	Aspect ratio : L/H
d	Characteristic length
g	Acceleration of gravity
h_c	Height of layer
H	Height of the tank
L	Width of the tank
N_S	Solutal Nusselt number at lower wall
N_T	Thermal Nusselt number at lower wall
Pr	Prandtl number : ν/κ_T
Ra	Thermal Rayleigh number based on η : $g\alpha\Delta\bar{T}\eta^3/(\kappa_T\nu)$
R_S	Solutal Rayleigh number : $g\beta\Delta\bar{S}d^3/(\kappa_T\nu)$
R_T	Thermal Rayleigh number : $g\alpha\Delta\bar{T}d^3/(\kappa_T\nu)$
r	Rising height ratio : η/L
S	Concentration
$\bar{S}_r$	Concentration in reference state
t	Time
T	Temperature
$\bar{T}_r$	Temperature in reference state
u	Horizontal velocity component
v	Vertical velocity component
$\mathbf{V}$	Velocity vector
x	Horizontal coordinate
y	Vertical coordinate
α	Thermal expansion coefficient
β	Solutal expansion coefficient
$\Delta\bar{S}$	Change in concentration : $\bar{S}_1 - \bar{S}_0$
$\Delta\bar{T}$	Change in temperature : $\bar{T}_1 - \bar{T}_0$
Δt	Time–step
κ_S	Solutal diffusivity
κ_T	Thermal diffusivity
η	Potential rising height : $\alpha\Delta\bar{T}/(\beta|\bar{\phi}_0|)$
ν	Kinematic viscosity
ϱ	Density
$\bar{\varrho}_r$	Density in reference state
τ	Diffusivity ratio : κ_S/κ_T
$\bar{\phi}_0$	Initial concentration gradient : $(\bar{S}_1 - \bar{S}_0)/H$

References

1. J.S. Turner, "Double–diffusive phenomena", Ann. Rev. Fluid Mech., **6**, 37–56 (1974).
2. J.S. Turner, "Multicomponent convection", Ann. Rev. Fluid Mech., **17**, 11–44 (1985).
3. J.S. Turner, "The behaviour of a stable salinity gradient heated from below", J. Fluid Mech., **33**, 168–200 (1968).
4. T.G.L. Shirtcliffe, "An experimental investigation of thermosolutal convection at marginal stability", J. Fluid Mech., **35**, 677–688 (1969).

5. H.E. Huppert and P.F. Linden, "On heating a stable salinity gradient from below", J. Fluid Mech., **95**, 431–464 (1979).

6. G. Veronis, "Effect of a stabilizing gradient of solute on thermal convection", J. Fluid Mech., **34**, 314–336 (1968).

7. E.K. Knobloch, D.R. Moore , J. Toomre and N.O. Weiss, "Transition to chaos in two–dimensional double–diffusive convection", J. Fluid Mech., **166**, 409–448 (1986).

8. E.K. Knobloch, M.R.E. Proctor and N.O. Weiss, "Heteroclinic bifurcations in a simple model of double–diffusive convection", J. Fluid Mech., **239**, 273–292 (1992).

9. H.E. Huppert and D.R. Moore, "Nonlinear double–diffusive convection", J. Fluid Mech., **78**, 851–854 (1976).

10. S.M. Chang, S.A. Korpela and Y. Lee, "Double–diffusive convection in the diffusive regime", Appl. Sci. Res., **39**, 301–319 (1982).

11. U. Ehrenstein and R. Peyret, "A Chebyshev collocation method for the Navier–Stokes equations with application to double–diffusive convection" , Int. J. Numer. Methods Fluids, **9**, 427–452 (1989).

12. B. Roux (ed), "Numerical Simulation of Oscillatory Convection in Lower–Pr Fluids", Proc. GAMM Workshop, Marseille (1988), Vieweg, Braunschweig, 1990.

13. P. Le Quéré, "Transition to unsteady natural convection in a tall water–filled cavity", Phys. Fluids, A **2**, 503-515 (1990).

14. J. Ouazzani, R. Peyret and A. Zakaria, "Stability of collocation–Chebyshev schemes with application to the Navier–Stokes equations", *in* D.Rues and W. Kordulla (eds), Proc. Sixth GAMM Conf. on Numerical Methods in Fluid Mechanics, Vieweg, Braunschweig, 1986, pp. 287–294.

15. J. Fröhlich, T. Gerhold, J.M. Lacroix and R. Peyret,"Fully implicit spectral methods for convection", *in* M. Durand and F. El Dabaghi (eds), High Performance Computing II, North–Holland, Amsterdam, 1991, pp. 585-596.

16. J.M. Vanel, R.Peyret and P. Bontoux, "A pseudo–spectral solution of vorticity-stream function equations using the influence matrix technique", *in* K.W. Morton and M.J. Baines (eds), Numerical Methods for Fluid Dynamics II, Clarendon Press, Oxford, 1986, pp.463–475.

17. D.B. Haidvogel and T. Zang, "The accurate solution of Poisson's equation by expansion in Chebyshev polynomials", J. Comput. Phys., **30**, 167–180 (1979).

18. U. Ehrenstein, "Méthodes spectrales de résolution des équations de Stokes et de Navier–Stokes. Application à des écoulements de convection double–diffusive", Thèse Doctorat, Mathématiques Appliquées, Université de Nice, 1986.

19. R. Bwemba and R. Pasquetti, "About the influence matrix used in the spectral solution of the 2D incompressible Navier–Stokes equations (vorticity-stream function formulation)", to appear.

20. S.A. Thorpe, P.K. Hutt ans R. Soulsby, "The effect of horizontal gradients on thermohaline convection", J. Fluid Mech., **38**, 375–400 (1969).

21. R. A. Wirtz, D.G. Briggs and C.F.Chen, "Physical and numerical experiments on layered convection in a density stratified fluid", Geophys. Fluid Dyn., **3**, 265–288 (1972).

22. V.N. Nekrasov, V.A. Popov and Yu.D. Chashechkin, "Formation of periodic convective–flow structure on lateral heating of a stratified liquid", Izv. Atmosph. Ocean. Phys., **12**, 1191-1200 (1976), English transl. pp.733-739.

23. Y. Suzukawa and U. Narusawa, "Structure of growing double–diffusive convection cells", J.Heat Transfer, **104**, 248-254 (1982).

24. J. Tanny and A.B. Tsinober, "The dynamics and structure of double–diffusive layers in sidewall-heating experiments", J. Fluid Mech., **196**, 135-156 (1988).

25. J. Tanny and A.B. Tsinober, "On the behavior of a system of double diffusive layers during its evolution", Phys. Fluids, **A1**, 606–609 (1989).
26. J.E. Hart, "On sideways diffusive instability", J. Fluid Mech., **49**, 279–288 (1971).
27. J.E. Hart, "Finite amplitude sideways diffusive convection", J. Fluid Mech., **59**, 47–64 (1973).
28. O. Kerr, "Heating a salinity gradient from a vertical sidewall : linear theory", J. Fluid Mech., **207**, 323–352 (1989).
29. O. Kerr, "Heating a salinity gradient from a vertical sidewall : nonlinear theory", J. Fluid Mech., **217**, 529–546 (1990).
30. C.F. Chen, "Onset of cellular convection in a salinity gradient due to a lateral temperature gradient", J. Fluid. Mech., **63**, 563-576(1974).
31. S. Thangam, A. Zebib and C. F. Chen, "Double–diffusive convection in an inclined fluid layer" , J. Fluid Mech., **116**, 363–378 (1982).
32. R. A. Wirtz and L. H. Liu, "Numerical experiments on the onset of layered convection in a narrow slot containing a stably stratified fluid", Int. J. Heat Mass Transfer, **18**, 1299-1305 (1975).
33. C.S. Reddy, "Cell merging and its effect on heat transfer in thermosolutal convection ", J. Heat Transfer, **102**, 172-174 (1980).
34. Y. Demay, J.M. Lacroix, R. Peyret and J.M. Vanel, "Numerical experiments on stratified fluids subject to heating", Proc. Third Inter. Symp. on Stratified Flows, Pasadena, 1987, E.J. List and G.H. Jirka (eds), American Society of Civil Engineers, New York, 1990, pp. 588-597.
35. H.C. Ku, R.S Hirsh and T.D. Taylor, "A numerical simulation of the effect of salinity on a thermally driven flow", *in* M. Deville (ed), Proc. Seventh GAMM Conf. on Numerical Methods in Fluid Mechanics, Vieweg, Braunschweig, 1988, pp. 151–158.
36. C.E. Mendenhall and M. Mason, "The stratified subsistence of fine particles", Proc. Nat. Acad. Sci. USA, **9**, 199-207 (1923).
37. C. Sabbah, "Etude numérique de la formation et l'évolution de cellules convectives dans un fluide stratifié chauffé latéralement", Projet de stage de DEA "Turbulence et Systèmes Dynamiques", Université de Nice–Sophia Antipolis, 1994.
38. J.P. Pulicani, "A spectral multi–domain method for the solution of 1–D Helmholtz and Stokes–type equations", Computers and Fluids, **16**, 207–215 (1988).
39. R. Peyret, "The Chebyshev multidomain approach to stiff problem in Fluid Mechanics", Comp. Meth. Appl. Mech. Eng., **80**, 129-145 (1990).
40. H. Guillard and R. Peyret," On the use of spectral methods for the numerical solution of stiff problems", Comp. Meth. App. Mech. Eng., **66**, 17–43 (1988).
41. J.M. Lacroix, R. Peyret and J.P. Pulicani, "A pseudospectral multidomain method for the Navier–Stokes equations with application to double–diffusive convection", *in* M. Deville (ed), Proc. Seventh GAMM Conf. on Numerical Methods in Fluid Mechanics, Vieweg, Braunschweig, 1988, pp. 167-174.
42. A. Mofid, "Application des méthodes spectrales à l'étude de l'effet d'une source de chaleur dans un fluide stratifié", Thèse de Doctorat, Sciences de l'ingénieur, Université de Nice-Sophia Antipolis, 1992.
43. J. P. Pulicani, "Modélisation isotherme d'une interface diffusive en convection de double–diffusion", Int. J. Heat Mass Transfer (to appear).

New Potentialities of Computational Experiment in Tsunami Problem

Yu.I. Shokin, G.S. Khakimsyanov, and L.B. Chubarov

Institute of Computational Technologies, Siberian Branch of the Russian Academy of Science, pr. Lavrentyeva 6, Novosibirsk 630090, Russia

1. Introduction

Computational experiment in tsunami problems is specified first of all by the multifactor physical process including the generation of the initial pertubation (Tsunami source) resulting from an underwater earthquake, volcano eruption or another similar large-scale hazard, propagation of the wave in deep ocean and its transformation in the coastal zone, interaction with floating and fixed objects and its running up on the shore.

The applied tsunami problems differ in content. One of them is associated with carrying out of apriori investigations, such as the coast zoning according to vulnerability, parameters calculation for local, regional and global tsunami warning systems, preparation of maps of inverse isochrones, scenario calculations. Another group of problems is determined by the needs of on-line forecasts and monitoring of tsunami and involves the development of mathematical models, algorithms and applied software for the reception and processing of continuously incoming information as well as for the expert assessment of the current situation in the ocean, development of the tsunami service regulations and designing the text forms of specially issued messages.

2. Basic Mathematical Models

Transformation modelling of long surface gravitational waves is mostly carried out within the scope of shallow water theory. Depending on the intervals of characteristic parameter change, different approximations of this theory are to be considered such as linear, non-linear and non-linearly dispersive.

Very important is the construction of a hybrid mathematical model which can describe the wave during its propagation from the pertubation source to the coast with automatic adjustment to a specific algorithm according to the current wave characteristics and the zone of the water area where it is propagating.

Such a model must include approximated hydrodynamic shallow water models and complete hydrodynamic equations fully allowing for vertical displacements of fluid.

The construction of an efficient hybrid mode requires the development of the selection criteria of its individual components, algorithms of joining the solutions and boundary conditions describing the entire variety of simulated physical phenomena (open boundaries, interaction with floating and fixed objects, running of the waves up the shore). At the same time we must proceed from the reliability requirements to the wave processes reproduction as well as from time and effort saving computational algorithms which must make it possible to carry out apriori and on-line modelling.

Dispersion effects and their interaction with non-linearity effects can be corectly allowed for by means of the models whose dispersive properties are retained within a wide range of wavelengths changes and the notation and type of the output assertions which enable one to employ efficient stable algorithms in a two-dimensional (planned) statement.

In parametric form the equations of the shallow water theory, in particular the non-linearly dispersive Aleshkov model, can be written as follows:

$$
\begin{aligned}
\eta_t + \nabla \cdot \left[(h + r\eta) u \right] &= s \left\{ (u \cdot \nabla h) \left| \nabla h^2 \right| + h \left[\left| \nabla h^2 \right| (\nabla \cdot u) + \right. \right. \\
&\quad + 2\nabla h \cdot \nabla (u \cdot \nabla h) + (u \cdot \nabla h) \Delta h \right] + \\
&\quad + h^2/2 \left[\Delta (u \cdot \nabla h) + 2\nabla h \cdot \nabla (\nabla \cdot u) \right] + \quad (2.1) \\
&\quad \left. + (h^3/6) (\Delta (\nabla \cdot u)) \right\}, \\
u_t + \nabla \left(g\eta + \frac{r}{2} u^2 \right) &= \frac{s}{2} \nabla \left[(u \cdot \nabla h)^2 + \nabla \cdot (h^2 u_t) \right],
\end{aligned}
$$

where η is the deviation of the free surface from its unpertubated level, h stands for the depth of unpertubated fluid level, u designates the vertically averaged velocity components, $(\Delta), (\nabla)$ are the two-dimensional Laplace and gradient operators in the (x, y)-plane respectively, g is the acceleration of gravity, r and s denote parameters such that:

- $r = 0$, $s = 0$ is a linear model;
- $r = 0$, $s = 1$ is a linearly dispersive model;
- $r = 1$, $s = 0$ is a non-linear model;
- $r = 1$, $s = 1$ is a non-linearly dispersive (NLD) model.

Let us dwell upon some unusual non-linearly dispersive models of shallow water expanding the class of solvable problems.

Thus *Urusov, Shokin* [1] suggested a NLD model within which the interaction of waves with a fixed partially submerged body can be described. Equations of this model are the consequences of the Eqs (2.1) under the assumption of the smallness of the wave amplitude:

$$
\begin{aligned}
\eta_t + \nabla (H\nabla\phi) &= \frac{\mu}{2} \nabla \cdot \left\{ \nabla \left(h^2 \nabla\phi \cdot \nabla h \right) + h^2 \nabla h \Delta\phi + \right. \\
&\quad \left. + \frac{h^3}{3g} \nabla \left[\frac{\phi_{tt} - \nabla h \cdot \nabla\phi}{h} \right] \right\} + O(\varepsilon\mu, \mu^2), \\
0 &= \left(\phi - \frac{\mu}{2} \nabla \cdot \left(h^2 \nabla\phi \right) \right)_t + g\eta + \quad (2.2)
\end{aligned}
$$

$$\frac{\varepsilon}{2}|\nabla\phi|^2 + O(\varepsilon\mu,\mu^2), \quad (x,y) \notin \omega,$$

$$\nabla \cdot (d\nabla\psi) = \mu\nabla \cdot \left\{ d\nabla d\,(\nabla\psi \cdot \nabla d) + \frac{d^2}{2}\left[(\nabla d\Delta\psi) + \nabla(\nabla\psi \cdot \nabla d)\right] + \frac{d^3}{6}\nabla(\Delta\psi)\right\} + O(\mu^2), \quad (x,y) \in \omega.$$

From here on we shall use mostly the same notation. We also take that ϕ designates the potential value at the bottom of the pool outside the partially submerged body, ψ denotes the same but in the area ω, i.e. under the body, $H = h + \eta$ is the total depth, $d = d(x,y)$ is the height of the gap between the bottom of the water area and the bottom of the body, A_0 and H_0 denote the characteristic wave amplitude and the depth of the water area, L_0 is the characteristic horizontal dimension, μ and ε parameters such that $\mu = (H_0/L_0)^2, \varepsilon = A_0/H_0$.

The conjunction conditions on the areas' interfaces result from the requirement of potential and pressure continuity.

The suggested equations ensure the approximation with an accuracy of $O(\varepsilon\mu,\mu^2)$. In model derivation we assumed also that the relation between the non-linearity and dispersion parameters was $\varepsilon/\mu = O(1)$.

The formulation of the Eqs (2.1) in terms of the wave height (η) and potential (ϕ,ψ) lowers the order of space derivatives entering into these equations and makes it easier to construct computational algorithms.

A one-dimensional model analogue of model (2.2) was successfully employed in the solutionof some test and simulation problems [2].

The means of mathematical simulation for on-line tsunami warning service deal mostly with the determination of some kinematic characteristics such as expected tsunami travelling time from the epicentre of the tsunamigenic earthquake to the protected points.

Appropriate mathematical models can be based either on the modified Huygen's principle or on the eikonal euation [3]. Algorithms implementing such a model were used for the calculation of the maps of tsunami propagation times in the Pacific Ocean, the error was obtained by the comparison of calculated and observed times and did not exceed 2-3 minutes per hour of the wave propagation.

At the same time it is vitally important to work out algorithms for the on-line calculation of the dynamic characteristics such as tsunami heights and velocities. Such algorithms must be very efficient, i.e. they must be able to simulate the wave propagations over transoceanic distances in the on-line operation mode ahead of the real wave propagation and allow for the non-linearity and dispersion effects which become the crucial factor under the description of the global-scale wave processes.

This accounts for the wide use of NLD models which, unfortunately, are realized by means of awkward and cumbersome computational algorithms which hinder the on-line operation. One of the possible ways out would be making use of an approximated NLD model suggested in [4]:

$$\eta_{tt} - g\nabla \cdot (h\nabla\eta) = -f^2\eta + \frac{1}{3}\nabla \cdot (h\nabla(h\eta_{tt})) + \frac{2}{3}g\nabla \cdot \left(h\nabla\left(\frac{\eta^2}{h}\right)\right), \quad (2.3)$$

where f is the Coriolis parameter.

In this work so called absorbing boundary conditions are suggested that enable one to simulate the wave propagation along the route from the source to the coast by means of the "running window". Making use of the model (2.3) is reduced to the solution of a single equation for η which, nevertheless, provides a means for taking into account various dispersion factors such as the effect of the rotation of the Earth, non-linear and frequency dispersion.

Note also the possibilities of the NLD model construction which allows for the influence of the water area bottom porosity on the wave process.

A one-dimensional analogon of such a model suggested in [5] has the following form:

$$u_t + \left(\frac{u^2}{2}\right)_x + g\eta_x = \left[\left(\frac{h^2 u}{2}\right)_{xx}\right]_t$$

$$\eta_t + [(h+\eta)u]_x = \frac{1}{\rho}[h_x u_t - (H_0 - h)u_{xt}] + \left(\frac{h^3}{6}u\right)_{xxx} \quad (2.4)$$

where $\rho = \nu/\aleph$, ν is the kinematic viscosity and $\aleph$ denotes the permeability of the porous medium.

Non-linearity of the wave propagation process is mostly manifested when the wave approach the shore. Therefore when carrying out computational experiments it is necessary to have also more accurate models for the coastal wave along with the already described mathematical models making it possible to obtain more accurate calculation results allowing for significant vertical displacements of the fluid, bottom topography, real water-line geometry and coastal structures as well as the presence of submerged or semi-submerged bodies. The correction is achieved through making use of non-linear boundary conditions on the free surface, allowing for three-dimensionality of the flow, curvilinear moving coordinate systems employed for the description of the flow.

The most general model we employed in the coastal zone was the mathematical model of vortex flows of ideal incompressible free-surface fluid [6]. Under additional assumptions one can also experiment with non-linear and linear models of potential flows. For each of these models there are program realizations of calculation algorithms for 3D, plane and axially symmetrical flows. On the side boundaries the conditions of impermeability, free passage of waves or the conditions of wave generation are imposed.

Under numerical solution the moving domain $\Omega(t)$ occupied by the fluid is mapped onto the stationary calculated domain Q of a simple form which is either a single cube or a part of a single cube composed of rectangular parallelepipeds (for two-dimensional flows Q is a single square or part of this square made up of rectangles). Such mapping is performed on the basis of one-to-one coordinate transformation consistent with the boundary $\Gamma = \partial\Omega$

$$t = q^0, \quad x^\alpha = x^\alpha(q^\beta, q^0), \quad \alpha, \beta = 1, 2, 3, \tag{2.5}$$

where x^α are initial coordinates and q^β denote curvilinear coordinates.

The mathematical statement of the problem of vortex fluid flows consists in the determination of the vector potential ψ and the vortex vector ω which must satisfy the imposed initial conditions, boundary conditions on ∂Q and in Q the equations

$$\sum_{\alpha,\gamma=1}^{3} \frac{\partial}{\partial q^\alpha} \left(k_{\alpha\gamma} \frac{\partial \psi^\beta}{\partial q^\gamma} \right) = -J\omega^\beta, \quad \beta = 1, 2, 3, \tag{2.6}$$

$$\sum_{\alpha,\gamma=1}^{3} \frac{\partial}{\partial q^\alpha} \left(k_{\alpha\gamma} \psi_\gamma \right) = 0, \tag{2.7}$$

$$\frac{\partial J\omega^\beta}{\partial t} + \sum_{\alpha=1}^{3} \frac{\partial}{\partial q^\alpha} J \left(v^\alpha \omega^\beta - v^\beta \omega^\alpha \right) = 0, \quad \beta = 1, 2, 3, \tag{2.8}$$

whose coeficcients $k_{\alpha\beta}$ are expressed in terms of the metric tensor components and Jacobian J of the transformation, v^α, ω^α are the contravariant components of the velocity vector and the vector $\omega, \psi^\beta, \psi_\beta$ are contravariant and covariant components of the vector potential.

In the case of potential flows instead of eqs. (2.6) - (2.8) we have one nonlinear equation for the velocity potential ϕ ([7])

$$\sum_{\alpha,\gamma=1}^{3} \frac{\partial}{\partial q^\alpha} \left(k_{\alpha\gamma} \frac{\partial \psi}{\partial q^\gamma} \right) = 0, \tag{2.9}$$

related to the covariant velocity components by the equalities

$$\nu_\gamma = \frac{\partial \phi}{\partial q^\gamma}, \quad \gamma = 1, 2, 3.$$

The considered problems can be solved by finite-difference method of calculation on a curvilinear grid adjusted to the flow interface and the singularities of the solutions. The developed numerical method was used for the study of the processes of the waves running up on the shore and on the coastal constructions of a complex geometrical configuration, for the investigation of waves transformation over irregularities of the bottom, wave interaction with floating or submerged objects.

3. Methodological Principles of Some Applied Problems of Tsunami by Computational Experiment

3.1 Apriori Zoning of the Coast

It is known that apriori zoning of the coast with respect to the vulnerability to hazardous waves is necessary to determine the location, intensity and main pecularities of the mechanism of pertubation sources, to analyse the quantitative characteristics of the waves at individual coastal points. The natural input data in this case will be historical data and the measurement results from the level whose interpretation may result in the long-term local forecast for the points of reference, i.e. for the given period of a forecast t_* the value of the level rise h_{t_*}:

$$h_{t_*}(x_i) \quad = \quad F_1(x_i; t_*, k(x_i), A, T),$$

where the coefficients A and T are "regional" parameters that determine the wave behavior in shallow water: A denotes the recurrence of major events in the studied region, T stands for the predominant wave period; the coefficient $k(x_i)$ is associated to the route of the pertubation propagation and characterizes the local singularities of the protected point x_i location. Then the problem of apriori zoning is reduced to the calculation of the function

$$H_{t_*}(x_i, x_j) \quad = \quad H_{t_*}(F_1, K(x_j))$$

along the investigated coast at the points x_j with respect to "reference" values $h_{t_*}(x_i)$ where $K(x_i)$ is the recount coefficient from point to point alomg the coast.

The computational experiment in this case consists in the assessment of the above mentioned fundamental parameters and obtaining the functions F_1, i.e. building up the model of apriori forecast.

The series of computational experiments based on the above considered mathematical models involves scenario calculations of the wave propagation from the reference and hypothetical (critical) source and simulation of real events. The finite result in this case is scarcely affected by the input conditions because in the prognostic model the so-called "effective" wave source is considered whose parameters in the deep water do not change much from point to point, i.e. are stable characteristics within the assumed prognostic model. The accuracy of the estimated coefficients $k(x_i)$ and $K(x_j)$ is directly related to the employed mathematical models and numerical algorithms.

3.2 On-line Assessment of the Hazardous Wave Parameters

The second group of problems connected with the calculation of the safety criteria of particular objects depends on the requirements of the on-line real-time monitoring of the hazardous process. The suggested approach involves the organization of a rational location scheme of deep-water hydrophysical sensors for early detection of dangerous waves, the development of special methods for reliable and timely analysis of the recorded data, their interpretation in prognostic terms.

Therefore, local on-line forecast constists in the estimated height of the level rise h_{t_*} at the protected point x which is obtained from the observation data of the wave characteristics on the local system sensor.

$$h_{t_\alpha}(x) = F_2(x; t_\alpha, k(x), \alpha, \tau),$$

where t_α is the so-called "good time" parameter which determines the minimum time required for descision-making and taking measures of protection. The parameters α and τ are the wave amplitude and period registered by the sensor respectively; $k(x)$ is the coefficient of the wave transformation from the sensor to the protected point at point x.

In solving this problem it is possible to avoid uncertainty associated with the pertubation source if we take the wave characteristics which are stable in the deep water (α, τ) from the sensor measurements. The coefficient $k(x)$ characterizes the local conditions and specific manifestations of the wave in the vicinity of the procted point.

The prognostic model F_2 built as described above whose reliability is determined by the algorithmic support of the numerical experiment and the completeness of the natural data includes nomograms for the estimate of the probable travelling time to the flooded zones at the protected point.

4. Computational Experiment Facilities Employed for the Training of the Tsunami Warning Service Personnel and of the Population in the Threatened Zones

The qualification requirements to the tsunami warning services and similar services intented for warning the population in the case of natural and man-made hazards as well as to their equipment for monitoring and investigation of hazardous processes are becoming greater and greater. This necessitated the development of software systems intented for scenario calculations illustrating the possible variants of catastrophic events depending on the characteristics of a physical phenomena, specific features of the protected points and the operation of the local administration; professional retraining of the special services personnel.

Training programs (TP) development in the considered fields involving rather complex phenomena and accordingly non-trivial mathematical models and computational algorithms are of special interest.

The TP structure involves the mathematical simulation facilities, interface, training, analysis of results and work with informational bases. TPs due to their specific functions require allowing for the psychological pecularities of the user, his interaction with TP which makes them significantly differ from the applied packages (AC) with respect to the selection of mathematical models, computational algorithms as well as in the selection of means and ways of their program implementation. A problem naturally is to be solved during a not too long PC session. This necessitates the selection of simple mathematical models retaining, at the same time, reliability of the reproduction of essential characteristics of a phenomenon, making use of computational algorithms which can operate within a wide range of input data and model parameters.

We must also name such specific features of the TPs as some forced limitation of the input data range due to the requirements of guaranteed stability and accuracy of calculations, adjustment to particular objects of the user's professional activity and the strict discipline of the control on order to train the necessary skills and habits of working with applied programs and obtaining meaningful results.

The authors made use of the above consideration in the development and operation of TP "Tsunami" which ensures successful study of hydrodynamic problems associated with tsunami, peculiar features of the wave behavior in the protected zones within the responsibility of specific services, possibilities of prevention and protection from the hazardous consequences of tsunami.

Acknowledgement

The investigation has been supported by the Foundation of Fundamental Research of the Russian Federation.

References

1. A.I. Urusov, Yu.I. Shokin: *On the Modelling of the interaction of Long Surface Water Waves with the Partly Submerged Body*, Proceedings of the All-Union Symposium on Numerical Methods in Wave Hydrodynamics, Krasnoyarsk, 1991 (in Russian), pp. 33-40
2. A.I. Urusov, Yu.I. Shokin: *Numerical Study of the Surface Water Waves Run-up on partly Submerged Obstacles Based on the Non-Linear-Dispersive Shallow-Water Model* , Proceedings of the All-Union Symposium on Numerical Methods in Wave Hydrodynamics, Krasnoyarsk, 1991 (in Russian), pp. 33-40
3. Y.I. Shokin, L.B. Chubarov, V.A. Novikov, A.N. Sudakov: *Calculation of Tsunami Travel Time Charts in the Pacific Ocean*, Vol. 5, No. 2, Models, Algorithms, Techniques, Results Tsunami Hazards, 1987, pp. 85-113

4. K.Y. Kim, R.O. Reid, R.E. Whitaker: *On an Open Radiational Boundary Condition for Weakly Dispersive Tsunami Waves*, J. Comput. Phys., Vol. 76, pp. 327-348, 1988
5. G. Flaten, O.B. Rygg: *Dispersive Shallow Water Waves over a Porous Sea Bed*, Coastal Engineering, Vol. 15, pp. 347-369, 1991
6. Y.I. Shokin, G.S. Khakimzyanov: *Finite-difference Method for Potential and Vortex Fluid Flows with a Free Surface*, Proc. of the 5th International Symposium on Computational Fluid Dynamics, Vol. 3, Sendai, Japan, 1993, pp. 121-128
7. Y.I. Shokin, G.S. Khakimzyanov: *On Numerical Calculation of Ideal Incompressible Fluids*, Proc. of the International Symposium on Computational Fluid Dynamics, Nagoya, Japan, 1989, pp. 408-417

General Balance Equations for a Fluid-Fluid Interface in Magnetofluiddynamics

Soubbaramayer

Département des Procédés d'Enrichissement CEA-Saclay
91191 Gif sur Yvette Cédex, France

Summary. The boundary conditions on the interface separating two magneto-fluids in motion are derived. As an application the surface depression in a high current arc weld pool is calculated and compared with experimental results.

1. Introduction

The purpose of this communication is to derive the general conservation laws (for quantities such as mass, momentum, energy, electric and magnetic fields) at a deforming interface separating two magnetofluids in motion. We mean by magnetofluid an electrically conducting fluid interacting with an electro-magnetic field, applied from exterior or self-induced. We refer to Cabannes (1970) for the theoretical mechanics of such fluids. In the current literature, the fluid-fluid interface equations are well studied in the case of neutral fluids, i.e. electrically non-conducting and not interacting with an electro-magnetic field (see for instance Gatignol (1986) which contains also a comprehensive bibliography). The present extension to MFD (MFD is the usual acronym of magnetofluiddynamics) is motivated by theoretical interest, but aims also at practical applications like arc-welding, electrowinning of aluminium, etc. ... The book of Szekely (1979) develops the industrial applications of MFD in metals processing. Another important application is pertinent to the devices of crystal growth from the melt using applied magnetic field (Müller 1988). In all these problems the boundary conditions on the interface play a key-role, and the full set of balance equations derived here are more complete than the simplified boundary conditions used by the authors in the current literature.

2. MFD Model: Integral Conservative Form

The basic MFD model includes the equations of dynamics, the Maxwell equations of electromagnetism, the phenomenological OHM's law and a thermodynamic state law. The whole set is presented and commented hereafter (Eqs. (2.1) to (2.8)).

The equations of dynamics (2.1) to (2.3) are written in the integral form as in Cabannes (1970). We denote by $\frac{d}{dt}$ the hydrodynamic derivative, (w) a control volume bounded by the surface (s), ρ the specific mass, $d\tau$ the element of volume, V the velocity vector of the fluid, J the electric current, B the magnetic field, T the stress tensor (including the pressure and the viscous stress), n the unit vector directed along the outward normal to the surface (s), ds the element of surface, ϵ_i the specific internal energy, E the electric field and θ the heat flux.

$$\frac{d}{dt}\iiint_{(w)} \rho \, d\tau = 0 \tag{2.1}$$

$$\frac{d}{dt}\iiint_{(w)} \rho V \, d\tau = \iiint_{(w)} J \times B \, d\tau + \iint_{(s)} T \cdot n \, ds \tag{2.2}$$

$$\frac{d}{dt}\iiint_{(w)} (\rho \, \epsilon_i + \frac{1}{2}\rho V^2) d\tau =$$

$$\iiint_{(w)} E \cdot J \, d\tau + \iint_{(s)} (T \cdot n) \cdot V \, ds - \iint_{(s)} \theta \cdot n \, ds \tag{2.3}$$

$$\Delta \cdot B = 0 \tag{2.4}$$

$$\Delta \times E = -\frac{\partial B}{\partial t} \tag{2.5}$$

$$\Delta \times B = 4\pi(J + J_D), \qquad J_D = \frac{1}{4\pi c^2}\frac{\partial E}{\partial t} \tag{2.6}$$

$$\Delta \cdot E = 4\pi c^2 q \tag{2.7}$$

$$J = \sigma(E + V \times B) \tag{2.8}$$

The equations of electromagnetism (2.4) to (2.7) are written in emu. The displacement current J_D is negligible in most of the practical cases. In Eq. (2.7) q denotes the net electrical charge density and is generally unknown. The system is completed by OHM's law (Eq. (2.8)), where σ is the electrical conductivity.

Our next step is to transform the preceding equations into the integral conservative form of the general type

$$\frac{d}{dt}\iiint_{(w)} A \, d\tau + \iint_{(s)} \alpha \cdot n \, ds = 0 \tag{2.9}$$

where A is a vector and α a tensor. Bringing Eqs. (2.2) and (2.3) to the form (2.9) requires the transformation (into surface integrals) of the two volume integrals

$$\iiint_{(w)} J \times B \, d\tau \quad and \quad \iint_{(w)} E \cdot J \, d\tau. \tag{2.10}$$

The transformation of the first one has been carried out by Cabannes (1970) and results in:

$$\iiint_{(w)} J \times B \, d\tau = \iint_{(s)} T_B \cdot n \, ds \tag{2.11}$$

$$T_B = \frac{1}{4\pi}(B,B) - \frac{B^2}{8\pi}\, U. \tag{2.12}$$

In Eq. (2.12) the dyadic (B,B) is the Maxwell stress, U the unit tensor and $B^2/(8\pi)$ the magnetic pressure. Regarding the second integral in Eq. (2.10), we will make use of the Poynting theorem

$$E \cdot J = -\Delta \cdot P_G - \frac{\partial \epsilon_m}{\partial t} \tag{2.13}$$

where P_G is the Poynting vector and ϵ_m the energy of the electromagnetic field

$$P_G = \frac{E \times B}{4\pi}, \qquad \epsilon_m = \frac{B^2}{8\pi} + \frac{E^2}{8\pi c^2}. \tag{2.14}$$

Integrating Eq. (2.13) results in:

$$\iiint_{(w)} E \cdot J \, d\tau = -\iint_{(s)} P_G \cdot n \, ds - \frac{d}{dt}\iiint_{(w)} \epsilon_m \, d\tau + \iint_{(s)} \epsilon_m V \cdot n \, ds. \tag{2.15}$$

The transformation of the Maxwell equations requires some algebra with no special difficulty. Finally, the integral conservative form of the MFD model follows (taking into account the symmetric nature of the shear stress tensor T).

$$\frac{d}{dt}\iiint_{(w)} \rho \, d\tau = 0 \tag{2.16}$$

$$\frac{d}{dt}\iiint_{(w)} \rho V \, d\tau = \iint_{(s)} (T_B + T) \cdot n \, ds \tag{2.17}$$

$$\frac{d}{dt}\iiint_{(w)} \left(\rho \epsilon_i + \frac{1}{2}\rho V^2 + \epsilon_m\right) d\tau =$$
$$\iint_{(s)} (-P_G + \epsilon_m V + T \cdot V - \theta) \cdot n \, ds \tag{2.18}$$

$$\iint_{(s)} B \cdot n \, ds = 0 \tag{2.19}$$

$$\frac{d}{dt}\iiint_{(w)} B \, d\tau + \iint_{(s)} (n \times E - (B,V) \cdot n) \, ds = 0 \tag{2.20}$$

$$\iint_{(s)} (J + J_D) \cdot n \, ds = 0 \tag{2.21}$$

3. Balance Equations for the Interface

The final result follows:

$$[\rho(V - V_i)] \cdot n = 0 \tag{3.1}$$

$$[(\rho V, V - V_i) - T_B - T] \cdot n = (U - (n, n)) \cdot \nabla\gamma - \gamma\nabla \cdot n \, n \tag{3.2}$$

$$\left[(\rho\epsilon_i + \frac{1}{2}\rho V^2)(V - V_i) + P_G - \epsilon_m V_i - T \cdot V + \theta\right] \cdot n = 0 \tag{3.3}$$

$$[B] \cdot n = 0 \tag{3.4}$$

$$[n \times E - (B, V_i) \cdot n] = 0 \tag{3.5}$$

$$[J + J_D] \cdot n = 0. \tag{3.6}$$

These equations are derived by starting from Eqs. (2.16) to (2.21) and using Kotchine's theorem (Flügge, 1960 p527), extended now currently to interfaces possessing material properties (see for instance Barrere and Prud'homme 1973). We restrict our case to only one material property of the interface, the surface tension γ, and we have written the balance equations with the notations V_i for the velocity of the interface and the brackets [ ] for the jump of a quantity across the interface. In Eq. (3.6), J and J_D may alternatively be replaced by their expressions from Eqs. (2.8) and (2.6).

4. Application to the Calculation of the Surface Depression in a High-Current Arc Weld Pool

The modelling of arc welding, (Choo et al., 1990), is a topic of active research in welding science and technology, Fig. 4.1 gives a schematic sketch of an arc welding operation. A high-current arc is struck between the cathode and the metallic workpiece, generating an energetic plasma (gaseous argon plasma in current operations). The plasma undergoes a flow, driven mainly by the Lorentz force. At the impact on the upper surface of the metal, the energy of the plasma is partially transferred to the metal, raises the surface temperature and causes local melting and formation of a pool. The melt in the pool undergoes a thermo-convective motion. The modelling aims at the calculations of both the arc plasma flow and the molten metal flow. The basic equations for both flows are MFD type and should be completed by appropriate boundary conditions. The common boundary (the interface plasma-melt) is a free boundary, i.e. a distorted surface whose shape is unknown. This difficulty is gotten around by Choo et al. by assigning an interface shape chosen on the basis of experimental measurements of Lin and Eager (1985). Now, the balance equations established in the present communication yields also a basis, a theoretical one, to calculate the shape of the interface. As a matter of fact, the projection of Eq. (3.2) on the normal to the interface results in a second-order differential equation governing the free surface geometry. We

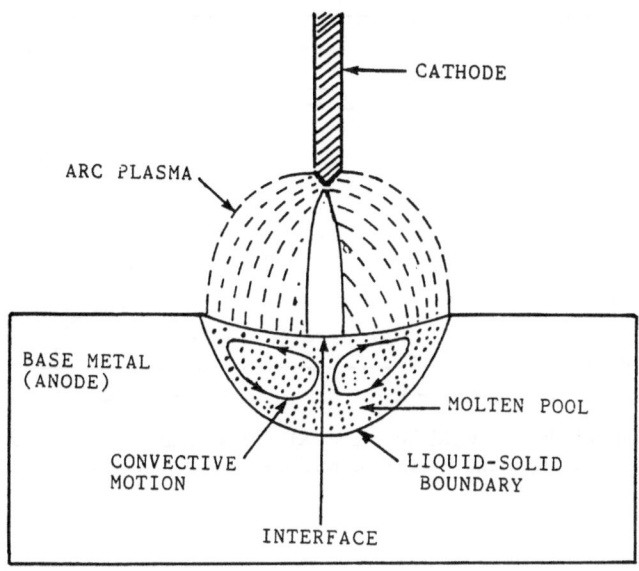

Fig. 4.1. Schematic of arc-welding operations

want to compare the measurements of Lin and Eager with the solution of the differential equation calculated for the same data: arc current 300 amps, argon plasma, and steel melt. In cylindrical symmetric geometry, if we denote $Z = f(r)$ the interface equation, the differential equation governing the function $f(r)$, after some simplifications, follows:

$$f'' + \frac{f'}{r}(1+f'^2) - \frac{(1+f'^2)^{\frac{3}{2}}}{\gamma}\left(\rho_L\, gf + p + \frac{B_\theta^2}{8\pi} + \frac{2f'\tau_{zr}}{1+f'^2}\right) = 0 \quad (4.1)$$

with boundary conditions $r = 0$, $f' = 0$, and $r = \infty$, $f = 0$.

In Eq. (4.1), ρ_L is the specific mass of the molten metal, g the gravity, p the pressure, B_θ the azimuthal magnetic field and τ_{zr} the shear stress, the latter quantities being pertinent to the plasma near over the interface. B_θ is related to the axial current J_z by the relation

$$B_\theta = \frac{4\pi}{r}\int_0^r J_z\, r\, dr.$$

Eq. (4.1) is integrated with a standard computer program for 2-point boundary differential equations. The following data are input:

$$\rho_L = 7200\ Kg/m^3, \qquad g = 9.81\ m/s^2, \qquad \gamma = 1.2\ N/m$$

$$p = p_{max}\ exp\left(-\frac{r^2}{r_0^2}\right), \qquad J_z = J_{max}\ exp\left(-\frac{r^2}{r_0^2}\right)$$

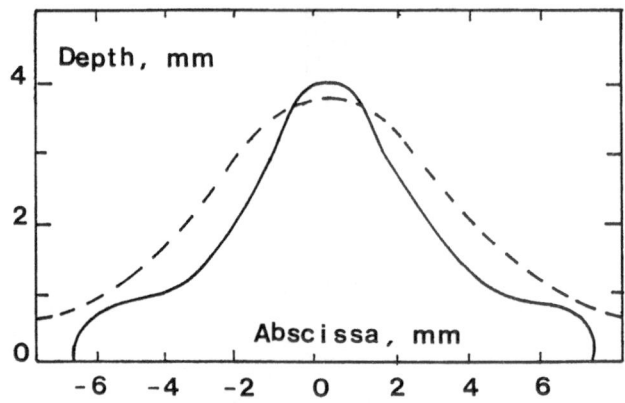

Fig. 4.2. Shape of the arc-weld interface: a) Experimental (Lin and Eager 1985), b) (−−) Calculated from the balance equations established in the present paper

$$\tau_{zr} = \tau_{max} \frac{r}{r_\tau} \exp\left(1 - \frac{r}{r_\tau}\right), \qquad p_{max} = 1140\,Pa, \qquad J_{max} = 6\,A/mm^2$$

$$\tau_{max} = 100\,Pa, \qquad r_0 = 4\,mm, \qquad r_\tau = 1.2\,mm$$

The spatial distributions for p and J_z are taken respectively from Stepanov et al. (1977) and Nestor (1962). The spatial distribution for τ_{zr} is a fitting function with Fig. 13 in Choo et al. (1990). The other constants are also taken from Choo. The results of the computations are plotted in Fig. 4.2 where we have also plotted the experimental shape of the interface. There is no significant discrepancy between experiments and theory.

5. Conclusion

The balance equations at a fluid-fluid interface in MFD, established in this communication are an indispensable tool to study the coupling between the fore and back MFD flows. They yield also the deformation of the interface.

References

1. Barrere, M., and Prud'homme, R. (1973). Equations fondamentales de l'aerothermochimie. Masson, Paris.
2. Cabannes, H. (1970). Theoretical Magnetofluiddynamics. Academic Press.
3. Choo, R.T.C., Szekely, J., and Westhoff, R.C. (1990). Modelling of High Current Arcs with Emphasis on Free Surface Phenomena in the Weld Pool. J. of Welding, **9**, 346-361.
4. Flügge, S. (1960). Handbuch der Physik, Vol. 3/1, 527. Springer–Verlag.

5. Gatignol, R. and Seppecher, P. (1986). Modelisation of Fluid-Fluid Interfaces with Material Properties. Journal de Mecanique Theorique et Appliquee., numro special, 225-247.
6. Lin, M.L. and Eager, T.W. (1985). Influence of Arc Pressure on Weld Pool Geometry. Welding Journal, **64** (6), 163s-169s.
7. Müller, G. (1988). Crystal Growth from the Melt. Springer–Verlag.
8. Nestor, O.H. (1962). Heat Intensity and Current Density Distributions at the Anode of High Current, Inert Gas Arcs. J. Appl. Phys., **33** (5), 1638-1648.
9. Stepanov, V.V. (1977). Distribution of Pressure in the Plasma Arc and Effects of the Rate of Flow of Plasma-Forming Gas on Weld Quality. Auto.Weld., **30** (6), 5-7.
10. Szekely, J. (1979). Fluid Flow Phenomena in Metals Processing. Academic Press.

Group-Invariant Solutions of Hydrodynamics

S.V. Coggeshall

Los Alamos National Laboratory, Los Alamos, NM 87545, USA

1. Introduction

The equations of hydrodynamics, being nonlinear, are in general difficult to solve analytically. A great deal of effort has therefore gone into the numerical solution of these equations, using a wide variety of algorithms. Issues associated with these numerical solutions include accuracy and stability of the algorithms and their associated solutions. Comparison to experiment is one basic way to address the validity of numerical solutions, but the issues of diagnostics and experimental error are always present. Further, there are regimes for which experimental results are either costly or impossible to obtain. Due to this, analytic solutions to such equations in relevant physical regimes have been sought. Such analytic, exact solutions can be used for three purposes: (1) benchmarks for numerical algorithms, (2) the basis for analytic models, (3) to provide insight into more general solutions.

One method of constructing analytic solutions that has been highly successful is the use of Lie group reductions. This technique was invented by Sophus Lie in the latter part of the 19th Century specifically to find solutions to differential equations, both ordinary and partial. Since their inception, Lie groups have been recognized as a powerful formalism applicable to a wide variety of physical and mathematical problems. Ironically, their utility as originally intended as tools to solve differential equations fell out of use until the middle of this century. In the past few decades there has been an explosion in the application of Lie groups to differential equations.[1-6]

The bottom line of the technique is this: invariance of a differential equation under the action of a Lie group allows (1) the reduction in order of ordinary differential equations (ODE's), (2) the reduction of the number of independent variables of partial differential equations (PDE's), and (3) construction of new solutions from existing ones. Each allowed nontrivial invariance group provides such a reduction. Further, chains of reduction are generally possible such that invariance of a differential equation under an n-parameter solvable Lie group allows n such reductions, taking and $n-$th order ODE into quadrature and a PDE with $n+1$ independent variables into an ODE. Systems of equations are reduced in the same fashion.

The first step in this procedure is to identify the Lie groups of transformations that leave a differential equation invariant, that is, the continuous transformations of the space of dependent and independent variables into new variables such that the differential equation written in terms of the new variables is identical to the old differential equation. Once these transformations with this property are found, two things can be done. New solutions can be found as transformations of old solutions, since the transformations take the solution surface into itself (the differential equation is invariant). Second, as a consequence of this invariance, new coordinates can be identified in terms of which the differential equation takes a simpler, reduced form. It is usually easier to obtain a solution in this reduced space, which can then be translated back into the original space to provide a solution to the original differential equation.

Starting with PDE's, one attempts to do reductions until ODE's are reached. Solutions of these ODE's are sought and then transformed back into the original variables where they are solutions of the original PDE's. These group-invariant solutions are only particular solutions to the PDE's, not general solutions. They exist and evolve only for particular initial/boundary conditions. Often the special initial/boundary conditions are physically relevant and the particular solutions are physically interesting. Some solutions are more artificial. All the particular solutions can be used as numerical benchmarks, although issues of stability should be considered.

2. Lie Groups Applied to Differential Equations

A Lie group is a collection of elements along with a binary operation such that the four group axioms, closure, associativity, existence of an identity and inverse are satisfied. A point transformation is a transformation of a point in space to a new point according to some defining relations, such as

$$\tilde{x} = f_1(x, y), \quad \tilde{y} = f_2(x, y).$$

(They are called point transformations because the transformation functions f_i depend only on the point (x, y) values and not on derivatives.) If we parameterize this transformation with a continuous parameter a such that the transformation functions f_1 and f_2 are continuous and continuously differentiable to all orders in x, y, and a, we have a collection of continuous point transformations

$$\tilde{x} = f_1(x, y; a), \quad \tilde{y} = f_2(x, y; a).$$

We can now consider the collection of all these transformations for all allowed values of the parameter a. We identify first an identity element a_0 that takes all points back into themselves:

$$x = f_1(x, y; a_0), \quad y = f_2(x, y; a_0).$$

Further, we consider a binary operation among this collection of transformations: a combined transformation that does the action of two consecutive transformations. That is, let the parameter a take the point (x, y) into $(\tilde{x}, \tilde{y})$ and the parameter b take the point $(\tilde{x}, \tilde{y})$ into another point $(\tilde{\tilde{x}}, \tilde{\tilde{y}})$:

$$\tilde{x} = f_1(x, y; a), \quad \tilde{y} = f_2(x, y; a), \qquad \tilde{\tilde{x}} = f_1(\tilde{x}, \tilde{y}; b), \quad \tilde{\tilde{y}} = f_2(\tilde{x}, \tilde{y}; b).$$

A binary operation would then be the transformation with parameter c that takes the point (x, y) directly into the point $(\tilde{\tilde{x}}, \tilde{\tilde{y}})$:

$$\tilde{\tilde{x}} = f_1(x, y; c), \quad \tilde{\tilde{y}} = f_2(x, y; a).$$

This can be written in short as $f(c) = f(b) * f(a)$.

A Lie groups of point transformations is simply a collection of all such transformations along with this binary composition function that satisfies the four group axioms, written as

1. For all $a, b \in D$, the domain of the group parameter, there exists an element $c \in D$ such that $f(c) = f(a) * f(b)$.
2. There exists an element $a_0 \in D$ such that for all $a \in D$, $f(a) * f(a_0) = f(a)$.
3. For all $a \in D$ there exists an element a^{-1} such that $f(a) * f(a^{-1}) = f(a_0)$.
4. For all $a, b, c \in D$, $[f(a) * f(b)] * f(c) = f(a) * [f(b) * f(c)]$.

There exist many common transformations that satisfy these four properties and are therefore Lie groups of transformations. These include translations, rotations, scale transformations, Galilean transformations, and Lorentz transformations.

One of the fundamental accomplishments of Lie was to show that for such transformations, all the information concerning the global action of the transformation is contained in the infinitesimal transformation around the identity element a_0. This is a consequence of the continuity properties required for the global transformation functions f_i. Because of this, all the invariance conditions required to construct such transformation groups become linear.

There is a one-to-one correspondence between the global transformation equations using the f_i's and the infinitesimal transformation equations around the identity. The infinitesimal transformation equations are found by expanding a Taylor series around the identity value

$$\tilde{x} = x + (a - a_0) \frac{\partial f_1}{\partial a}\bigg|_{a=a_0} + O[(a-a_0)^2], \tilde{y} = y + (a-a_0)\frac{\partial f_2}{\partial a}\bigg|_{a=a_0} + O[(a-a_0)^2]$$

Special symbols and names are given to these first derivatives of the the transformation equations evaluated at the identity element. They are called the coordinate functions,

$$\frac{\partial f_1}{\partial a}\bigg|_{a=a_0} = \xi(x, y), \quad \frac{\partial f_2}{\partial a}\bigg|_{a=a_0} = \eta(x, y),$$

and they contain complete information about the global transformations since

$$\tilde{x} = e^{(a-a_0)(\xi\partial_x + \eta\partial_y)}x, \quad \tilde{y} = e^{(a-a_0)(\xi\partial_x + \eta\partial_y)}y.$$

Here the exponentiation of an operator A formally means

$$e^{(a-a_0)A}x = \left(1 + (a-a_0)A + \frac{(a-a_0)^2}{2!}(A)(A) + \dots\right)x.$$

We can also ask how a general function F of x and y changes under the action of the group. We expand F in a Taylor series around the identity as

$$F(\tilde{x},\tilde{y}) = F(x,y) + (a-a_0)\left[\frac{\partial f_1}{\partial a}\bigg|_{a=a_0}\frac{\partial F}{\partial x} + \frac{\partial f_2}{\partial a}\bigg|_{a=a_0}\frac{\partial F}{\partial y}\right] + O\left[(a-a_0)^2\right],$$

$$F(\tilde{x},\tilde{y}) - F(x,y) = (a-a_0)\left[\xi\frac{\partial F}{\partial x} + \eta\frac{\partial F}{\partial y}\right] \equiv (a-a_0)UF,$$

where the differential operator

$$U = \xi\frac{\partial}{\partial x} + \eta\frac{\partial}{\partial y}$$

is called the *generator* of the group action. This operator describes how functions change infinitesimally under the action of the group, and it's exponentiation generates the global transformation action,

$$F(\tilde{x},\tilde{y}) = e^{(a-a_0)U}F(x,y),$$

which is just the Taylor series.

Using this differential operator U we define two types of invariance. A function $F(x,y)$ is said to be an invariant function under the action of the group if $UF = 0$ identically. An equation $F(x,y) = 0$ is an invariant equation if $UF = 0$ whenever $F = 0$. This is written either as $UF|_{F=0} = 0$ or $UF = \lambda(x,y)F$ for some function λ. We now have all the machinery to understand how Lie group theory is used to reduce the space of independent variables for PDE's.

Recall the solution of quasilinear PDE's using the method of characteristics. Given the PDE

$$g_1(\mathbf{x})\frac{\partial F}{\partial x_1} + g_2(\mathbf{x})\frac{\partial F}{\partial x_2} + \dots + g_n(\mathbf{x})\frac{\partial F}{\partial x_n} = 0,$$

the general solution is $F = F(c_1, c_2, \dots, c_{n-1})$, where the c_i's are the $n-1$ integration constants of the characteristic equations

$$\frac{dx_1}{g_1} = \frac{dx_2}{g_2} = \dots = \frac{dx_n}{g_n}.$$

Since the invariance condition $UF = 0$ is this type of PDE, we know that if a function F is invariant under a group with generator $U = \xi\partial x + \eta\partial y$, we can rewrite this function as another function $G(c)$, where c is the integration constant of the characteristic equations

$$\frac{dx}{\xi} = \frac{dy}{\eta},$$

and the dimensionality of the space has been reduced by one.

A simple example is rotation in xy space, where

$$\tilde{x} = x \cos a - y \sin a, \quad \tilde{y} = x \sin a + y \cos a,$$

and the group parameter a is seen to be the angle of rotation. Simple calculation gives $\xi = -y$ and $\eta = x$, so the generator is $U = -y\partial_x + x\partial_y$. (A good excercise is to recover the global equations for $\tilde{x}$ and $\tilde{y}$ from the infinitesimal generator U through $\tilde{x} = e^{aU}x$. $\tilde{y} = e^{aU}y$.) Solving the characteristic equations for this case

$$\frac{dx}{-y} = \frac{dy}{x} \text{ gives } c = x^2 + y^2,$$

and we note any function $F(c)$ is invariant under this group. Therefore, any function $F(x^2 + y^2)$ can be written in terms of this new variable c, which is seen to be the square of the radius. In other words, any function which is rotationally symmetric can be written in terms of the radius only.

The concept of symmetry can be generalized. By definition, if an object is invariant under a transformation it posses a symmetry with respect to that transformation (e.g., reflection and rotation). Differential equations define surfaces (solution surfaces) in space. If this solution surface is invariant under a transformation (it is mapped back into itself), it possesses a symmetry with respect to that transformation. Invariance of the differential equation under the generator U implies a symmetry with respect to the transformation e^{aU}.

With this technique we find we can reduce the number of variables, given invariance under this special linear differential operator U. For a system of PDE's with n independent variables, we can invoke this mechanism $n - 1$ times to reduce the system to ODE's, which are then easier to solve. Any solution of the ODE's will provide a particular solution to the original PDE's. Additionally, since the system of equations is invariant, the transformations take solutions into solutions. Often we can use the global transformations to generate a new solution from a given one. Specific examples will be given for the hydrodynamics equations.

3. Hydrodynamics Model

The equations of hydrodynamics can be written

$$\rho_t + \mathbf{u} \cdot \nabla\rho + \rho\nabla \cdot \mathbf{u} = 0,$$

$$\mathbf{u}_t + \mathbf{u} \cdot \nabla \mathbf{u} + \frac{1}{\rho} \nabla P = 0,$$

$$E_t + \mathbf{u} \cdot \nabla E + \frac{1}{\rho} P \nabla \cdot \mathbf{u} + \frac{1}{\rho} \nabla \cdot F = S.$$

Here ρ is the material density, $\mathbf{u}$ the 3–dimensional velocity vector whose components can be written (u, v, w), E and P the specific energy and pressure, F is the heat flux, and S is a general energy source term. The equations must be closed with the equation of state, a relationship between energy and pressure, which can be written in terms of a material temperature as

$$E = E(\rho, T), \quad P = P(\rho, T).$$

The heat flux F can be chosen to represent normal material conduction or a nonlinear conduction typical for radiation diffusion, creating Marshak heat fronts. The energy and pressure terms can also be chosen to allow radiation energies and pressures in various forms, such as in a black-body equilibrium form

$$P = \frac{aT^4}{3} \quad \text{and} \quad E = \frac{aT^4}{\rho},$$

a the radiation constant. Radiation conduction can be written

$$F = -\frac{c\lambda}{3} \nabla aT^4.$$

Here c is the speed of light and $\lambda(\rho, T)$ is the radiation mean-free-path.

The Lie group properties of these equations in 1–D including the radiation diffusion terms for an arbitrary material equation of state are listed in Reference 7 along with special cases of reductions to ODE's. For the remainder of this chapter the model will be restricted to a perfect gas EOS with no radiation energy and pressure terms included:

$$\rho_t + \mathbf{u} \cdot \nabla \rho + \rho \nabla \cdot \mathbf{u} = 0,$$

$$\mathbf{u}_t + \mathbf{u} \cdot \nabla \mathbf{u} + \frac{1}{\rho} \nabla (\Gamma \rho T) = 0, \tag{3.1}$$

$$T_t + \mathbf{u} \cdot \nabla T + (\gamma - 1) T \nabla \cdot \mathbf{u} - \frac{\gamma - 1}{\Gamma \rho} \nabla \cdot \kappa \nabla T - \frac{\gamma - 1}{\Gamma} S = 0.$$

Here the material energy and pressure are written in terms of the temperature as

$$P = \Gamma \rho T, \quad E = \frac{\Gamma}{\gamma - 1} T,$$

Γ the gas constant and γ the adiabatic exponent. The heat flux F is written in a general diffusion approximation $F = -\kappa(\rho, T)\nabla T$, and can include radiation Marshak behavior.

The Lie groups which leave Eqs. (3.1) invariant are generated by the differential operators

$$U_x = \frac{\partial}{\partial x} \equiv \partial_x,$$
$$U_y = \partial_y,$$
$$U_z = \partial_z,$$
$$U_t = \partial_t,$$
$$U_{Gx} = t\partial_x + \partial_u,$$
$$U_{Gy} = t\partial_y + \partial_v,$$
$$U_{Gz} = t\partial_z + \partial_w,$$
$$U_{xy} = -y\partial_x + x\partial_y - v\partial_u + u\partial_v,$$
$$U_{yz} = -z\partial_y + y\partial_z - w\partial_v + v\partial_w,$$
$$U_{zx} = -x\partial_z + z\partial_x - u\partial_w + w\partial_u,$$
$$U_{st} = t\partial_t - u\partial_u - v\partial_v - w\partial_w - 2T\partial_T,$$
$$U_{ss} = x\partial_x + y\partial_y + z\partial_z - q\rho\partial_\rho + u\partial_u + v\partial_v + w\partial_w + 2T\partial_T,$$
$$U_{s\rho} = \rho\partial_\rho,$$
$$U_p = xt\partial_x + yt\partial_y + zt\partial_z + t^2\partial_t - q\rho t\partial_\rho + (x - ut)\partial_u + (y - vt)\partial_v$$
$$+ (z - wt)\partial_w - 2Tt\partial_T \tag{3.2}$$

with the following conditions for the conductivity κ and energy source S:

$$\hat{\kappa} = \kappa[a_{s\rho} - a_{st} + (2 - q)a_{ss} - qa_p t],$$

$$US - S(2a_{ss} - 3a_{st} - 4ta_p) - qa_p \frac{\Gamma T}{\gamma - 1}\left(\gamma - \frac{q+2}{q}\right) = 0. \tag{3.3}$$

Here we have written

$$\hat{\kappa} \equiv U\kappa = \eta^\rho \frac{\partial \kappa}{\partial \rho} + \eta^T \frac{\partial \kappa}{\partial T},$$

and expressed the operator U as a linear combination of the 14 separate generators,

$$U = a_x U_x + a_y U_y + ... + a_{st} U_{st} + a_{ss} U_{ss} + a_{s\rho} U_{s\rho} + a_p U_p.$$

The first four groups are translations in x, y, z, t. The groups generated by U_{Gi} are Galilean boosts in the x, y, z directions. The next three groups are rotations. Following those are three separate scaling groups in time, space and density. The final group is a projective group in the space-time plane. Note that for the case of no conduction or source S, the conditions (3.3) become

$$a_p = 0 \quad \text{unless} \quad \gamma = \frac{q+2}{q}.$$

The parameter q takes the value $q = \text{Max}(N, k + 1)$, where N is the number of spatial independent variables in the problem and k is a geometry factor equal to 0, 1 or 2 for planar, cylindrical, or spherical geometry, respectively (see Reference 8).

4. One-Dimensional Solutions

In one-dimensional arbitrary coordinates, Eqs. (3.1) become

$$\rho_t + u\rho_r + \rho u_r + \frac{k\rho u}{r} = 0,$$

$$u_t + uu_r + \frac{1}{\rho}\Gamma T\rho_r + \Gamma T_r = 0, \tag{4.1}$$

$$\frac{\Gamma}{\gamma - 1}(T_t + uT_r) + \Gamma T u_r + \Gamma T\frac{ku}{r} + \frac{1}{\rho}\left(F_r + \frac{kF}{r}\right) = 0.$$

We let r be the single spatial coordinate for general geometry (planar, cylindrical, spherical), with geometry factor $k = 0, 1$, or 2. Again, the heat flux can be related to the temperature through a (nonlinear) diffusion approximation $F = -\kappa(\rho, T)T_r$. For the time being we consider the case of no conduction and set $F = 0$.

The groups allowed in one-dimensional coordinates are represented by the generators U_r, U_t, U_{Gr}, U_{st}, U_{ss}, $U_{s\rho}$, and U_p. For the one-dimensional generators we replace x with r and set all terms with y, z, v, w to zero. An additional condition in 1–D occurs for space translation, $ka_r = ka_{Gr} = 0$, which says that translations and Galilean boosts are only allowed in planar geometry ($k = 0$). The condition (3.3) on γ to keep the projective group in 1–D with no source is $\gamma = (k + 3)/(k + 1)$.

4.1 Traditional Similarity Solutions

We note that the "classic" similarity solutions found through dimensional analysis and/or the use of the Buckingham Pi Theorem[9] are exactly the result of the use of scaling groups. The general scaling group allowed for Eqs. (4.1) is a linear combination of the three generators U_{st}, U_{ss} and $U_{s\rho}$,

$$U = U_{st} + aU_{ss} + (b + qa)U_\rho = ar\partial_r + t\partial_t + b\rho\partial_\rho + (a - 1)u\partial_u + 2(a - 1)T\partial_T,$$

a and b free parameters. A group generator can always be arbitrarily scaled, so in the linear combination we can choose one of the multipliers as 1. The global equations for the action of this group are

$$\tilde{r} = e^a r, \quad \tilde{t} = e^1 t, \quad \tilde{\rho} = e^b \rho, \quad \tilde{u} = e^{a-1} u, \quad \tilde{T} = e^{2(a-1)} T.$$

When these tilded variables are substituted into Equations (4), we find the equations remain the same as (4.1) and are therefore invariant under this scale transformation. Since this U operating on the system (4.1) is equal to zero on the solution surface of (4.1), we know that the system can be rewritten solely in terms of the integration constants of the characteristic equations

$$\frac{dr}{ar} = \frac{dt}{t} = \frac{d\rho}{b\rho} = \frac{du}{(a-1)u} = \frac{dT}{2(a-1)T}.$$

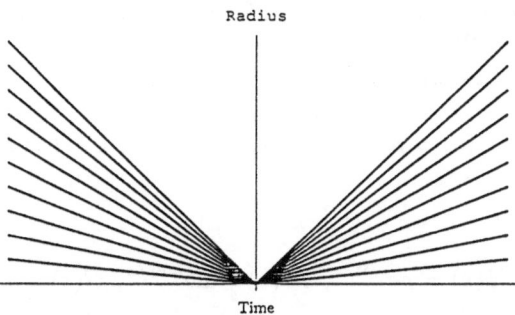

Radius

Time

Fig. 4.1. Material trajectories for a solution $u(r,t) = u_0 r/t$

Integrating the first relation gives $c_1 = rt^{-a}$, which is identified as the new independent variable s. Integrating the other relations provides the three new dependent variables

$$(c_2 =)H(s) = \frac{\rho}{t^b}, \quad M(s) = \frac{u}{t^{a-1}}, \quad G(s) = \frac{\Gamma T}{t^{2(a-1)}}, \quad \left(s = \frac{r}{t^a}\right).$$

To transform Eqs. (4.1) into these new variables the required derivatives are calculated using the chain rule. For example,

$$\rho = H(s)t^b, \quad \rho_t = H'\frac{\partial s}{\partial t}t^b + bHt^{b-1} = -aH'st^{b-1} + bHt^{b-1}.$$

When these are substituted into Eqs. (4.1), they become the ODE's

$$bH + HM' + (M - as)H' + \frac{kHM}{s} = 0$$

$$(a-1)M + (M - as)M' + \frac{H'}{H}G + G' = 0 \qquad (4.2)$$

$$2(a-1)G + (M - as)G' + (\gamma - 1)G\left[M' + \frac{kM}{s}\right] = 0$$

Many well-known solutions can be constructed from these ODE's, for instance the Sedov and Taylor point explosion,[10] imploding shocks (e.g., Guderley[11]), and a reflecting shock (e.g., Noh[12]). A characteristic of these types of solutions with $M \sim s$ is that the velocity can be written $u(r,t) = u_0 r/t$, which gives material trajectories as $r = r_0|t|^{u_0}$, shown in Fig. 4.1 for $u_0 = 1$. Solutions 0, 1, 2, and 16 in Section 5 are generated using these scaling groups.

4.2 Exponential Solutions

If the translation groups are added to the above scaling groups the solution of the characteristic equations takes two new branches, depending on the choice of parameters. The generator is written

$U = cU_{st} + aU_{ss} + (b + aq)U_{sp} + dU_r + eU_t = (ar + d)\partial_r + (ct + e)\partial_t + bp\partial_\rho + (a - c)u\partial_u + 2(a - c)T\partial_T.$

The two new branches are

$$s = \frac{e^{cr/d}}{ct + e}, \ H(s) = \rho e^{-br/d}, \ M(s) = ue^{cr/d}, \ G(s) = \Gamma T e^{2cr/d}$$

for $\quad d, c \neq 0, a = 0,$

$$s = \frac{e^{at/e}}{ar + d}, \ H(s) = \rho e^{-bt/e}, \ M(s) = ue^{-at/e}, \ G(s) = \Gamma T e^{-2at/e}$$

for $\quad c = 0, e, a \neq 0.$

The use of scaling groups in conjunction with translation groups gives rise to solutions which are exponential in either time (time translation with space scaling) or space (space translation with time scaling). These solutions have also been found through inspection by many authors.

4.3 Projective Group Solutions

The projective group generator

$$U_p = rt\partial_r + t^2\partial_t - qpt\partial_\rho + (r - ut)\partial_u - 2Tt\partial_T$$

creates the global transformations

$$\tilde{r} = \frac{r}{1 - a_p t}, \ \tilde{t} = \frac{t}{1 - a_p t}, \ \tilde{\rho} = \rho(1 - a_p t)^{k+1}, \ \tilde{u} = u(1 - a_p t) + a_p r, \quad (4.3)$$

$$\tilde{T} = T(1 - a_p t)^2.$$

This is a projective group because these transformation equations take a straight line in $r - t$ space into another straight line, $t = br + c \rightarrow \tilde{t} = b\tilde{r}/(1 - a_p c) + c$. This interesting group is allowed only for the special value of the adiabatic exponent $\gamma = (q + 2)/q$, which for spherical coordinates is $\gamma = 5/3$.

Keeping the projective group along with density scaling and time translation, we write the generator

$$U = U_p + bU_\rho - a^2 U_t = rt\partial_r + (a^2 - t^2)\partial_t + (b - qt)\rho\partial_\rho + (r - ut)\partial_u + 2Tt\partial_T.$$

The integration constants of the characteristic equations become the new variables

$$s = \frac{r}{(a^2 - t^2)^{1/2}}, H(s) = \rho(a^2 - t^2)^{q/2}\left(\frac{t + a}{t - a}\right)^{b/2a}, M(s) = u(a^2 - t^2)^{1/2} - st,$$

$$G(s) = \Gamma T(a^2 - t^2).$$

Written in terms of these new variables, the PDE's (4.1) become the ODE's

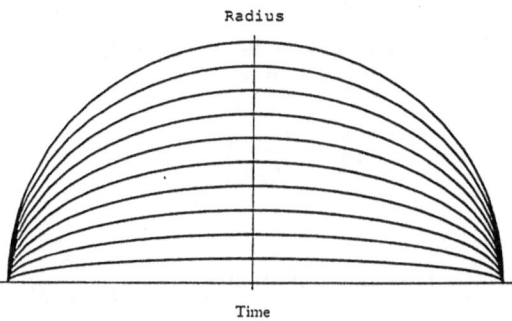

Radius

Time

Fig. 4.2. Material trajectories for Solution 3

$$bH + H'M + HM' + \frac{kHM}{s} = 0$$

$$-a^2s + M'M + \Gamma\frac{GH'}{H} + \Gamma G' = 0,$$

$$G'M + (\gamma - 1)G\left(M' + \frac{kM}{s}\right) = 0.$$

One can solve these equations with a power law assumption,

$$H(s) = H_0s^h, \quad M(s) = M_0s^m, \quad G(s) = G_0s^g,$$

but the only nontrivial ($M_0 \neq 0$) solution reduces to a special case solution of the ODE's for the scaling groups, so nothing new is obtained. However, making the assumption $M = 0$, the first and third equations reduce to $b = 0$, and the middle equation can be solved in general with only the assumption of isentropic flow (see Reference 13, Appendix C for details) to construct Solution 4. Here the constants R_i and R_o are locations such that at initial time, $\rho(r \leq R_i, 0) = T(r \leq R_i, 0) = 0$, so the solution is for a hollow shell of material initially extending from R_i to R_o. When $R_i = 0$, the solution is no longer hollow and can be written more simply as Solution 3. For the special case of $d = 2\gamma$, this solution becomes Solution 5.

These solutions using the projective group have the interesting property of two singularities on time axis at $t = \pm a$. The material trajectories for Solution 3 are shown in Fig. 4.2. At time $t = -a$, a point explosion occurs and all material expands out from the point $r = 0$ like a "Big Bang". At time $t = a$ all the material collapses again in a "Big Crunch." At the middle point $t = 0$ all material has stopped expanding and the velocity is zero everywhere. This type of solution is also described by Sedov[10] as a pulsating periodic solution. The projective group can also be used to generate interesting new solutions from known solutions. For instance, Solution 0 is a trivial stationary pressure balance solution. This solution can be transformed to another, nontrivial solution using the global transformation Eqs. (4.3). We write Solution 0 in tilded variables, $\tilde{\rho}(\tilde{r}, \tilde{t}) = \rho_0 \tilde{r}^b$, $\tilde{u} = 0$, and $\tilde{T}(\tilde{r}, \tilde{t}) = T_0 \tilde{r}^{-b}$, and use the transformation (4.3) to generate the new solution

$$\tilde{\rho} = \rho(1 - a_p t)^{k+1} = \rho_0 \tilde{r}^b = \rho_0 \left(\frac{r}{1 - a_p t} \right)^b \Rightarrow \rho(r, t) = \rho_0 r^b (1 - a_p t)^{-b-k-1}.$$

Similarly,

$$\tilde{u} = u(1 - a_p t) + a_p r = 0 \Rightarrow u(r, t) = -\frac{a_p r}{1 - a_p t}, \quad \text{and}$$

$$\tilde{T} = T(1 - a_p t)^2 = T_0 \tilde{r}^{-b} = T_0 \left(\frac{r}{1 - a_p t} \right)^{-b} \Rightarrow T(r, t) = T_0 r^{-b} (1 - a_p t)^{b-2}.$$

This solution is identical with Solution 1 after the time translation $\tilde{t} = t - 1/a_p$ is used.

The projective group can often be used to generate new solutions in this fashion. If we have a solution for which $u(r, t) = u_0 r/t$, the material trajectories are found by integrating $\dot{r} = u$ which provides $r(t) = r_0 |t|^{u_0}$. Under the action of the projective group with parameter a, these relations become

$$u(r, t) = \frac{r(u_0 + at)}{t(1 + at)}, \quad r(t) = r_0 |t|^{u_0} |1 + at|^{1-u_0}.$$

Figs. 4.3 – 4.5 show the action of the projective group for various values of u_0. It is seen that when $u_0 = 1$ there is no change of the flow. The trivial solution $u_0 = 0$ (Solution 0) goes into a linear imploding/exploding trajectory (Solution 1). Solution 2 has the value $u_0 = 1/2$ for the required γ value for the projective group $\gamma = (k + 3)/(k + 1)$. This solution is transformed by the projective group into Solution 3 (see Fig. 4.5). It should be noted that the process of generating a new solution from a given one can only be performed once. A second similar transformation with a different group element gives only a single transformation with a combined composition function value of the two group elements, as is known from the closure property. Secondly, when a particular solution has been obtained as a solution to the ODE's constructed using a certain group, the use of the group again to generate a new solution gives an identity transformation. This is described as follows:

When a subgroup H of the multiparameter invariance group G is used to generate a solution, that solution is called an H-invariant solution. This means that it is invariant under the subgroup H, and any use of the global transformations of H go back into the original solution. To generate a new solution from an H-invariant solution one must use a different transformation from G not in H.

Since a solution can be used to generate new solutions using different groups, it would be nice to identify the minimum collection of subgroups that will generate all possible group invariant solutions. Such a collection is

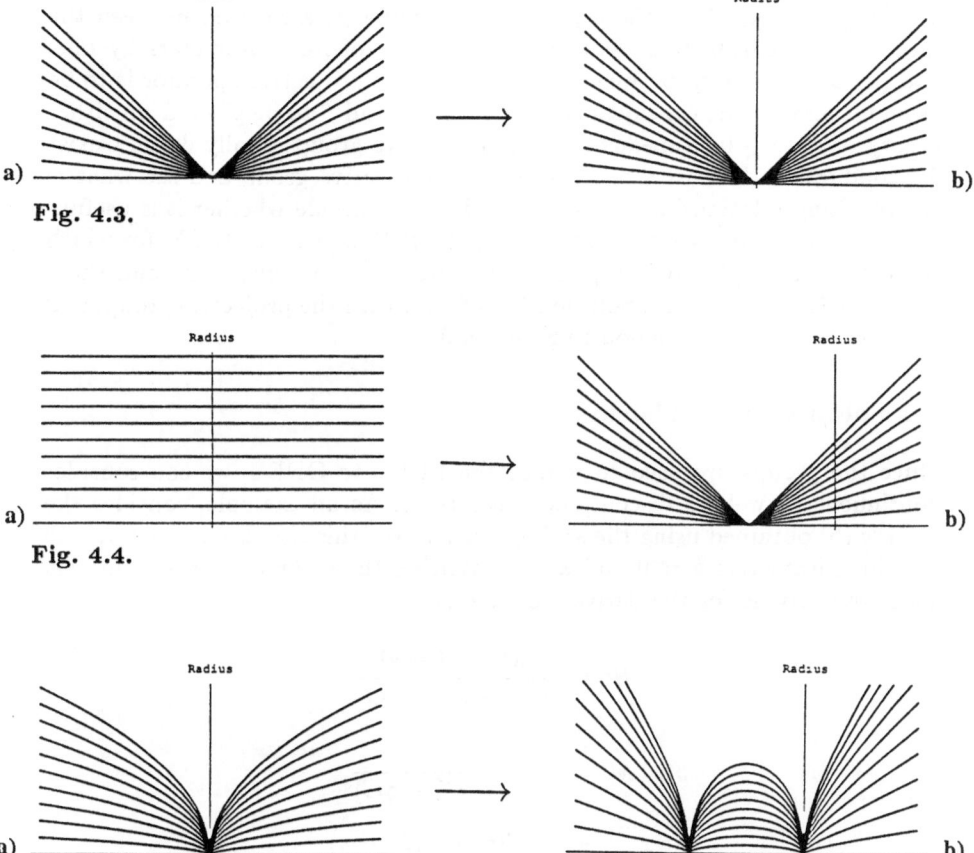

Fig. 4.3.

Fig. 4.4.

Fig. 4.5. (a) show material trajectories for solutions $u(r,t) = u_0 r/t$, with values $u_0 = 1$, 0, and 1/2 respectively. (b) show the effect of the projective group action on these solutions

called an optimal system,[1,2,8] and it is constructed by examining the ways in which group invariant solutions transform among themselves through the adjoint operation.

4.4 Solutions Including Conduction

Group reduction of the PDE's (4.1) can be performed including the diffusive conduction term provided the first condition of (3.3) is met. This condition becomes two relations

$$\rho\kappa_\rho(a_{s\rho} - qa_{ss}) + 2T\kappa_T(a_{ss} - a_{st}) = \kappa[a_{s\rho} - a_{st} + (2-q)a_{ss}], \quad \text{and}$$

$$q\rho\kappa_\rho + 2T\kappa_T = \kappa q,$$

83

since it is an identity in the variable t. For a power law form of the conductivity, $\kappa = \kappa_0 \rho^\alpha T^\beta$, these conditions become (i) a relation between the three scaling parameters, reducing the number of free parameters by one, and (ii) a relationship between α, β, and q if the projective operator is to be used. Generally, one considers the exponents α and β to be given from either theory or data, which means that the projective group usually doesn't exist for conduction. However, one can use the projective group and see what is the resulting relationship between α and β, and decide whether it is useful.

Again, the groups are used to reduce the PDE's (4.1) to ODE's, for which solutions are sought. It is typical that power law solutions exist, and these generate Solutions 7 – 15. Solution 15 is found with the projective group, and is the conduction companion to Solution 3.

4.5 Solutions with Shocks

Once the groups are used to reduce the PDE's to ODE's, we can consider continuous as well as discontinuous solutions. As an example, consider the ODE's (5) obtained using the scaling groups. For this simple example we will set the parameters $b = 0$ and $a = 1$. Writing these three ODE's in matrix form and solving for the derivatives, we get

$$H' = \frac{kMH(M-s)}{s[(M-s)^2 - \gamma G]},$$

$$M' = \frac{\gamma kMG}{s[(M-s)^2 - \gamma G]},$$

$$G' = \frac{-kMG(\gamma - 1)(M-s)}{s[(M-s)^2 - \gamma G]}.$$

Note for $a = 1$ and $b = 0$, $s = r/t$, $H = \rho$, $M = u$, and $G = T$.

A simple solution with a shock can be found by solving these ODE's separately in two regions and connecting the regions with the standard shock jump conditions

$$\rho_1(\dot{R} - u_1) = \rho_2(\dot{R} - u_2) \equiv m_s,$$

$$\Gamma \rho_1 T_1 - \Gamma \rho_2 T_2 = m_s(u_1 - u_2),$$

$$\frac{\Gamma}{\gamma - 1}T_1 + \Gamma T_1 + \frac{1}{2}(\dot{R} - u_1)^2 = \frac{\Gamma}{\gamma - 1}T_2 + \Gamma T_2 + \frac{1}{2}(\dot{R} - u_2)^2,$$

where R is the shock location and the quantities are evaluated at the shock surface $\pm\epsilon$. The shock location will be stationary in the s coordinate system (required by self-similarity). and the shock location in s space can be called s_1, so $R = s_1 t$, $\dot{R} = s_1$.

We look for a solution describing cold material flowing into the origin with a velocity $u_0(< 0)$, causing a shock to move from the origin outward. There are then two regions: (1), the central region that has experienced the shock, and (2) the outer region with velocity u_0, that the shock is propagating into.

In Region 2, we integrate the ODE's analytically to the shock location, s_1. At $s = \infty$, the density is a constant ρ_0, the velocity is a constant u_0 and the temperature is zero ($T = G = 0$). Since $G = 0$ the last ODE says that G stays zero for all of Region 2. This information in the second ODE says that $M' = 0$, or $M = u_0$ for all Region 2. The first ODE can then be integrated

$$H' = \frac{ku_0 H(u_0 - s)}{s(u_0 - s)^2} \quad \Rightarrow \quad H(s) = H_0 \left(\frac{s - u_0}{s} \right)^k.$$

The condition $H(\infty) = \rho_0$ gives $H_0 = \rho_0$. So the solution in Region 2 is

$$H = \rho_0 \left(\frac{s - u_0}{s} \right)^k, \quad M = u_0, \quad G = 0.$$

The jump conditions are now used to connect the solution at the Region 2 side of the shock to just inside Region 1, where the integration will be continued. We find

$$\rho_1 = \rho_2 \frac{s_1 - u_0}{s_1 - u_1}, \quad \Gamma T_1 = (s_1 - u_1)(u_1 - u_0),$$

$$u_1 = 2s_1 - u_0(1 - \gamma)$$

$$= -u_0(\gamma + 1)$$

There are two possibilities for u_1 since the jump condition is quadratic. To find out which one is the correct value, we look at the M' equation with the requirement that at the origin the velocity is zero, $M(0) = 0 \quad \Rightarrow \quad M = 0$ for all of Region 1. This fact in the ODE's demands that G and H are constant in Region 1, where they are just $G = T_1$, $H = \rho_1$. Also, since $u_1 = 0$, $s_1 = u_0(1 - \gamma)/2$ (the top condition for u_1 is used; the bottom requires $u_0 = 0$) and therefore $\rho_1 = \rho_2(\gamma + 1)/(\gamma - 1)$. So for Region 1, the solution is

$$H = \rho_1 = \rho_2 \left(\frac{\gamma + 1}{\gamma - 1} \right), \quad M = 0, \quad G = -s_1 u_0.$$

The value of ρ_2 is $H(s_1 + \epsilon)|_{\epsilon=0} = \rho_0[(s_1 - u_0)/s_1]^k = \rho_0[(\gamma + 1)/(\gamma - 1)]^k$. Putting this all together gives Solution 16. It is interesting that this solution can be projected to produce Solution 17 in the same way that Solution 0 projected produced Solution 1.

More complicated shock solutions are certainly possible. One interesting extension is the addition of conduction, which can be then represented by four first order ODE's (one more being the relationship between the heat flux and the temperature). These equations are generally too difficult to solve analytically, but the ODE's can easily be solved numerically to get similarity solutions including heat conduction. An example of such a solution is in Reference 7.

4.6 Boundary Conditions

The analytic solutions as given in Section 7 contain no boundary conditions, which must be taken into account for numerical solutions. There are two approaches to this concern. The first and simplest is to consider a finite region initialized with the properties of any of the analytic solutions with no consideration of special boundary conditions. In this approach the evolution at the boundary immediately deviates from the analytic solution, and a rarefaction wave propagates into the material of interest. The solution is valid only in the region which has not felt this rarefaction wave. This approach, while simplest to implement, causes the region of validity to shrink as the problem evolves.

The second approach is to apply the correct boundary conditions at the edge of the problem. This is immediate if the calculation is Eulerian. For a Lagrangian calculation one must calculate the location of the boundary at all times and the appropriate material properties for that location must be imposed. One concern with this approach is that small errors made in these boundary conditions can propagate into the problem and confuse the investigation of internally generated errors. For this reason we generally use the first and simplest approach. More general discussions of the treatment of boundary conditions can be found in References 7 and 13.

5. Two-Dimensional Solutions

5.1 Multiple Reductions Using Lie Groups

In two dimensions we consider the case of axisymmetry, rotational symmetry around the z axis. This coordinate system can be described in either spherical (r, θ) or cylindrical (R, z) coordinates. There are seven allowed groups in this geometry, which form a 7-parameter group G^7 generated by the operators $(U_z, U_{Gz}, U_{ss}, U_t, U_{st}, U_{s\rho}, U_p)$.

In two dimensions we need to perform two successive group reductions in order to reduce the PDE's to ODE's. We could perform one transformation, reduce the space to 2 independent variables, and then seek the group invariance properties of these reduced equations in order to reduce them a second time to ODE's. Alternatively, it is known that successive reductions are facilitated by looking at the structure of the Lie group via the Lie algebra. Details in the Lie algebra provide information so that invariance properties of the original equations can be inherited by the reduced equations, and the double reduction can be done in one step.

Recall, an algebra is a vector space with a bilinear composition operation $[X, Y]$. A Lie algebra is an algebra for which $[X, Y] = -[Y, X]$. The Lie algebra associated with a Lie group is simply the collection of group generators. The composition operation is the commutator of two generators $[X, Y] = XY - YX$. Lie's principal theorem states that for an n-parameter Lie group with generators U_1, U_2, ..., U_n, the commutator of any two generators is a linear combination of the generators,

$$[U_i, U_j] = \sum_{k=1}^{n} C_{ijk} U_k.$$

The constants C_{ijk} are called the structure constants of the algebra. There is a one-to-one correspondence between the Lie group and the Lie algebra.

A table of commutators can be constructed for any Lie group that allows the identification of subgroups, that is, subsets of generators that are closed under commutation. For this 7-parameter group G^7, the commutator table is shown in Table 1. From this table we can find examples of subgroups. Any collection of generators that commute ($[U_i, U_j] = 0$) form a subgroup, such as $(U_{s\rho}, U_z, U_{Gz})$. This is a commutative or Abelian subgroup, since all commutators are zero. Examples of non-Abelian subgroups are (U_z, U_{Gz}, U_{ss}), (U_z, U_{Gz}, U_p), (U_t, U_{ss}, U_{st}), $(U_t, U_{ss}, U_{st}, U_p)$, etc. Each single generator forms a 1-parameter subgroup.

Table 1. Table of commutators for the subgroup allowed in 2-D axisymmetric geometry

	$U_{s\rho}$	U_z	U_{Gz}	U_{ss}	U_t	U_{st}	U_p
$U_{s\rho}$	0	0	0	0	0	0	0
U_z	0	0	0	U_z	0	0	U_{Gz}
U_{Gz}	0	0	0	U_{Gz}	$-U_z$	$-U_{Gz}$	0
U_{ss}	0	$-U_z$	$-U_{Gz}$	0	0	0	0
U_t	0	0	U_z	0	0	U_t	$U_{ss} + 2U_{st}$
U_{st}	0	0	U_{Gz}	0	$-U_t$	0	U_p
U_p	0	$-U_{Gz}$	0	0	$-U_{ss} - 2U_{st}$	$-U_p$	0

A special kind of subgroup occurs when the commutator with an element from the subgroup and one from the larger group always goes back into the subgroup. These are called normal subgroups and their corresponding algebras are called ideals.

Definition: Consider a group G and a subset of its elements H along with their associated algebras $\mathcal{G}$ and $\mathcal{H}$. If for any two elements h_1 and h_2 in $\mathcal{H}$, $[h_1, h_2]$ is a linear combination of elements in $\mathcal{H}$, then H is a *subgroup* of G and $\mathcal{H}$ is a *subalgebra* of $\mathcal{G}$. If for any element h of $\mathcal{H}$ and any element g of $\mathcal{G}$,

$[h, g]$ is a linear combination of elements in $\mathcal{H}$, then H is a *normal subgroup* of G and $\mathcal{H}$ is an *ideal* of $\mathcal{G}$.

All subgroups whose commutators with the larger group are zero are normal subgroups. Examples of normal subgroups are $(U_{st}) \subset_n (U_{st}, U_{ss})$, $(U_{Gz}, U_t) \subset_n (U_{Gz}, U_t, U_z)$, $(U_t) \subset_n (U_t, U_{ss}, U_{st}, U_{s\rho})$, etc.

For any subgroup H of G we can identify the collection of all operators v_i whose commutation with any element in $\mathcal{H}$ goes back into $\mathcal{H}$. This collection, which must contain all of H, is called the *normalizer* $\mathrm{Nor}_G(H)$ of the subgroup H of G. It is the largest subgroup of G with the property that H is a normal subgroup of $\mathrm{Nor}_G(H)$. The corresponding algebra $\mathcal{N}_g(\mathcal{H})$ is the collection $v_i : [v_i, v_j] \in \mathcal{H} \; \forall \; v_j \in \mathcal{H}, v_i \in \mathcal{G}$.

The theorem pertaining to sequential reductions can now be stated:

Theorem   Consider a system of differential equations E invariant under a multiparameter group G with subgroup H. The system E/H obtained by reducing E with the subgroup H will be invariant under the quotient group $Q = \mathrm{Nor}_G(H)/H$.

The quotient group can be formed by simply removing one of the elements of H from $\mathrm{Nor}_G(H)$. A simple proof of this theorem can be found in Ovsiannikov[1982].

To summarize the procedure for multiple reductions, first choose a subgroup H for a reduction. Next, form the normalizer $\mathrm{Nor}_G(H)$ of this collection by finding all the other generators that, when commuted with the collection H, go back into H. Finally, remove any one of the generators of H from the set $\mathrm{Nor}_G(H)$. The resulting collection is the quotient group $Q = \mathrm{Nor}_G(H)/H$. Now use any element from H for the first reduction. One is then guaranteed to be able to do a second reduction using any element of Q.

Therefore, for any two generators U_1 and U_2 which could be linear combinations of single generators, when $[U_1, U_2] = U_1$, use any part of U_1 for the first reduction and then U_2 for the second reduction. It is best to use a part of U_1 that is not included in U_2. More detailed discussion on this process can be found in Reference 8. In that paper is listed the 2–Dimensional optimal system Θ_2, which is a collection of possible ways to do such double reductions. The collection is "optimal" in the sense that all possible double group reduction solutions can be found from the members of this list. Each member provides a path for reducing the 2–D hydro PDE's to ODE's.

5.2 Reductions to ODE's

Tables 2 and 3 list the new variables for each entry of Θ_2 for the 2–D form of Eqs. (3.1) with no conduction or source S. Each entry of Tables 2 and 3 reduces the 2–D PDE's to ODE's. These variables are sometimes easier to write in spherical coordinates (Table 2) and sometimes in cylindrical (Table 3).

As an example, consider the similarity variables found using the reduction path $\mathcal{H}_7$ from Table 2, which uses first the combination $U_1 = U_{st} + \alpha U_{s\rho}$,

Table 2. Members of the 2-dimensional optimal system Θ_2 that are easier to write in spherical coordinates. Each entry in Tables 2 and 3 show the generators U_1, U_2 used for the double reduction and the resulting new variables in which the 2-D hydro equations become ordinary differential equations.

Group	U_1	U_2	new ind. var.	ρ	ΓT	u	v
$\mathcal{H}_{1+}$	$a^2 U_t + U_p + \alpha U_{sp}$	$U_{ss} + \beta U_{sp}$	θ	$Hr^{\beta-3}(a^2+t^2)^{-\beta/2}\exp\left(\frac{\alpha}{a}\tan^{-1}\frac{t}{a}\right)$	$G\frac{r^2}{(a^2+t^2)^\tau}$	$\frac{U+t}{a^2+t^2}r$	$V\frac{r}{a^2+t^2}$
$\mathcal{H}_{1-}$	$-a^2 U_t + U_p + \alpha U_{sp}$	$U_{ss} + \beta U_{sp}$	θ	$Hr^{\beta-3}(a^2-t^2)^{-\beta/2}\left(\frac{a-t}{a+t}\right)^{\alpha/2a}$	$G\frac{r^2}{(a^2-t^2)^\tau}$	$\frac{U-t}{a^2-t^2}r$	$V\frac{r}{a^2-t^2}$
$\mathcal{H}_2$	U_p	$U_{ss} + (1-\frac{1}{a})U_{st} + \frac{\beta}{a}U_{sp}$	θ	$Hr^{-3}\left(\frac{r}{t}\right)^\beta$	$\frac{G}{t^2}\left(\frac{r}{t}\right)^{2a}$	$\frac{U}{t}\left(\frac{r}{t}\right)^a + \frac{r}{t}$	$\frac{V}{t}\left(\frac{r}{t}\right)^a$
$\mathcal{H}_3$	$U_p + \alpha U_{sp}$	$U_{ss} + \beta U_{sp}$	θ	$Hr^{\beta-3}t^{-\beta}e^{-\alpha/t}$	Gr^2/t^4	$\frac{U+t}{t^2}r$	Vr/t^2
$\mathcal{H}_5$	U_p	$U_{st} + \beta U_{sp}$	θ	$Ht^{-3}\left(\frac{r}{t}\right)^{-\beta}$	G/t^2	$\frac{U+r}{t}$	V/t
$\mathcal{H}_7$	$U_{st} + \alpha U_{sp}$	$U_{ss} + (\beta+3)U_{sp}$	θ	$Ht^\alpha r^\beta$	Gr^2/t^2	Ur/t	Vr/t
$\mathcal{H}_{13}$	$U_t + \alpha U_{sp}$	$U_{ss} + (\beta+3)U_{sp}$	θ	$Hr^\beta e^{\alpha t}$	Gr^2	Ur	Vr
$\mathcal{H}_{14}$	U_t	$U_{ss} + (\alpha+1)U_{st} + (\beta+3)U_{sp}$	θ	Hr^β	Gr^{2a}	Ur^a	Vr^a

Table 3. Members of Θ_2 written in cylindrical coordinates (R, z). Each entry reduces the 2-D partial differential equations to ordinary differential equations.

Group	U_1	U_2	newind. var.	ρ	$\Gamma\Upsilon$	u_R	u_z
$\mathcal{H}_4$	U_ρ	$U_{Gz} + a(U_{ss} + U_{st}) + \beta U_{s\rho}$	R/t	$Ht^{-3}e^{\beta z/t}$	$e^{2az/t}G/t^2$	$(Ae^{az/t} + R)/t$	$(Be^{az/t} + z)/t$
$\mathcal{H}_6$	$U_\rho + \alpha U_{s\rho}$	$U_{Gz} + \beta U_{s\rho}$	R/t	$Ht^{-3}e^{(\beta z-\alpha)/t}$	G/t^2	$(A+R)/t$	$(B+z)/t$
$\mathcal{H}_8$	$U_{ss} + U_{st} + (\alpha+3)U_{s\rho}$	$U_{Gz} + \beta U_{s\rho}$	R/t	$Ht^\alpha e^{\beta z/t}$	G	A	$B+z/t$
$\mathcal{H}_9$	$U_{st} + \alpha U_{s\rho}$	$U_z + \beta U_{s\rho}$	R	$Ht^\alpha e^{\beta z}$	G/t^2	A/t	B/t
$\mathcal{H}_{10}$	$aU_{Gz} + U_{ss} + bU_t + (\alpha+3-\beta b)U_{s\rho}$	$aU_z - U_t + \beta U_{s\rho}$	$R/(z+at+b)$	$Hx^a e^{-\beta t}$	Gx^2	Ax	$Bx-a$
$\mathcal{H}_{11}$	$2aU_{Gz} + U_t$	$2U_{ss} + U_{st} + 2(\beta+3)U_{s\rho}$	$R/(z-at^2)$	Hx^β	Gx	$Ax^{1/2}$	$Bx^{1/2}+2at$
$\mathcal{H}_{12}$	$2aU_{Gz} + U_t + \alpha U_{s\rho}$	$U_z + \beta U_{s\rho}$	R	$H\exp(\alpha t + \beta z - a\beta t^2)$	G	A	$B+2at$
$\mathcal{H}_{16}$	U_t	$U_z + aU_{st} + \beta U_{s\rho}$	R	$He^{\beta z}$	Ge^{-2az}	Ae^{-az}	Be^{-az}
$\mathcal{H}_{18}$	U_{Gz}	$aU_{st} + U_\rho + (\beta+3a)U_{s\rho}$	$\frac{R}{t}e^{a/t}$	$Ht^{-3}\exp(\alpha z/t - 2az/t^2 + \beta/t)$	$e^{-2a/t}G/t^2$	$(Ae^{-a/t} + R)/t$	$(Be^{-a/t} + z)/t$
$\mathcal{H}_{19}$	$U_{Gz} + \alpha U_{s\rho}$	$aU_z + U_\rho + \beta U_{s\rho}$	R/t	Ht^β	G/t^2	$(A+R)/t$	$(B+z)/t+a/t^2$
$\mathcal{H}_{20}$	U_{Gz}	$aU_z + U_{st} + \beta U_{s\rho}$	R	Ht^β	G/t^2	A/t	$(B+z-a\ln t)/t$
$\mathcal{H}_{22}$	U_{Gz}	$aU_{ss} + U_{st} + (\beta+3a)U_{s\rho}$	R/t^a	Ht^β	$Gt^{2(a-1)}$	At^{a-1}	$(Bt^a+z)/t$
$\mathcal{H}_{23}$	U_z	$aU_{Gz} + U_{ss} + U_{st} + (\beta+3)U_{s\rho}$	R/t	Ht^β	G	A	$B+a\ln t$
$\mathcal{H}_{24}$	U_z	$aU_{ss} + U_{st} + (\beta+3a)U_{s\rho}$	R/t^a	Ht^β	$Gt^{2(a-1)}$	At^{a-1}	Bt^{a-1}
$\mathcal{H}_{25}$	U_z	$aU_t + U_t + (\beta+3a)U_{s\rho}$	Re^{-at}	$He^{\beta t}$	Ge^{2at}	Ae^{at}	Be^{at}

and next $U_2 = U_{ss} + (\beta + 3)U_{s\rho}$, α and β arbitrary constants. To perform a multiple reduction we could first reduce by U_1, write out the new PDE's (which now only have 2 independent variables), and then reduce again with U_2 to ODE's. We know the reduced, intermediate PDE's in two independent variables will be invariant under U_2 since we chose U_1 and U_2 by the above theorem to have that property. We find that it is not necessary to write out the intermediate PDE's; we can go directly from the original PDE's in three independent variables to ODE's in one transformation step as follows.

In spherical coordinates, we find

$$U_1 = t\partial_t + \alpha\rho\partial_\rho - u\partial_u - v\partial_v - 2T\partial_T, \quad \text{and}$$

$$U_2 = r\partial_r + \beta\rho\partial_\rho + u\partial_u + v\partial_v + 2T\partial_T.$$

We first calculate the group invariants of the first generator U_1 by solving the characteristic equations

$$\frac{dr}{0} = \frac{d\theta}{0} = \frac{dt}{t} = \frac{d\rho}{\alpha\rho} = \frac{du}{-u} = \frac{dv}{-v} = \frac{dT}{-2T}$$

for the integration constants $a_1 = r$, $a_2 = \theta$, $a_3 = \rho t^{-\alpha}$, $a_4 = ut$, $a_5 = vt$, $a_6 = Tt^2$. We now write the second generator in terms of these new variables (the a_i's) as

$$U_2 = f_1\partial_{a_1} + f_2\partial_{a_2} + f_3\partial_{a_3} + f_4\partial_{a_4} + f_5\partial_{a_5} + f_6\partial_{a_6},$$

where the f_i's are as of yet unknown functions of the a_i's. The f_i's are found by noticing that $f_i = U_2 a_i$. We calculate

$$f_1 = U_2 a_1 = U_2 r = r = a_1, \quad f_2 = U_2 a_2 = 0,$$

$$f_3 = U_2 a_3 = U_2(\rho t^{-\alpha}) = \beta\rho t^{-\alpha} = \beta a_3, \quad \text{etc.}$$

Continuing, we find we can write U_2 completely in terms of the new variables as

$$U_2 = a_1\partial_{a_1} + \beta a_3\partial_{a_3} + a_4\partial_{a_4} + a_5\partial_{a_5} + 2a_6\partial_{a_6}.$$

We now solve the characteristic equations for this operator

$$\frac{da_1}{a_1} = \frac{da_2}{0} = \frac{da_3}{\beta a_3} = \frac{da_4}{a_4} = \frac{da_5}{a_5} = \frac{da_6}{2a_6}$$

for the integration constants $b_1 = a_2 = \theta$, $b_2 = a_3 a_1^{-\beta} = H(\theta)$, $b_3 = a_4/a_1 = M(\theta)$, $b_4 = a_5/a_1 = V(\theta)$, $b_5 = a_6/a_1^2 = G(\theta)$. The integration constants b_i are group invariants of both U_1 and U_2, and are therefore the new variables we are looking for that will reduce the PDE's to ODE's. These new variables H, M, V, and G are the new dependent variables and the new independent variable θ is itself an invariant under both U_1 and U_2. In the same manner as the single reduction in Section 4A, we calculate the derivatives of the old variables in terms of the new variables

$$\rho = H(\theta)r^\beta t^\alpha, \quad \rho_t = \alpha H r^\beta t^{\alpha-1}, \quad \rho_r = \beta H r^{\beta-1}t^\alpha, \quad \rho_\theta = H'r^\beta t^\alpha, \ldots$$

When these are substituted into the PDE's (3.1) they become the ODE's

$$\alpha + M(\beta + 3) + V\frac{H'}{H} + V' + \frac{V}{\tan\theta} = 0,$$

$$M^2 - M + VM' - V^2 + (\beta + 2)G = 0,$$

$$-V + 2MV + VV' + G\frac{H'}{H} + G' = 0,$$

$$M(2 + 3\gamma - 3) - 2 + \frac{VG'}{G} + (\gamma - 1)\left(V' + \frac{V}{\tan\theta}\right) = 0.$$

Any technique can be attempted to solve the reduced ODE's, which then provides specific solutions for the hydro PDE's. Solutions 18 – 24 are some solutions to the 2–D hydro equations found by making some ansatz in the solution of the associated reduced ODE's.

We find that we can transform Solution 19 with the projective group to get Solution 25. Just as occurred the 1–D solutions, a solution with one pole on the $r - t$ time axis is transformed into one with two poles (see Fig. 4.5). Solution 25 is an explosion/collapse ellipsoidal 2–D solution.

6. 3–D Solutions

The 2–dimensional optimal system Θ_2 was worked out in Reference 8, which yields the minimal reduction paths for the 2–D PDE's into ODE's for which all possible group-invariant solutions may be found. The corresponding 3–D optimal system Θ_3 for the paths for the 3–D PDE's into ODE's has not been worked out. However, guided by the multiple reduction theorem, we can choose a few reduction paths and look for particular solutions. One such path is $U_1 = U_{st} + c_1 U_{ss} + c_2 U_{xy} + c_3 U_{s\rho}$, $U_2 = U_{st} + c_4 U_{ss} + c_5 U_{s\rho}$, $U^3 = U_{st} + c_6 U_{s\rho}$. This gives the similarity variables

$$\rho = H(\theta)t^a r^b e^{c\phi}, \quad u = M(\theta)\frac{r}{t}, \quad v = V(\theta)\frac{r}{t}, \quad w = W(\theta)\frac{r}{t}, \quad \Gamma T = G(\theta)\frac{r^2}{t^2}.$$

When these new variables are substituted into the PDE's (3.1), they become the ODE's

$$a + M(b + 3) + V\frac{H'}{H} + V' + \frac{V}{\tan\theta} + c\frac{W}{\sin\theta} = 0$$

$$-M + M^2 + VM' - V^2 - W^2 + G(b + 2) = 0$$

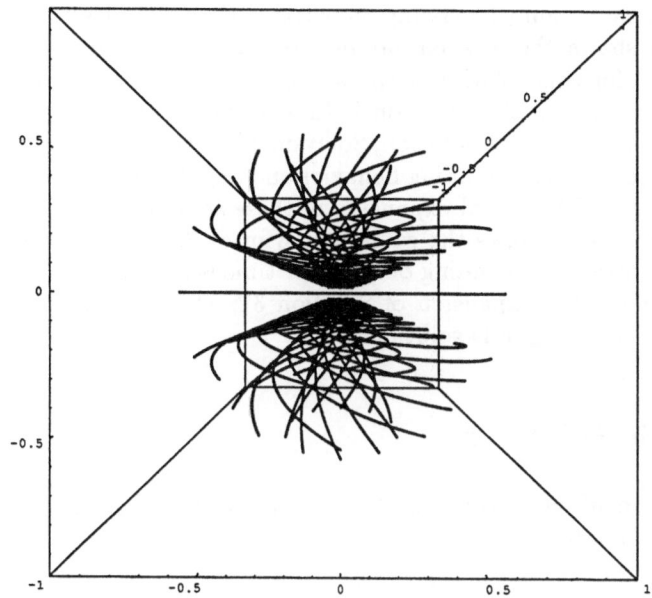

Fig. 6.1. Material trajectories for Solution 30 with a = 1 and b = 2

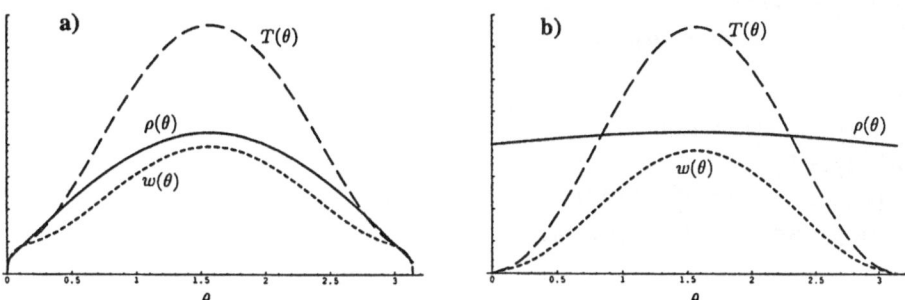

Fig. 6.2. Angular profiles for Solution 30 and 31 with a=1, b=2. (a) has $\rho_0 = 1, \rho_1 = 0.1$ and (b) uses $\rho_0 = 0.1, \rho_1 = 1$

$$-V + 2MV + VV' - \frac{W^2}{\tan\theta} + G\frac{H'}{H} + G' = 0$$

$$-W + 2MW + VW' + \frac{VW}{\tan\theta} + c\frac{G}{\sin\theta} = 0$$

$$-2 + M(2 + 3\gamma - 3) + V\frac{G'}{G} + (\gamma - 1)V' + (\gamma - 1)\frac{V}{\tan\theta} = 0.$$

A particular solution for these equations can be found through the ansatz $V = 0$, $H = H_0 + H_1(\sin\theta)^d$, $c = 0$, $\gamma = 5/3$, and yields Solution 30. Fig. 6.1 shows material trajectories for Solution 30, which has flow spinning either into or out of the origin (depending on whether time is positive or negative). Figs. 6.2a, b shows the angular profiles for the density, temperature, and ϕ-velocity (w) for this solution with two choices of the free parameters. An interesting property of this solution is that it can be transformed with the projective group into a new solution, Solution 31, in the same way as Solution 19 gave Solution 25. The spinning behavior and angular profiles are similar to Figs. 6.1 and 6.2. The radial expansion of these spinning solutions is shown in Fig. 4.5. Fig. 4.5a shows the radial behavior of Solution 30, which either expands or contracts, depending on whether time is positive or negative. Fig. 4.5b shows the radial expansion of Solution 31, which is a point explosion, point collapse spinning 3–D solution.

7. Analytic Solutions

A. 1–D General coordinates, $k = 0, 1, 2$ for planar, cylindrical or spherical coordinates

0.
$\rho(r,t) = \rho_0 r^b$,
$u(r,t) = 0$,
$T(r,t) = T_0 r^{-b}$.
Free parameters: b, ρ_0, T_0

1.
$\rho(r,t) = \rho_0 r^b t^{-b-k-1}$,
$u(r,t) = \frac{r}{t}$,
$T(r,t) = T_0 r^{-b} t^{b-(\gamma-1)(k+1)}$.
Free parameters: b, k, ρ_0, T_0

2.
$\rho(r,t) = \rho_0 r^b t^{\frac{-2(b+k+1)}{2+(\gamma-1)(k+1)}}$,
$u(r,t) = \frac{2}{2+(\gamma-1)(k+1)} \frac{r}{t}$,
$T(r,t) = \frac{2(\gamma-1)(k+1)}{\Gamma(b+2)[(\gamma-1)(k+1)+2]^2} \frac{r^2}{t^2}$.
Free parameters: b, k, ρ_0

3.
$\rho(r,t) = \rho_0 \frac{r^b}{(\tau^2-t^2)^{(k+1+b)/2}}$,
$u(r,t) = \frac{-rt}{\tau^2-t^2}$,
$T(r,t) = \frac{\tau^2 r^2}{\Gamma(b+2)(\tau^2-t^2)^2}$,
with
$\gamma = \frac{k+3}{k+1}$.
Free parameters: b, k, τ, ρ_0

94

4.

$$\rho(r,t) = \frac{R_o^{b/\gamma} \tau^{\frac{(k+1)\gamma-1-b}{\gamma-1}} r^{-k-1}}{\left[R_o^{2-b/\gamma} - R_i^{2-b/\gamma}\right]^{\frac{1}{\gamma-1}}} \left(\frac{r}{(\tau^2-t^2)^{1/2}}\right)^{k+1-b/\gamma} .$$

$$\cdot \left[\left(\frac{r}{(\tau^2-t^2)^{1/2}}\right)^{2-b/\gamma} - \left(\frac{R_i}{\tau}\right)^{2-b/\gamma}\right]^{\frac{1}{\gamma-1}},$$

$$u(r,t) = \frac{-rt}{\tau^2-t^2},$$

$$T(r,t) = \frac{\tau^2(\gamma-1)r^{-2}}{\Gamma(2\gamma-b)}\left(\frac{r}{(\tau^2-t^2)^{1/2}}\right)^{2+b/\gamma}\left[\left(\frac{r}{(\tau^2-t^2)^{1/2}}\right)^{2-b/\gamma} - \left(\frac{R_i}{\tau}\right)^{2-b/\gamma}\right],$$

with $\gamma = \frac{k+3}{k+1}$.

Free parameters: b, k, τ, R_o, R_i

5.

$$\rho(r,t) = \frac{(a^2-t^2)^{(1-k)/2}}{r^2}\left[\rho_0 + \frac{a^2(\gamma-1)}{T_0\gamma}\log\frac{r}{(a^2-t^2)^{1/2}}\right]^{1/(\gamma-1)},$$

$$u(r,t) = \frac{-rt}{a^2-t^2},$$

$$T(r,t) = T_0\frac{r^2}{(a^2-t^2)^2}\left[\rho_0 + \frac{a^2(\gamma-1)}{T_0\gamma}\log\frac{r}{(a^2-t^2)^{1/2}}\right].$$

Free parameters: a, k, ρ_0, T_o

6. With conduction

$$\rho(r,t) = \rho_0 r^{\frac{k-1}{\beta-\alpha+4}} t^{-k-1-\frac{k-1}{\beta-\alpha+4}},$$

$$u(r,t) = \frac{r}{t},$$

$$T(r,t) = T_0 r^{\frac{1-k}{\beta-\alpha+4}} t^{(1-\gamma)(k+1)+\frac{k-1}{\beta-\alpha+4}}.$$

Free parameters: α, β, k, ρ_0, T_0

7. With conduction

$$\rho(r,t) = \rho_0 r^{-(2\beta+k+7)/\alpha} t^{\frac{-2[\alpha(k+1)-2\beta-k-7]}{\alpha[2+(\gamma-1)(k+1)]}},$$

$$u(r,t) = \frac{2}{2+(\gamma-1)(k+1)}\frac{r}{t},$$

$$T(r,t) = \frac{2\alpha(\gamma-1)(k+1)}{\Gamma[2+(\gamma-1)(k+1)]^2(2\alpha-2\beta-k-7)}\frac{r^2}{t^2}.$$

Free parameters: α, β, k, ρ_0

8. With conduction

$$\rho(r,t) = \rho_0 r^{-k},$$

$$u(r,t) = \frac{4c\lambda_0 a}{3}\frac{\gamma-1}{\Gamma\gamma}k\rho_0^{\alpha-1}T_0^{\beta+3},$$

$$T(r,t) = T_0 r^k,$$

with $\alpha = \beta + 4 - \frac{1}{k}$, $k \neq 0$.

Free parameters: β, k, ρ_0, T_0

9. With conduction

$$\rho(r,t) = \rho_0 r^{(\gamma-1)(k+1)-2} t^{1-k-(\gamma-1)(k+1)},$$

$$u(r,t) = \frac{r}{t},$$

$$T(r,t) = T_0 r^{2-(\gamma-1)(k+1)} t^{-2},$$

with $\alpha = \beta + 4 + \frac{k-1}{2-(\gamma-1)(k+1)}$.

Free parameters: k, ρ_0, T_0, either α or β

10. With conduction

$\rho(r,t) = \rho_0 r^{\frac{2}{\alpha-\beta-4}} t^{\frac{-2}{\alpha-\beta-4}-k-1}$,

$u(r,t) = \frac{r}{t}$,

$T(r,t) = T_0 r^{\frac{-2}{\alpha-\beta-4}} t^{\frac{\alpha(k+1)-k-2}{\beta+3}+\frac{2(\alpha-1)}{(\beta+3)(\alpha-\beta-4)}}$,

with

$T_0 = \left[\frac{3\Gamma}{4c\lambda_0 a(\gamma-1)} \frac{\alpha-1+(\beta+3)(\gamma-1)}{\beta+3} \rho_0^{1-\alpha} \frac{\beta+4-\alpha}{2} \right]^{\frac{1}{\beta+3}}$.

Free parameters: α, β, k, ρ_0

11. With conduction

$\rho(r,t) = \rho_0 r^{-k-b}$,

$u(r,t) = r^b \sqrt{\Gamma T_0 \frac{k-b}{b}}$,

$T(r,t) = T_0 r^{2b}$,

with

$b = \frac{k-1-\alpha k}{2+\alpha-2(\beta+4)}$, and

$T_0 = \left\{ \frac{b}{\Gamma(k-b)} \left(\frac{4c\lambda_0 a}{3} \frac{\gamma-1}{\Gamma} \right)^2 \frac{\rho_0^{2\alpha-2} 16 b^4}{[2b+(\gamma-1)(k+b)]^2} \right\}^{\frac{-1}{5+2\beta}}$.

Free parameters: α, β, k, ρ_0

12. With conduction

$\rho(r,t) = \rho_0 r^{\frac{2}{\alpha-\beta-4}} t^{-k-1-\frac{2}{\alpha-\beta-4}}$,

$u(r,t) = \frac{r}{t}$,

$T(r,t) = T_0 r^{\frac{-2}{\alpha-\beta-4}} t^{-2}$,

with

$\frac{2}{\alpha-\beta-4} = \frac{k+4-\alpha(k+1)-2(\beta+4)}{\alpha-1}$, and

$T_0 = \left[\frac{3}{4c\lambda_0 a} \frac{\Gamma}{2(k+1)(\gamma-1)} \rho_0^{1-\alpha} \left\{ 2 + [2-(\gamma-1)(k+1)](\alpha-\beta-4) \right\} \right]^{\frac{1}{\beta+3}}$.

Free parameters: k, ρ_0, either α or β

13. With conduction

$\rho(r,t) = \rho_0 r^{-k-b}$,

$u(r,t) = u_0 r^b$,

$T(r,t) = \frac{u_0^2 b}{\Gamma(k-b)} r^{2b}$,

with

$\alpha = 1 - \frac{1}{k}$, $\beta = \frac{1}{2}\alpha - 3$, $k \neq 0$, and

$\rho_0 = \left[\frac{16c\lambda_0 a}{3}(\gamma-1) \right]^k \frac{b^{(5k-1)/2}}{u_0(k-b)^{(k-1)/2}\Gamma^{(3k-1)/2}[2b+(\gamma-1)(k+b)]^k}$.

Free parameters: b, k, u_0

14. With conduction

$\rho(r,t) = \rho_0 r^{\frac{2\beta-4}{1-\alpha}} t^{\frac{2\beta+5}{\alpha-1}}$,

$u(r,t) = u_0 \frac{r}{t}$,

$T(r,t) = T_0 \frac{r^2}{t^2}$,

with $u_0 = \frac{2\beta+5}{2\beta-4+(1-\alpha)(k+1)}$,

$$T_0 = \frac{(\alpha-1)(2\beta+5)[9-(1-\alpha)(k+1)]}{\Gamma[2\beta-4+(1-\alpha)(k+1)]^2[2\beta-4+2(1-\alpha)]},$$

$$\rho_0 = \left[\frac{-2+u_0[2+(\gamma-1)(k+1)]}{2\left[\frac{\alpha}{1-\alpha}(2\beta-4)+2\beta+k+7\right]} \frac{3}{4c\lambda_0 a} \frac{\Gamma}{\gamma-1} T_0^{-\beta-3}\right]^{\frac{1}{\alpha-1}}.$$

Free parameters: α, β, k

15. With conduction

$$\rho(r,t) = \rho_0 r^{-\frac{1}{\alpha}(2\beta+k+7)}(\tau^2 - t^2)^{-\frac{1}{2}(k+1)+\frac{1}{2\alpha}(2\beta+k+7)},$$

$$u(r,t) = \frac{-rt}{\tau^2-t^2},$$

$$T(r,t) = \frac{\alpha\tau^2}{\Gamma(2\alpha-2\beta-k-7)}\frac{r^2}{(\tau^2-t^2)^2},$$

with $\gamma = \frac{k+3}{k+1}$.

Free parameters: α, β, k, τ, ρ_0

16. Shock, no conduction

Region 1:

$$\rho(r,t) = \rho_0 \left(\frac{\gamma+1}{\gamma-1}\right)^{k+1},$$

$$u(r,t) = 0,$$

$$T(r,t) = \frac{u_0^2(\gamma-1)}{2\Gamma},$$

Region 2:

$$\rho(r,t) = \rho_0 \left(\frac{r-u_0 t}{r}\right)^k,$$

$$u(r,t) = u_0(<0),$$

$$T(r,t) = 0.$$

The shock location is $R = -\frac{1}{2}(\gamma-1)u_0 t$.

Free parameters: k, u_0, ρ_0

17. Shock, no conduction

Region 1:

$$\rho(r,t) = \rho_0 \left(\frac{\gamma+1}{\gamma-1}\right)^{k+1}(1-at)^{-k-1},$$

$$u(r,t) = -\frac{ar}{1-at},$$

$$T(r,t) = \frac{u_0^2(\gamma-1)}{2\Gamma}(1-at)^{-2},$$

Region 2:

$$\rho(r,t) = \rho_0(1-at)^{-k-1}\left(\frac{r-u_0 t}{r}\right)^k,$$

$$u(r,t) = \frac{u_0-ar}{1-at},$$

$$T(r,t) = 0.$$

The shock location is $R = \frac{u_0(\gamma-1)}{4a}\frac{t(1-2at)}{1-at}$ with $\gamma = \frac{k+3}{k+1}$.

Free parameters: a, k, ρ_0, u_0

B. 2–D axisymmetric flow, (r,θ) or (R,z): Solutions 18 – 21 come from $\mathcal{H}_7$ with the ansatz $M = M_0(a + b\cos^2\theta)$, $V = V_0\sin\theta\cos\theta$.

18. $\beta \neq -2$, $bM_0 + V_0 = 0$, $aM_0 = 0$:

$$\rho(r,\theta,t) = \rho_0 t^{-2(\beta+1)/(\gamma+1)}(r\cos\theta)^\beta$$

$$u(r,\theta,t) = \frac{2}{(\gamma+1)}\frac{r}{t}\cos^2\theta$$

$$v(r,\theta,t) = -\frac{2}{(\gamma+1)}\frac{r}{t}\sin\theta\cos\theta$$
$$T(r,\theta,t) = \frac{2(\gamma-1)}{\Gamma(\gamma+1)^2(\beta+2)}\left(\frac{r}{t}\right)^2\cos^2\theta$$
Free parameters: β, ρ_0

19. $\beta \neq -2$, $bM_0 + V_0 = 0$, $aM_0 = 1$:
$$\rho(r,\theta,t) = \rho_0 t^{-\beta-3+3(\beta+1)(\gamma-1)/(\gamma+1)}\left(r\cos\theta\right)^\beta$$
$$u(r,\theta,t) = \frac{r}{t}\left(1 - 3\frac{\gamma-1}{\gamma+1}\cos^2\theta\right)$$
$$v(r,\theta,t) = 3\frac{\gamma-1}{\gamma+1}\frac{r}{t}\sin\theta\cos\theta$$
$$T(r,\theta,t) = \frac{6(\gamma-1)(2-\gamma)}{\Gamma(\gamma+1)^2(\beta+2)}\left(\frac{r}{t}\right)^2\cos^2\theta$$
Free parameters: β, ρ_0

20. $\beta = -2$, $aM_0 = 0$:
$$\rho(r,\theta,t) = \rho_0 t^{2-\gamma}r^{-2}\left(\sin\theta\right)^{1-\gamma}\left(\cos\theta\right)^{\gamma-3}$$
$$u(r,\theta,t) = \frac{r}{t}\cos^2\theta$$
$$v(r,\theta,t) = -\frac{r}{t}\sin\theta\cos\theta$$
$$T(r,\theta,t) = T_0\left(\frac{r}{t}\right)^2\left(\sin\theta\right)^{\gamma-1}\left(\cos\theta\right)^{3-\gamma}$$
Free parameters: ρ_0, T_0

21. $\beta = -2$, $aM_0 = 1$:
$$\rho(r,\theta,t) = \rho_0 t^{2(1-\gamma)}r^{-2}\left(\sin\theta\right)^{2\gamma-4}\left(\cos\theta\right)^{2-2\gamma}$$
$$u(r,\theta,t) = \frac{r}{t}(1-\cos^2\theta)$$
$$v(r,\theta,t) = \frac{r}{t}\sin\theta\cos\theta$$
$$T(r,\theta,t) = T_0\left(\frac{r}{t}\right)^2\left(\sin\theta\right)^{4-2\gamma}\left(\cos\theta\right)^{2\gamma-2}$$
Free parameters: ρ_0, T_0

22. $\mathcal{H}_{12}$ with the ansatz $A = 0$:
$$\rho(R,z,t) = \rho_0\exp\left(\alpha t + \beta z - \alpha\beta t^2\right)$$
$$u^R(R,z,t) = 0,$$
$$u^z(R,z,t) = -\frac{\alpha}{\beta} + 2at$$
$$T(R,z,t) = -\frac{2a}{\Gamma\beta}$$
Free parameters: ρ_0, α, β, a

23. $\mathcal{H}_{14}$ with the ansatz $M = 0$:
$$\rho(r,\theta,t) = \rho_0 r^{2[-2a+\gamma(a+1)]/(2-\gamma)}\left(\sin\theta\right)^{-2/(\gamma+1)}$$
$$u(r,\theta,t) = 0$$
$$v(r,\theta,t) = v_0 r^a\left(\sin\theta\right)^{(1-\gamma)/(1+\gamma)}$$
$$T(r,\theta,t) = v_0^2\frac{1-\gamma}{2\Gamma\gamma}r^{2a}\left(\sin\theta\right)^{(2-2\gamma)/(\gamma+1)}$$
Free parameters: ρ_0, a, v_0

24. $\mathcal{H}_{14}$ with the ansatz $M = M_0\cos\theta$, $V = V_0\sin\theta$:
$$\rho(r,\theta,t) = \rho_0 r^{-2\gamma/(2\gamma-1)}\left(\sin\theta\right)^{(2-2\gamma)/(2\gamma-1)}$$

$$= u_0\cos\theta$$
$$(\;\lor\;,t) = u_0\frac{1-\gamma}{\gamma}\sin\theta$$
$$T(r,\theta,t) = u_0^2\frac{(\gamma-1)(2\gamma-1)}{2\Gamma\gamma^3}\sin^2\theta$$
Free parameters: $\rho_0,\ v_0,\ u_0$

25. (Solution 19, projected)
$$\rho(r,\theta,t) = \rho_0(r\cos\theta)^\beta(t-\tau)^{-\beta-3+3(\beta+1)(\gamma-1)/(\gamma+1)}$$
$$[1+s(t-\tau)]^{-3(\beta+1)(\gamma-1)/(\gamma+1)}$$
$$u(r,\theta,t) = \frac{r}{s\{[t-\tau+1/(2s)]^2-1/(4s^2)\}}\left(1-3\frac{\gamma-1}{\gamma+1}\cos^2\theta\right) + \frac{sr}{1+s(t-\tau)}$$
$$v(r,\theta,t) = \frac{3(\gamma-1)r\sin\theta\cos\theta}{s(\gamma+1)\{[t-\tau+1/(2s)]^2-\frac{1}{4s^2}\}}$$
$$T(r,\theta,t) = \frac{6(\gamma-1)(2-\gamma)}{\Gamma(\gamma+1)^2(\beta+2)}\cos^2\theta\left[\frac{r}{s\{[t-\tau+1/(2s)]^2-1/(4s^2)\}}\right]^2,$$
Free parameters: $\rho_0,\ \beta,\ s$. This solution reduces to Solution 19 when $\gamma \neq 5/3$.

C. 2–D cylindrical solutions (R,ϕ):

Here the groups $\langle U_{ss}+c_1U_{st}+c_2U_{xy}+c_3U_{s\rho},\ U_{xy}+c_4U_{ss}+c_5U_{s\rho}\rangle$ were used to generate the similarity variables
$$\lambda = R^a t^b e^{-c\phi},\ \rho = H(\lambda)\left(\frac{R}{t}\right)^d t^e,\ u^R = M(\lambda)\frac{R}{t},$$
$$u^\phi = V(\lambda)\frac{R}{t},\ T = G(\lambda)\left(\frac{R}{t}\right)^2.$$

26. $\gamma = 2,\ e = 0$:
$$\rho(R,\phi,t) = \rho_0\left(\frac{R}{t}\right)^2$$
$$u^R(R,\phi,t) = \frac{R}{2t}$$
$$u^\phi(R,\phi,t) = v_0\frac{R}{t}$$
$$T(R,\phi,t) = \frac{4v_0^2+1}{16\Gamma}\left(\frac{R}{t}\right)^2$$
Free parameters: $\rho_0,\ v_0$

27. $\gamma = 2,\ e = 0$:
$$\rho(R,\phi,t) = \rho_0 R^{-2}\left(\frac{R}{t}\right)^a$$
$$u^R(R,\phi,t) = \frac{R}{t}$$
$$u^\phi(R,\phi,t) = v_0 R^{-1}\left(\frac{R}{t}\right)^b$$
$$T(R,\phi,t) = \frac{v_0^2}{\Gamma(a+2b-4)}R^{-2}\left(\frac{R}{t}\right)^{2b}$$
Free parameters: $\rho_0,\ v_0,\ a,\ b$

28.
$$\rho(R,\phi,t) = \rho_0\left(\frac{R}{t}\right)^{d-bf}R^{f(a+b)}e^{-c\phi}$$
$$u^R(R,\phi,t) = \frac{R}{\gamma t}$$
$$u^\phi(R,\phi,t) = \frac{-d+(2+d+af)/\gamma}{c}\frac{R}{t}$$

$$T(R,\phi,t) = \frac{(2-\gamma)(1-\gamma)R^2}{\Gamma\gamma^2[(\gamma-1)(af+2)+fb\gamma-d]t^2}$$
with $c^2 = \frac{\gamma}{1-\gamma}[f(b+a/\gamma) + (d+2)/\gamma) - d][(\gamma-1)(af+2) + fb\gamma - d]$
Free parameters: ρ_0, a, b, d, f

29. $V = 0$:
$$\rho(R,\phi,t) = \rho_0 \left(\frac{R}{t}\right)^a t^{-2} e^{c\phi}$$
$$u^R(R,\phi,t) = \frac{R}{t}$$
$$u^\phi(R,\phi,t) = 0$$
$$T(R,\phi,t) = T_0 \left(\frac{R}{t}\right)^{-a} t^{2(1-\gamma)} e^{-c\phi}$$
Free parameters: ρ_0, a, T_0

D. 3–D solution, spherical coordinates (r,θ,ϕ): **30.** $V = 0$, $H = H_0 + H_1(\sin\theta)^a$, $c = 0$, $\gamma = 5/3$:
$$\rho(r,\theta,\phi,t) = t^{-(b+3)/2} r^b (\rho_0 + \rho_1 \sin^a\theta)$$
$$u(r,\theta,\phi,t) = \frac{r}{2t}$$
$$v(r,\theta,\phi,t) = 0$$
$$w(r,\theta,\phi,t) = \pm\frac{r}{4t}\left[\frac{4\Gamma T_0(b+2)(\sin\theta)^{b+2}}{\rho_0+\rho_1\sin^a\theta} - \frac{a\rho_0}{(\rho_0+\rho_1\sin^a\theta)(b+2-a)} + \frac{a}{(b+2-a)}\right]^{1/2}$$
$$T(r,\theta,\phi,t) = \left(\frac{r}{4\Gamma t}\right)^2 \left[\frac{4\Gamma T_0(\sin\theta)^{b+2}}{\rho_0+\rho_1\sin^a\theta} - \frac{a\rho_0}{(\rho_0+\rho_1\sin^a\theta)(b+2)(b+2-a)} + \frac{1}{(b+2-a)}\right]$$
Free parameters: ρ_0, ρ_1, T_0, a, b

31. (Solution 30, projected; $\gamma = 5/3$)
$$\rho(r,\theta,\phi,t) = [(t-\tau)(1+s(t-\tau))]^{-(b+3)/2} r^b (\rho_0 + \rho_1 \sin^a\theta)$$
$$u(r,\theta,\phi,t) = \frac{r}{2s\{[t-\tau+1/(2s)]^2-1/(4s^2)\}} + \frac{sr}{1+s(t-\tau)}$$
$$v(r,\theta,\phi,t) = 0$$
$$w(r,\theta,\phi,t) = \pm\frac{1}{4}\left[\frac{4\Gamma T_0(b+2)(\sin\theta)^{b+2}}{\rho_0+\rho_1\sin^a\theta} - \frac{a\rho_0}{(\rho_0+\rho_1\sin^a\theta)(b+2-a)} + \frac{a}{b+2-a}\right]^{1/2}$$
$$\times \frac{r}{s\{[t-\tau+1/(2s)]^2-1/(4s^2)\}}$$
$$T(r,\theta,\phi,t) = \frac{1}{4\Gamma}\left[\frac{4\Gamma T_0(\sin\theta)^{b+2}}{\rho_0+\rho_1\sin^a\theta} - \frac{a\rho_0}{(\rho_0+\rho_1\sin^a\theta)(b+2)(b+2-a)} + \frac{1}{(b+2-a)}\right]$$
$$\times \left[\frac{r}{s\{[t-\tau+1/(2s)]^2-1/(4s^2)\}}\right]^2 .$$
Free parameters: ρ_0, ρ_1, T_0, a, b, s

References

1. P.J. Olver, Application of Lie Groups to Differential Equations, Springer, NY 1986.
2. L.V. Ovsiannikov, Group Analysis of Differential Equations, Academic Press, NY 1982.
3. G.W. Bluman and J.D. Cole, Similarity Methods for Differential Equations, Springer-Verlag, 1974.

4. N.H. Ibragimov, Transformation Groups Applied to Mathematical Physics, D. Reidel, Dordrecht, Holland 1985.
5. G.W. Bluman and S. Kumei, Symmetries and Differential Equations, Springer-Verlag, NY 1989.
6. H. Stephani, Differential equations: Their solutions using symmetries, Cambridge University Press, Cambridge 1989.
7. S.V. Coggeshall and Roy A. Axford, "Lie group-invariance properties of radiation hydrodynamic equations and their associated similarity solutions", Phys. Fluids 29, 2398 (1986).
8. S.V. Coggeshall and J. Meyer-ter-Vehn, "Group-invariant solutions and optimal systems for multidimensional hydrodynamics", J. Math. Physics 33, 3585 (1992).
9. G.A. Birkhoff, Hydrodynamics; a Study in Logic, Fact and Simultude, Princeton Press, 1950.
10. L.I. Sedov, Similarity and Dimensional Methods in Mechanics, Academic Press, NY, 1959.
11. V.G. Guderley, Luftfahrtforschung 19, 302 (1942); J. Meyer-ter-Vehn and C. Schalk, Z. Naturforschung Teil A 37, 955 (1982).
12. W.F. Noh, "Artificial Viscosity (Q) and Artificial Heat Flux Errors for Spherically Divergent Shocks", Lawrence Livermore Report UCRL 89623, (1983).
13. S.V. Coggeshall, "Analytic solutions of hydrodynamic equations," Phys. Fluids A 3, 757 (1991).

Survey on Exact Solutions for Discrete Models of the Boltzmann Equation

Henri Cabannes

Lab. Modélisation en Mécanique, Université Pierre et Marie Curie
4 Place Jussieu, 75005 Paris, France

Summary. In Discrete Kinetic Theory the evolution of a gas is governed by a system of semi-linear partial differential equations. This paper is a review of exact solutions which have been obtained for some of those systems.

1. Introduction

The study of rarefied gas dynamics is based on the famous equation obtained by Boltzmann in 1872. That equation, integro-partial derivative, is very complex and several simplified models have been written and studied. Among those are discrete models obtained by the assumption that the velocities of the molecules of the gas can belong only to a finite set of vectors. The first discrete model has been introduced by Carleman [1] for a gas with two velocities; in 1964 two models, one with six velocities, another one with eight velocities have been introduced by Broadwell [2], [3]. The kinetic equations for the most general discrete model have been written by R. Gatignol [4], [5]. The works in Discrete Kinetic Theory can be set in two groups: 1. The study of flows, 2. the study of evolution of gas from given initial data. The authors of the works in the second group tried to prove the global existence of the solution of the initial value problem with extended conditions: one-dimensional flows, multi-dimensional flows, small initial densities, then bounded initial densities, binary collisions, multiple collisions, positive initial densities, then partially negative initial densities. Concurrently other authors, after Cornille [6], [7], succeeded in building exact solutions, more and more interesting and for more and more complex models.

Those exact solutions, beyond the esthetic interest, are useful because they have allowed on one hand to test the value of numerical integrations, on the other hand to observe that positivity of initial densities is not necessary to provide the global existence of the solution of the initial value problem. This last result was the starting point of several new researchs. In this paper we will give a survey of the main exact solutions obtained for various equations, which appear in Discrete Kinetic Theory.

2. Broadwell Equations

In Discrete Kinetic Theory, the simplest kinetic equations are the Broadwell equations.

$$\frac{\partial N_1}{\partial t} + c\frac{\partial N_1}{\partial x} = \frac{2cS}{3}\left(N_2^2 - N_1 N_4\right) \tag{2.1}$$

$$\frac{\partial N_4}{\partial t} - c\frac{\partial N_4}{\partial x} = \frac{2cS}{3}\left(N_2^2 - N_1 N_4\right) \tag{2.2}$$

$$\frac{\partial N_2}{\partial t} = \frac{cS}{3}\left(N_1 N_4 - N_2^2\right) \tag{2.3}$$

The three unknown functions N_1, N_4 and N_2 depend on two variables t, time and x, abscissa. The quantities c and S denote a constant velocity and a constant area respectively.

2.1 Self-Similar Solutions

The self-similar solutions are solutions for which the system (2.1, 2.2, 2.3) can be replaced by a system of ordinary differential equations. To simplify the writing we will assume from now that one has $c = 1$ and $2cS = 3$; one introduces the variable $\sigma = (x/t)$ and one looks, for the Broadwell equations, for solutions in the form (2.4). The three new functions $U(\sigma), V(\sigma)$ and $W(\sigma)$, satisfy the system (2.5).

$$N_1(t, x) = \frac{U(\sigma)}{t}, \qquad N_4(t, x) = \frac{V(\sigma)}{t}, \qquad N_2(t, x) = \frac{W(\sigma)}{t} \tag{2.4}$$

$$\frac{d}{d\sigma}\{(1-\sigma)U\} = -\frac{d}{d\sigma}\{(1+\sigma)V\} = \frac{d}{d\sigma}\{2\sigma W\} = W^2 - UV \tag{2.5}$$

Eqs. (2.5) possess two first integrals, which allow to explicit functions U and V with W and two integration constants a and b. The function $\varphi(\sigma) = 2\sigma W(\sigma)$ satisfies Eq. (2.7), a Ricatti equation.

$$U(\sigma) = \frac{a + b + 4\sigma W(\sigma)}{2(1 - \sigma)}, \qquad V(\sigma) = \frac{a - b - 4\sigma W(\sigma)}{2(1 + \sigma)} \tag{2.6}$$

$$4(1 - \sigma)\frac{d\varphi}{d\sigma} = \frac{1 + 3\sigma^2}{\sigma^2}\varphi^2 + 4b\varphi + b^2 - a^2 \tag{2.7}$$

For Eq. (2.7) the physically valid solutions are those for which densities are positive. As we are limited to the positive values of time, the three functions U, V and W must be positive. In the σ, φ plane, the straight lines $\varphi = 0, \pm 1$ and $\sigma = 0, (\pm a - b)/2$ divide the plane in 16 domains. Assuming $-b < a < b$ the condition of positivity of densities is satisfied only in the two following domains:

$$\mathcal{R}_1 : -1 \leq \sigma \leq 0, \quad -(a+b) < 2\varphi < 0; \quad \mathcal{R}_2 : 0 \leq \sigma \leq 1, \quad 0 < 2\varphi < (a-b).$$

In those domains Eq. (2.5) possesses three singular points, two nodes: origin and point $A(\sigma = -1, \varphi = -(a+b)/2)$, and a saddle point: $B(\sigma = 1, (a-b)/2)$. The integral curve AOB, joining the three singular points represents a solution of Eqs. (2.1, 2.2, 2.3), valid in the angular domain $-t \leq x \leq t$; outside that domain the solution cannot be longer self-similar because the densities would not be all positive; but another solution could be valuable as the points A (where $\sigma = -1$, $x = -t$) and B (where $\sigma = 1$, $x = t$) correspond to characteristics. Otherwise the other integral curves cannot represent a solution with positive densities, as they leave the domains $\mathcal{R}_1$ and $\mathcal{R}_2$ outside singular points, and as before, at least one of the densities becomes negative. The curve AOB is represented in Fig. 2.1; the values of U, V, and W, corresponding to the AOB curve, are represented in Figs. 2.2a,b,c.

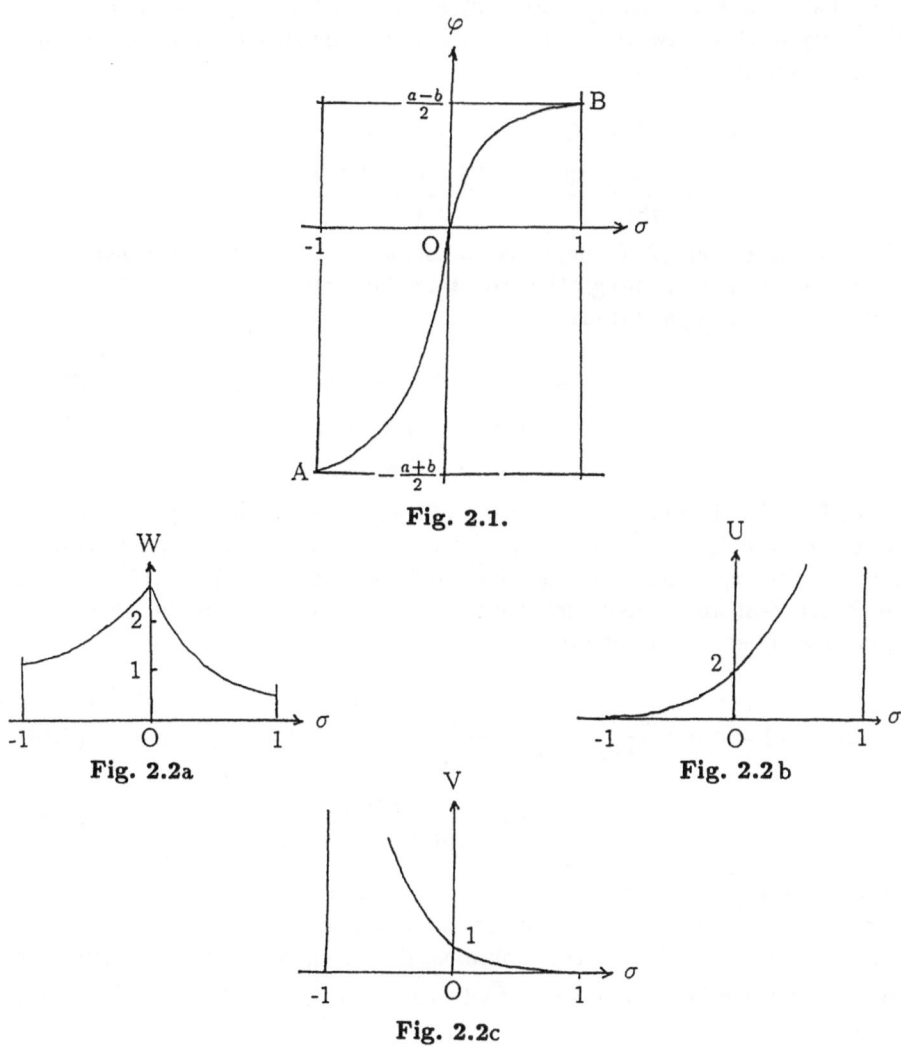

Fig. 2.1.

Fig. 2.2a

Fig. 2.2 b

Fig. 2.2c

One also obtains self-similar solutions for all one-dimensional discrete models of the Boltzmann equation with binary collisions: Eqs. (2.8), and also with multiple collisions, under the condition that all multiple collisions are of the same order.

$$\frac{\partial N_j}{\partial t} + u_j \frac{\partial N_j}{\partial x} = \sum_{lm} A_{jk}^{lm}(N_l N_m - N_j N_k), \qquad (j = 1, 2, ..., p) \quad (2.8)$$

Quantities A_{jk}^{lm} are constants positive or zero.

2.2 Cornille's Solutions

It is always possible to look for solutions of the Broadwell equations as solitons, that means to put: $N_j(t, x) = F_j(\xi)$, $j = 1, 2, 4$, $\xi = \sigma x + it$, $(i = \sqrt{-1})$. By putting always $c = 1$ and $S = (3/2)$ the Broadwell equations are reduced to a quadrature:

$$(\sigma + i)F_1 + 2iF_2 = a, \quad (\sigma - i)F_4 - 2iF_2 = -b \qquad (2.9)$$

$$2i\frac{dF_2}{d\xi} = \frac{(3 - \sigma^2)F_2^2 + (a + b)2iF_2 - ab}{1 + \sigma^2} \qquad (2.10)$$

The solution of Eq. (2.10) depends on coefficient σ, integration constants a and b and on a third integration constant. One obtains, for Eqs. (2.9) and (2.10) the following solutions:

$$2F_2 = \gamma - i\,\alpha\,\tan\xi \qquad (2.11)$$

$$2F_1 = \alpha + (\sigma - i)\,\gamma\,\tan\xi \qquad (2.12)$$

$$2F_4 = \alpha - (\sigma + i)\,\gamma\,\tan\xi \qquad (2.13)$$

As the Broadwell equations have real coefficients, one obtains again a solution by replacing $F_j(\xi)$ by $\overline{F}_j(\overline{\xi})$. Cornille has observed the possibility to choose the constants α, γ, and σ in order that the functions $N_j(t, x) = F_j(\xi) + \overline{F}_j(\overline{\xi})$, which are real and called bisolitons, are also a solution of the Broadwell equations. These solutions are:

$$N_1(t, x) = \alpha + \gamma\frac{sh(2t) + \sigma\sin(2\sigma x)}{ch(2t) + \cos(2\sigma x)}, \quad N_4(t, x) = N_1(t, -x) \qquad (2.14)$$

$$N_2(t, x) = \gamma + \alpha\frac{sh(2t)}{ch(2t) + \cos(2\sigma x)} \qquad (2.15)$$

with $\alpha = -4\frac{\sqrt{3}}{3}$, $\gamma = \frac{2(3+\sqrt{3})}{3}$, $\sigma^2 = 2\sqrt{3} - 3$ The solution (2.14, 2.15), Cornille's solution, corresponds to an explosion at the time $t = 0$, at the point $2\sigma x = \pi$. The three functions, densities, $N_j(t, x)$ are positive for all x, if one has $2t \geq Log\,(2 + \sqrt{3} + \sqrt{10 + 6\sqrt{3}}) = 2.10995....$ Therefore, if one chooses as initial time a value t_0 positive, but smaller than 1.0549, the functions (2.14, 2.15) represent, for the initial value problem, a global solution in time which corresponds to partially negative initial densities.

The densities (2.14, 2.15) are periodic functions in x, with period π/σ. As one has:
$N_1(t,0) = N_4(t,0)$ and $N_1(t,\pi/2\sigma) = N_4(t,\pi/2\sigma)$, the Cornille solution represents the motion of a gas between two parallel planes ($x = 0$ and $x = \pi/2\sigma$) on which the molecules rebound following the laws of specular reflection. Putting $T = 2t, X = 2\sigma x$, the Fourier's expansion of functions $N_j(t,x)$ are the following

$$N_1(t,x) = \alpha + \gamma \left\{ 1 + 2\sum_{n=1}^{\infty}(-1)^n e^{-nT}[\cos(nX) - \sigma\sin(nX)]\right\} \qquad (2.16)$$

$$N_2(t,x) = \gamma + \alpha \left\{ 1 + 2\sum_{n=1}^{\infty}(-1)^n e^{-nT}\cos(nX) \right\} \qquad (2.17)$$

3. Models with 14 Velocities

The Broadwell equations correspond to a six-velocity model with some symmetries. In fact, as a consequence of symmetries one has obtained three partial differential equations for three unknown functions, depending on two variables; the system consists, in fact, of two linear equations and one nonlinear. It was natural to try to use the initial Cornille idea to found exact solutions for kinetic equations corresponding to more complex discrete models. This has been made for models with 14 velocities related to the cube [8].

Kinetic equations corresponding to the first model (model for which the ratio of moduli of velocities is equal to $\sqrt{3}$) are the following:

$$\frac{\partial N_1}{\partial t} - \frac{\partial N_1}{\partial x} = \frac{\sqrt{6}}{2}(N_2 N_4 - N_1 N_3) \qquad (3.1)$$

$$\frac{\partial N_2}{\partial t} + \frac{\partial N_2}{\partial x} = \frac{\sqrt{6}}{2}(N_1 N_3 - N_2 N_4) \qquad (3.2)$$

$$\frac{\partial N_3}{\partial t} + \frac{\partial N_3}{\partial x} = 2\sqrt{6}(N_2 N_4 - N_1 N_3) + \frac{4}{3}(N_5^2 - N_3 N_4) \qquad (3.3)$$

$$\frac{\partial N_4}{\partial t} - \frac{\partial N_4}{\partial x} = 2\sqrt{6}(N_1 N_3 - N_2 N_4) + \frac{4}{3}(N_5^2 - N_3 N_4) \qquad (3.4)$$

$$\frac{\partial N_5}{\partial t} = \frac{2}{3}(N_3 N_4 - N_5^2) \qquad (3.5)$$

From Eqs. (3.1,..,3.5) one can deduce three linear equations, and the system (3.1,..,3.5) is, in fact a system of three linear equations and only two nonlinear ones. When densities $N_1(t,x)$ and $N_2(t,x)$ are null, one again finds the Broadwell equations.

The method used to obtain exact solutions is general and can be applied to the general system (2.8); unknowns are functions $N_j(t, x)$, the quantities u_j are constant, the quantities A_{jk}^{lm} are constant, positive or zero. The method consists to try to satisfy equations (3.6) with functions in the form (3.7); the method can, however, only succeed if the number of discrete velocities is not too large.

$$\frac{\partial N_j}{\partial t} + u_j \frac{\partial N_j}{\partial x} = \sum_{lm} A_{jk}^{lm}(N_l N_m - N_j N_k), \qquad (j = 2, 2, ..., p) \qquad (3.6)$$

$$N_j(t, x) = \alpha_j + 2\mathcal{R}e\left\{a_j \tan(\sigma x + it)\right\}, \qquad (i = \sqrt{-1}) \qquad (3.7)$$

Coefficients α_j and σ are real numbers; coefficients a_j are complex numbers, the notation $\mathcal{R}e\ z$ denotes the real part of the complex number z. The number of real parameters to be determined is also $3p+1$. From Eqs. (3.6) one deduces q linear equations; $q = 2$ in the case of Broadwell equations, $q = 3$ in the case of Eqs. (3.1,..,3.5). To write that functions (3.7) satisfy Eqs. (3.6) one must identify constant terms and terms in $\tan(\sigma x + it), \tan^2(\sigma x + it)$ and $|\tan(\sigma x + it)|^2$, which leads to write six relations for each of the Eqs. (3.6). In the case of linear equations deduced from system (3.6) the number of relations to write is only 2. Finally, in order that functions (3.7) represent a particular solution of system (3.6), it is enough that the $3p + 1$ parameters satisfy a system of $6p - 4q$ algebraic equations of second degree.

Coming back to system (3.1,..,3.5), $p = 5$, $q = 3$ one obtains a system of 18 equations for 16 unknowns. As a consequence of symmetries the system possesses a solution which has been obtained first by hand [8], then by computer with the help of Macsyma software [9]. The coefficient σ is the solution of the equation $\sigma^2 = 2\sqrt{3} - 3$ (this corresponds to the particular case of the Broadwell system), or of the equation:

$$5(3 - 2\sqrt{3})\sigma^4 - 2(3 + 8\sqrt{6})\sigma^2 + 3(5 - 2\sqrt{6}) = 0 \qquad (3.8)$$

Introducing the parameter τ so that $16\sigma\tau = (10 - 3\sqrt{6})\sigma^2 - 3(2 - \sqrt{6})$, the values of the other coefficients are the following:

$$\alpha_1 = \alpha_2 = \frac{1 + \sigma^2}{4(\tau - \sigma)\sqrt{6}}, \quad \alpha_3 = \alpha_4 = -\frac{1 + \sigma^2}{\sqrt{6}}, \quad \alpha_5 = \frac{2(1 + \sigma\tau)}{\sqrt{6}} \qquad (3.9)$$

$$a_1 = -\bar{a}_2 = -\frac{\tau(1 - i\sigma)}{4\sqrt{6}}, \quad a_3 = -\bar{a}_4 = \frac{\sigma - \tau - i(1 + \sigma^2)}{\sqrt{6}}, \quad a_5 = \frac{i(1 + \sigma^2)}{\sqrt{6}}$$
$$(3.10)$$

Functions $N_j(t, x)$ are periodic in x, with the period $2\pi/\sigma$; furthermore, one has $N_2(t, x) = N_1(t, -x)$ and $N_4(t, x) = N_3(t, -x)$. As a result the exact solution, defined by formula (3.7) and (3.9, 3.10) represents the motion of a gas between two parallel plane walls; the walls are the planes $x = 0$ and

$x = \pi/\sigma$; on the walls the molecules rebound with the law of specular reflection. When the time increases indefinitely functions $N_j(t, x)$ tend to constant positive values, and the gas is at rest in a maxwellian state; those functions are positive for $t \geq 0.22$; for $0 < t \leq 0.22$ they are partially negative. The negative values of the time are less interesting because there is an explosion (infinite densities) at time $t = 0$ at point $x = \pi/(2\sigma)$. Also, as the Cornille solution, the solution (3.7) - (3.10), represents, for Eqs. (3.6, 3.7), a global solution in time, corresponding to partially negative initial data.

A second 14-velocity model exists, [10], for which the equation for σ, similar to Eq. (3.9, 3.10) is:

$$2184\, \sigma^6 + 4264\, \sigma^4 + 2306\, \sigma^2 - \sqrt{11} \left\{ 696\, \sigma^6 + 1562\, \sigma^4 + 640\, \sigma^2 + 71 \right\} = 0 \tag{3.11}$$

That last equation, with degree three in σ^2, possesses only one real root ($\sigma^2 = -7.565...$); that real root being negative, Eq. (3.11) possesses no real root, and the kinetic equations of the second 14-velocity model do not possess exact solutions in the form of (3.7).

4. Models with Triple Collisions

In the study of evaporation and condensation problems we have have to consider models with ten collisions obtained by cancellation of four velocities in each of the models studied in the former section. In relation with the physical origin of the question one assumes some symmetries; those hypotheses have, as a new consequence, for one of the models, the disparition of terms corresponding to binary collisions; and for this model it is necessary to consider triple collisions. We have also obtained two ten-velocity models with triple collisions. Kinetic equations corresponding to one of those models are the following:

$$\frac{\partial N_1}{\partial t} + \frac{\partial N_1}{\partial x} = -\frac{\partial N_3}{\partial t} + \frac{\partial N_3}{\partial x} = \beta(N_3^2 N_2 - N_1^2 N_4) \tag{4.1}$$

$$\frac{\partial N_2}{\partial t} + 2\frac{\partial N_2}{\partial x} = -\frac{\partial N_4}{\partial t} + 2\frac{\partial N_4}{\partial x} = \beta(N_1^2 N_4 - N_3^2 N_2) \tag{4.2}$$

Exact solutions of system (4.1, 4.2) have been obtained by Tiem [11], by a generalization of results obtained in the former sections. By introducing always the complex variable $z = \sigma x + it$ Tiem tried, for system (4.1, 4.2), to obtain solutions in the form of (4.3), similar to the form (3.7), but in which one has replaced function $\tan(z)$, which satisfies the differential equation $(df/dz) = 1 + f^2$, by a function $f(z)$ solution of differential equation (4.4).

$$N_j(t, x) = \alpha_j + 2\mathcal{R}e\ \{a_j f(z)\} \tag{4.3}$$

$$(df/dz) = f^3(z) + pf(z) + q \tag{4.4}$$

Coefficients p and q, as coefficients a_j, are complex numbers to be determined; coefficients α_j are real. The introduction of functions (4.3) in Eqs. (4.1, 4.2) and the identification of terms of the same degree in $f(z)$ and $\overline{f}(\overline{z})$ leads to a system of algebraic equations for unknowns α_j, a_j, p, and q. The function $f(z)$ is determined by the integration of Eq. (4.4) and the densities $N_j(t,x)$ by formulas (4.3). One of the solutions obtained by Tiem is the following:

$$N_1(t,x) = 5\alpha\left\{1 + F + G\right\}, \qquad N_3(t,x) = 5\alpha\left\{1 + F - G\right\} \qquad (4.5)$$

$$N_2(t,x) = 4\alpha\left\{1 - 2F - G\right\}, \qquad N_4(t,x) = 4\alpha\left\{1 - 2F + G\right\} \qquad (4.6)$$

$$F = \sqrt{\frac{3}{2}\left\{\frac{1}{\sqrt{D}} + \frac{1 + e^T \cos \pi X}{D}\right\}}, \qquad G = \sqrt{\frac{3}{2}\left\{\frac{1}{\sqrt{D}} - \frac{1 + e^T \cos \pi X}{D}\right\}} \qquad (4.7)$$

$$D = 1 + 2e^T \cos \pi X + e^{2T}, \qquad \pi X = 120\,\alpha^2\beta x, \qquad T = 120\,\alpha^2\beta t \qquad (4.8)$$

The functions $N_j(T,X)$ being densities must be positive, which is the case for large values of time, as functions F and G tend to zero when the time increases indefinitely. The function $N_1(T,X)$ is always positive; the regions of the X,T plane, in which the other functions $N_j(T,X)$ are positive are shown in Fig. 4.1. The function $N_3(T,X)$ is negative inside the closed curve on which it is null; the two other functions $N_j(T,X)$, $(j = 2,4)$ are positive for values of time greater that values for which they are null. Therefore the four functions $N_j(X,T)$ are positive for all X, if one has $T \geq 2.672....$ The macroscopic variables: densitiy n, velocity $\mathbf{u}$ and temperature θ are given (with a constant factor) by relations:

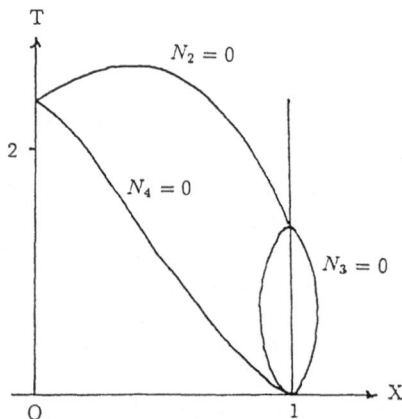

Fig. 4.1.

$$n(T, X) = 4(N_1 + N_3) + N_2 + N_4 = 48\,\alpha + 24\,\alpha F(T, X) \qquad (4.9)$$

$$(n\mathbf{u})(T, X) = 32\,\alpha G(T, X)\mathbf{x} \qquad (4.10)$$

$$\theta = 12(N_1 + N_3) + N_2 + N_4 = 128\,\alpha + 96\,\alpha F(T, X) \qquad (4.11)$$

Even when the microscopic densities are negataive, macroscopic variables n and θ are always positive, which agrees with their physical significations. Furthermore, if one chooses, as initial time, a positive but less than 2.67 value of T, the Tiem solution, as exact solution obtained in former sections, is a global solution in time corresponding to partially negative initial data.

5. Two-Dimensional Semi-Continuous Model

The simplest semi-discrete model of the Boltzmann equation is two–dimensional and obtained when the velocity vectors have all the same modulus but an arbitrary direction. On the orthogonal axis those vectors $\mathbf{u}(\theta)$ have components: $c\cos\theta$ and $c\sin\theta$. The density $N(t, x; \theta)$ of particles with velocity $\mathbf{u}(\theta)$ satisfies the Eq. (5.1), called the semi-discrete, or semi-continuous, two-dimensional Boltzmann equation.

$$\frac{\partial N}{\partial t} + \cos\theta\frac{\partial N}{\partial x} = \frac{1}{2\pi}\int_0^{2\pi} \{N(\varphi)N(\varphi + \pi) - N(\theta)N(\theta + \pi)\}\,d\varphi \qquad (5.1)$$

The function $N = N(t, x; \theta)$ is a periodic function, with period 2π of the variable θ. The right-hand side of Eq. (5.1) can be written in the form (5.2), and the function $F(t, x; \theta)$ satisfies relation (5.3).

$$F(t, x; \theta) = \frac{1}{2\pi}\int_0^{2\pi} N(\varphi)N(\varphi + \pi)d\varphi - N(\theta)N(\theta + \pi) \qquad (5.2)$$

$$\int_\theta^{\theta+\pi} F(t, x; \varphi)d\varphi = \frac{1}{2}\int_0^{2\pi} N(\varphi)N(\varphi + \pi)d\varphi - \int_\theta^{\theta+\pi} N(\varphi)N(\varphi + \pi)d\varphi \qquad (5.3)$$

As the function $N(\theta)$ is periodic, with period 2π, the second integral on the right-hand side of the relation (5.3) is independent of θ, and one can write:

$$\int_\theta^{\theta+\pi} \left\{\frac{\partial N}{\partial t} + \cos\theta\frac{\partial N}{\partial x}\right\}\,d\theta = \int_\theta^{\theta+\pi} F(t, x; \theta)d\theta = 0 \qquad (5.4)$$

The identity (5.4), linear with respect to derivatives of the function $N(t, x; \theta)$, is a conservation equation. As the angle θ is arbitrary, they are an infinity of conservation equations, corresponding to an infinity of summational invariants.

For the two-dimensional 4p-velocity model, symmetric with respect to the two axis Ox and Oy, one obtains, starting from the same ideas as in former sections, the following solution:

$$N_j = \frac{n_0}{1+c_j}\left(1 + \frac{c_j^2 - 1}{1 + e^z}\right), \qquad for \qquad 1 \le j \le p \quad (5.5)$$

$$N_j = n_0(1 + c_{2p+1-j})\left(1 - \frac{1}{1 + e^z}\right), \qquad for \qquad p+1 \le j \le 2p \quad (5.6)$$

with $c_j = \rho \cos \alpha_j$, $\alpha_j = (\overrightarrow{Ox}, \mathbf{u}_j)$, $\rho = \sqrt{p/(\cos^2 \alpha_1 + ... + \cos^2 \alpha_p)}$, and $z = k(ct - \rho x)$, $k =$ positive and arbitrary; $N_{4p-j} = N_j$. When the number $4p$ of velocities increases indefinitely, one obtains, for Eq. (5.1), the solution (5.7), where $z = ct - x\sqrt{2}$.

$$N(t, x; \theta) = (1 + \sqrt{2} \cos \theta)\frac{1}{1 + e^z} + M(\theta) \qquad (5.7)$$

$$M(\theta) = 0, \qquad for \qquad -\frac{\pi}{2} \le \theta \le \frac{\pi}{2} \qquad (5.8)$$

$$M(\theta) = \frac{1 + \cos 2\theta}{1 - \sqrt{2}\cos \theta}, \qquad for \qquad \frac{\pi}{2} \le \theta \le \frac{3\pi}{2} \qquad (5.9)$$

The function $N(t, x; \theta)$ is always positive, continuous and also its derivatives with respect to z (therefore with respect to t and x); its first derivative with respect to θ is also continuous. For this function $N(t, x; \theta)$ the two sides of Eq. (5.1) are equal to $-c \cos(2\theta)e^z/(1 + e^z)^2$. To the conservation Eq. (5.4) corresponds the shock Eq. (5.10), [13], where ξ is the speed of propagation of shock wave.

$$\int_\theta^{\theta+\pi} (\xi - c \cos \theta)[N(t, x; \theta)]d\theta \qquad (5.10)$$

A continuous solution will represent a structure of shock if the relation (5.10) is satisfied when the discontinuity $[N(t, x; \theta)]$ is replaced by the difference $N(t, +\infty; \theta) - N(t, -\infty, \theta)$. In fact, for the solution (5.7) the discontinuity $[N(t, x; \theta)]$ is given by relation (5.11), and the condition of shock (5.10) becomes relation (5.12).

$$[N(t, x; \theta)] = (1 - \sqrt{2}\cos \theta)[1/(1 + e^z)] \qquad (5.11)$$

$$[1/(1 + e^z)](\pi - \sqrt{8}\sin \theta)(\xi - c/\sqrt{2}) = 0 \qquad (5.12)$$

As a consequence the solution (5.7) of Eq. (5.1) represents the structure of a shock wave, propagating with the speed $\xi = c/\sqrt{2}$. The macroscopic variables density $n(t, x)$, mean velocity $u(t, x)$, temperature $T(t, x)$, and pressure $p(t, x)$ are defined by the following formulas

$$n(t,x) = 2 \int_0^\pi N(t,x;\theta)d\theta = 2\pi(a_0 + \frac{1}{1+e^z}) \qquad (5.13)$$

$$(nu)(t,x) = 2c \int_0^\pi N(t,x;\theta)\cos\theta \, d\theta = c\pi(a_1 - \frac{\sqrt{2}}{1+e^z}) \qquad (5.14)$$

$$p(t,x) + \tilde{k}\,nT = \frac{\mu}{3}n(c^2 - u^2) \qquad (5.15)$$

with $a_0 = \frac{\sqrt{2}+\log(1+\sqrt{2})}{\pi} - \frac{1}{2}$, $a_1 = \frac{1-2a_0}{\sqrt{2}}$, μ denotes the mass of molecules, $\tilde{k}$ the constant of perfect gases. The variation of pressure versus $= -ct+xz/\sqrt{2}$, that means versus time t at a given point, or versus the abscissa x at a given time, is represented in Fig. 5.1a. The similar variation of temperature is represented in Fig. 5.1b.

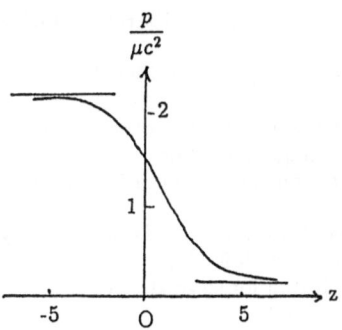
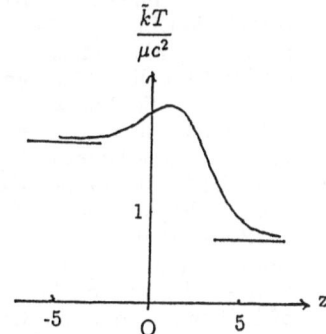

Fig. 5.1. a) - Pressure variation   b) - Temperature variation

6. Conclusion

The possibility to obtain exact solutions for equations modelling physical phenomena is rare as early as equations are non-linear. Those exact solutions are interesting since they allow to test numerical methods. In the case of the Discrete Kinetic Theory, the Cornille solutions and others have showed that the positivity of initial data was not necessary to prove the global existence of the solution of the initial problem. That remark was the starting point of new researchs which led to the global existence with partially negative initial data [10], [14], [15]. The last example concerns no more a system of partial differential equations, but an integro-differential equation, issuing from the fact that if the moduli of velocities are discretized, the directions vary continuously. This last model is closer to the original Boltzmann equation than the classical discrete models, and we hope that the construction of an exact solution could be continued in the future.

Acknowledgements

The results presented in this paper have been obtained in several works supported by the Ministry of Defense, contracts D.R.E.T. from 1986 to 1993.

References

1. CARLEMAN T., 1957, Problémes mathématiques de la théorie cinétique des gaz. *Publications Scientifiques de l'Institut Mittag-Leffler*, Uppsala
2. BROADWELL J., 1964-a Shock Structure in a simple discrete velocity gas. The Physics of Fluids **7**, 1243-1247.
3. BROADWELL J., 1964-b, Study of a rarefied shear flow by discrete velocity method. J. Fluid Mech., **19**, 367-370.
4. GATIGNOL R:, 1970 Théorie cinétique d'un gaz á répartition discréte de vitesses. Zeit- für Flugwissenschaften, **18**, 93-97.
5. GATIGNOL R., 1975 Théorie cinétique d'un gaz á répartition discréte de vitesses. Lecture Notes in Physics, **36**, Springer-Verlag, Heidelberg.
6. CORNILLE H., 1987-a, Exact solutions of the Broadwell model in 1+1 dimensions. J. Phys. A: Math. Gen., **20**, 1973-1988.
7. CORNILLE H., 1987-b, Exact (1+1)-Dimensional Solution of Planar Velocity Boltzmann Models. Journal of Statistical Physics, **48**, 789-811.
8. CABANNES H. and TIEM D.H., 1987, Exact solution for some discrete models of Boltzmann equation. Complex Systems, **1**, 575-584.
9. CABANNES H. and DURUISSEAU J.P., 1990, Construction, using Macsyma, of exact solutions for some equations of discrete kinetic theory. Symbolic Computations and their Impact on Mechanics, 277-284, ASME, New York.
10. CABANNES H., 1991, On the initial-value problem in discrete kinetic theory. Eur. J. Mech., B/Fluids, **10**, 207-224.
11. TIEM D.H., 1991, Solutions exactes pour certains modéles discrets de l'équation de Boltzmann avec collisions multiples. C.R. Acad. Sc. Paris, **II**, 995-998.
12. CORNILLE H., 1989, Solutions exactes pour deux modéles cinétiques discrets tridimensionnels á dix vitesses. C.R. Acad. Sc. Paris, **313, II**, 743-747.
13. TIEM D.H. and CABANNES H. 1990, Solution exacte pour le modéle plan semi-continu de l'équation de Boltzmann. C.R. Acad. Sc. Paris, **315, II**, 1175-1179.
14. BONY J.M., 1991, Existence globale á données de Cauchy petites pour les modéles discrets de l'équation de Boltzmann. Communications in Partial Differential Equations, **16**, 533-545.
15. CABANNES H., 1992, Solution of a Discrete Boltzmann Equation with Triple Collisions, for Partially Negative Initial Data. Transport Theory and Statistical Physics, **21**, 437-450.

Boundary Conditions for Discrete Models of Gases and Applications to Couette Flows

Amah D'Almeida and Renée Gatignol

Laboratoire de Modélisation en Mécanique Associé au CNRS, Université Pierre et Marie Curie, 4 Place Jussieu, 75252 Paris Cédex 05, FRANCE

Summary. Boundary conditions are investigated for discrete models of gases in order to study gas flows along walls described as usual continuum media. To avoid difficulties due to the existence of some macroscopic variables other than mass, momentum and energy related to spurious invariants, a particular class of models is considered. Then the steady Couette flow between two parallel plates is studied. The thermophoresis phenomenon is analysed; it is remarkable when the difference between the velocities of the plates is small and the difference between their temperatures is large.

1. Introduction

The main idea of the discrete kinetic theory of gases is that the velocities of the particles belong to a *given finite set of vectors*.

The theory for a general model with discrete velocities has been given for the first time in 1970 [10]. The Boltzmann equation is replaced by a system of partial differential equations with a very interesting mathematical structure. Many papers concern the proof of the existence of the solutions with given initial data ([14],[16],...), and the trends of the kinetic equations to the hydrodynamical equations ([1],[7],...). Alike for particular models, some exact solutions have been found ([8],...).

But very few papers discuss the flow problems where we have to take into account the interactions of the gas molecules with a solid wall. Sure, there is the pionner work of Broadwell [2]. Some extensions have been made ([3],[12]) and more recently a few papers concern the mathematical theory of the system of kinetic equations with boundary conditions [4].

One of the difficulties is in the formulating of the well-adapted boundary conditions. This is due to the existence of some nonphysical macroscopic variables associated with the spurious invariants. The boundary conditions on an impermeable wall for any discrete gases have been previously investigated [12]; all the macroscopic variables with physical or nonphysical meaning were taken into account. But we have some trouble. The physical understanding of the results is obviously confused. Furthermore, we have to give a meaning to the extra macroscopic variables *inside* the wall. In the way of the last remark, it is not possible to compare the descriptions of the flow given by the discrete models with those given by the usual Boltzmann equation.

In this paper we consider a class of "good" discrete models for which the summational invariants are linearly independent and reduced to the physical ones (mass, momentum and energy). After a short presentation of the discrete kinetic theory and of the selected models (Section 2), the interaction of the particles with a solid wall is investigated (Section 3). Motivated by similar studies in classical kinetic theory, some boundary conditions are suggested. Then, we study the gas flow for the Couette flow problem between two parallel plates (Section 4). By using a simple model with ten velocities, we can obtain an exact solution. It is to notice that the walls are described as continuum media with their own velocities and their own temperatures. At last (Section 5), the results are presented and discussed. An emphasis is given to the thermophoresis phenomenon in which the velocity of the gas near the plate with the lowest temperature is greater than the velocity of the plate.

2. Discrete Kinetic Theory

The gas is composed of identical particles of mass m, whose velocities are restricted to a *given finite* set of p vectors: $\mathbf{u}_1, \mathbf{u}_2, ..., \mathbf{u}_p$. We denote by $N_i(\mathbf{r}, t)$ the number density of particles with velocity $\mathbf{u}_i$ (also called particles "i") at the point $\mathbf{r}$ and at time t.

The first theories ([10],[11]) with binary collisions only have been generalized to multiple collisions, with the aim to reduce the number of summational invariants and in favourable cases to obtain only the physical ones ([6],[7]).

By definition an r-collision ($r \geq 2$) involves r particles. Let $I_r = (i_1, ..., i_r)$ and $J_r = (j_1, ..., j_r)$ be two elements of $\mathcal{E}_r$, the set of the r-not arranged numbers taken in the set $\{1, ..., p\}$. Then if the velocities of the r particles are $\mathbf{u}_{i_1}, ..., \mathbf{u}_{i_r}$ before the collision and $\mathbf{u}_{j_1}, ..., \mathbf{u}_{j_r}$ after, the collision is written $I_r \Rightarrow J_r$. The collision must bear out the conservation of momentum and energy (the conservation of mass being borne out automatically).

A "transition probability" denoted by $A_{I_r}^{J_r}$ is associated to each r-collision $I_r \Rightarrow J_r$. So the number of r-collisions $I_r \Rightarrow J_r$ per unit volume and unit time is $A_{I_r}^{J_r} N_{I_r}$, where N_{I_r} denotes the product $N_{i_1} N_{i_2} ... N_{i_r}$. The quantities $A_{I_r}^{J_r}$ are positive or zero, and as usual it is convenient to assign a zero transition probability to an unrealizable collision. It will be also assumed that the transition probabilities satisfy the hypothesis of microreversibility: $A_{I_r}^{J_r} = A_{J_r}^{I_r}$.

Now, a balance equation should be written for the number density of particles "k". Through an r-collision, several particles "k" can be created or destroyed. Let $\delta(k, I_r)$ be the number of indices k present in the r-set I_r. Obviously, $\delta(k, I_r)$ is positive or zero. If we put $\delta(k, J_r, I_r) = \delta(k, J_r) - \delta(k, I_r)$, then $\delta(k, J_r, I_r)$ is *the algebraic number of particles "k" created through the collision $I_r \Rightarrow J_r$*. With these notations, the balance equation for the number density of particles "k" is:

$$\frac{\partial}{\partial t} N_k + \mathbf{u}_k \cdot \nabla N_k = \sum_{r=2}^{R} \sum_{I_r \in \mathcal{E}_r} \sum_{J_r \in \mathcal{E}_r} \delta(k, J_r, I_r) A_{I_r}^{J_r} N_{I_r}. \qquad (2.1)$$

In equation (2.1), the r-collisions with $2 \le r \le R$ are taken into account. The set of equations (2.1) with $1 \le k \le p$ can be written in a short form:

$$\frac{\partial}{\partial t} \mathbb{N} + \mathcal{A} \mathbb{N} = \sum_{r=2}^{R} \mathcal{F}^r (\mathbb{N}, ..., \mathbb{N}) \equiv \mathcal{C}(\mathbb{N}). \tag{2.2}$$

The r-collision operator $\mathcal{F}^r$ is an r-linear symmetrical operator from $\mathbb{R}^p \times ... \times \mathbb{R}^p$ into $\mathbb{R}^p$, which has all the properties of the collision operator of the Boltzmann equation.

Now we give the definition of the *summational invariants*. The summational invariants are quantities attached to the conservation properties through all the collisions. They are defined as the p-component vectors $\Phi = (\varphi_1, ..., \varphi_p)$ satisfying the following conditions:

$$A_{I_r}^{J_r} \sum_{k=1}^{p} \delta (k, J_r, I_r) \varphi_k = 0, \quad \forall I_r, J_r \in \mathcal{E}_r, \quad \forall r, \; 2 \le r \le R. \tag{2.3}$$

In particular, the vectors Φ such that φ_i are equal to m, $m\mathbf{u_i}$ or $m \mid \mathbf{u}_i\mid^2$ are summational invariants. They are called physical invariants. In constrast to the classical kinetic theory of monoatomic gases, the geometric character of the set of the given velocities may allow other summational invariants which are called *spurious invariants* or cause the physical invariants to be linearly dependent. The summational invariants generate a subspace $\mathbb{F}$ of $\mathbb{R}^p$. We denote by q the dimension of $\mathbb{F}$ and we introduce orthonormal bases in $\mathbb{F}$ and $\mathbb{R}^p$: $\mathcal{V}^1, ..., \mathcal{V}^q$ in $\mathbb{F}$, $\mathcal{W}^{q+1}, ..., \mathcal{W}^p$ in $\mathbb{F}^\perp$. We thus can write:

$$\mathbb{N} = \sum_{\alpha=1}^{q} a_\alpha \mathcal{V}^\alpha + \sum_{\beta=q+1}^{p} b_\beta \mathcal{W}^\beta. \tag{2.4}$$

The quantities a_α are called macroscopic state variables of the gas. Among them there are the number density n, the mean velocity $\mathbf{u}$ and the total energy E defined by:

$$n = \sum_{i=1}^{p} N_i, \quad n\mathbf{u} = \sum_{i=1}^{p} N_i \mathbf{u}_i, \quad nE = \frac{1}{2} \sum_{i=1}^{p} N_i \mid \mathbf{u}_i\mid^2. \tag{2.5}$$

The presence of multiple collisions reduces the dimension of the invariant space $\mathbb{F}$. In favourable cases the spurious invariants are removed ([6],[7]). From now, we assume that all the discrete models in consideration in this paper are tridimensional and have only physical linearly independent summational invariants. So we have the usual balance laws for the mass, momentum and energy:

$$\frac{\partial}{\partial t}\rho + \nabla.(\rho\mathbf{u}) = 0$$

$$\frac{\partial}{\partial t}(\rho\mathbf{u}) + \nabla.(\rho\mathbf{u}\mathbf{u}) + \nabla.\mathbb{P} = 0 \qquad (2.6)$$

$$\frac{\partial}{\partial t}(\rho E) + \nabla.(\rho E\mathbf{u} + \mathbb{P}.\mathbf{u} + \mathbf{q}) = 0$$

where:

$$\rho = nm, \quad \mathbb{P} = m\sum_{i=1}^{p} N_i(\mathbf{u}_i-\mathbf{u})(\mathbf{u}_i-\mathbf{u}), \quad \mathbf{q} = \frac{m}{2}\sum_{i=1}^{p} N_i(\mathbf{u}_i-\mathbf{u})^2(\mathbf{u}_i-\mathbf{u}).$$
$$(2.7)$$

Owing to the discretization, some definitions of the macroscopic variables, such as the temperature, given by the classical kinetic theory are no longer valid [5]. Thus we introduce the "kinetic" temperature T defined by the relation:

$$\frac{3kT}{2m} + \frac{1}{2}|\mathbf{u}|^2 = E, \qquad (2.8)$$

where k is the Boltzmann constant. This definition of the temperature has been used before ([10],[7]). In fact, to study the evolution of the "kinetic" temperature amount to study that of the internal energy $E - \frac{1}{2}|\mathbf{u}|^2$.

The Boltzmann H-function is defined by $H = \sum_{i=1}^{p} N_i Log N_i$. We denote by $Log\mathbb{N}$ the p-component vector $(Log N_1, ..., Log N_p)$. For an uniform gas, the computations of dH/dt yields the H-theorem:

$$\frac{dH}{dt} = \langle Log\mathbb{N}, \mathcal{C}(\mathbb{N})\rangle \leq 0. \qquad (2.9)$$

where $\langle .,.\rangle$ denotes the scalar product in $\mathbb{R}^p$. As usual the *Maxwellian state* is defined by $dH/dt = 0$ or $Log\mathbb{N} \in \mathbb{F}$. As in the discrete kinetic theory for binary collisions [11], the densities N_i of a Maxwellian state depend only on the mean macroscopic variables. Moreover, for a gas having given values of the macroscopic variables, there exists *one and only one associated Maxwellian state*. For a model having only linear independent physical invariants, the densities in a Maxwellian state associated with the macroscopic variables n, $\mathbf{u}$ and E are:

$$N_i = exp\left(\alpha + \boldsymbol{\beta}.\mathbf{u}_i + \gamma \mid \mathbf{u}_i \mid^2\right), \qquad (2.10)$$

the parameters α, $\boldsymbol{\beta}$ and γ are given by the implicit relations (2.5) .

3. Boundary Conditions

We investigate in this section the conditions which must satisfy the densities N_i on a wall $\partial\Omega$. Let M be a point on the surface $\partial\Omega$. At time t its position is $\mathbf{r}$. We denote by $\mathbf{n}(\mathbf{r},t)$ and $\mathbf{u}_w(\mathbf{r},t)$, respectively, the inward-pointing (i.e., into the gas) unit vector normal to the wall and the velocity of the wall at

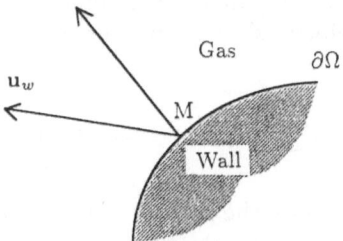

Fig. 3.1.

M. We arrange the p velocities into two groups, corresponding to impinging molecules and emerging molecules. This leads to the partition of the set of velocity numbers into two sets:

$$I = \{i; (\mathbf{u}_i - \mathbf{u}_w) . \mathbf{n} < 0\}, \quad R = \{j, (\mathbf{u}_j - \mathbf{u}_w) . \mathbf{n} > 0\}.$$

Doing so, we assume that the models in consideration do not possess any velocity parallel to the wall.

We assume that the macroscopic state variables of the wall at M are its velocity $\mathbf{u}_w$ and its total energy E_w which is defined as the sum of its kinetic energy and the internal energy of a gas having its temperature T_w:

$$E_w = \frac{3kT_w}{2m} + \frac{|\mathbf{u}_w|^2}{2}. \tag{3.1}$$

We define a discrete gas in *Maxwellian equilibrium* with the wall as the discrete gas whose microscopic densities are the microscopic densities N_{iw} of the Maxwellian state associated with 1, $\mathbf{u}_w$ and E_w. Then we introduce the positive coefficients α_i such that at point M and at time t, the microscopic densities at the wall can be written as:

$$N_i(\mathbf{r}, t) = \alpha_i(\mathbf{r}, t) N_{iw}(\mathbf{r}, t). \tag{3.2}$$

The coefficients α_i express the accomodation of the discrete gas to the wall quantities.

Now we introduce the probability density $B(\mathbf{u}_i \to \mathbf{u}_j; \mathbf{r}, t) = B_{ij}$ such that a molecule striking the wall at the point M and at time t with velocity $\mathbf{u}_i$, re-emerges at the same point and at the same time with velocity $\mathbf{u}_j$. It is natural to assume that [12] :

$$B_{ij} \geq 0, \quad \sum_{j \in R} B_{ij} = 1. \tag{3.3}$$

The second condition means that the wall re-emits all the incident gas molecules.

We can easily write the boundary conditions for an impermeable wall by the following argument: the number of molecules emerging with velocity $\mathbf{u}_j$

from an unit surface element per unit time is equal to the total number of incident molecules (of all velocities) that will change their velocities to $\mathbf{u}_j$; that is [12],

$$|(\mathbf{u}_j - \mathbf{u}_w).\mathbf{n}|\, N_j = \sum_{i\in I} B_{ij}\, |(\mathbf{u}_i - \mathbf{u}_w).\mathbf{n}|\, N_i, \quad j \in R. \qquad (3.4)$$

As consequences of (3.3) and (3.4), the normal component of the macroscopic velocity at the wall is zero, or equivalently, per unit surface element and per unit time the number of emerging molecules is equal to the number of impinging molecules. One has:

$$n\,(\mathbf{u} - \mathbf{u}_w).\mathbf{n} = \sum_{i\in I} N_i\,(\mathbf{u}_i - \mathbf{u}_w).\mathbf{n} + \sum_{j\in R} N_j\,(\mathbf{u}_j - \mathbf{u}_w).\mathbf{n}$$

$$= \sum_{i\in I} \left(-\alpha_i + \sum_{j\in R} \alpha_i B_{ij} \right) |(\mathbf{u}_i - \mathbf{u}_w).\mathbf{n}|\, N_{iw} = 0.$$

By similar computations the total density near the wall is:

$$n = \sum_{i\in I} N_i + \sum_{j\in R} N_j = \sum_{j\in R}\sum_{i\in I} \alpha_i B_{ij} \left(1 + \frac{|(\mathbf{u}_i - \mathbf{u}_w).\mathbf{n}|}{|(\mathbf{u}_j - \mathbf{u}_w).\mathbf{n}|} \right) N_{iw}. \qquad (3.5)$$

Now we have to specify B_{ij}. As in the classical theory this is not an easy problem, since the situation at the wall is very complicated, due to the complex phenomena of adsorption and evaporation which can take place. So we shall adopt the simple rules of a perfectly reflecting wall and of a perfectly absorbing wall, i.e specular reflection and pure diffuse reflection ([15],[12]). The specular reflection is possible only for discrete models having their velocities symmetrical with respect to the normal at any point of the wall. It is easy to see that specular reflection boundary conditions can be applied to regular models with adequate symmetrical properties. However, these boundary conditions are unrealistic unless the wall is adiabatic.

For pure diffuse reflection boundary conditions, the properties of the reflected molecules are independent of their properties before the impact. In other words the re-emitted stream has completely lost its memory of the incoming stream, except for conservation of the number of molecules. For this reason, B_{ij} does not depend on the index i. We can thus assume:

$$B_{ij} = B_j, \quad \forall i \in I. \qquad (3.6)$$

Moreover, we impose the following condition: "the wall locally behaves as a thermostat"; that is, the Maxwellian gas in equilibrium with the wall satisfies the boundary conditions (3.4). So we have:

$$|(\mathbf{u}_j - \mathbf{u}_w).\mathbf{n}|\,N_{jw} = \sum_{i \in I} B_j\,|(\mathbf{u}_i - \mathbf{u}_w).\mathbf{n}|\,N_{iw}, \quad j \in R,$$

$$B_j = \frac{|(\mathbf{u}_j - \mathbf{u}_w).\mathbf{n}|\,N_{jw}}{\sum_{i \in I}|(\mathbf{u}_i - \mathbf{u}_w).\mathbf{n}|\,N_{iw}}, \quad \forall j \in R. \tag{3.7}$$

Due to (3.2), (3.4), (3.6) and (3.7), we can see that α_j, $(j \in R)$, does not depend on j. So at the wall we have:

$$N_j(\mathbf{r},t) = \lambda(\mathbf{r},t)N_{jw}(\mathbf{r},t), \quad \forall j \in R. \tag{3.8}$$

Then letting:

$$n_{Rw} = \sum_{j \in R} N_{jw}, \quad n_I = \sum_{i \in I} \alpha_i N_{iw}, \quad n_{Rw}\mu_{Rw} = \sum_{j \in R} N_{jw}\,|(\mathbf{u}_j - \mathbf{u}_w).\mathbf{n}|,$$

$$n_I\mu_I = \sum_{i \in I} \alpha_i N_{iw}\,|(\mathbf{u}_i - \mathbf{u}_w).\mathbf{n}|, \quad \theta_d = \frac{\mu_{Rw}}{\mu_I},$$

and using the definition of n and the relation $(\mathbf{u} - \mathbf{u}_w).\mathbf{n} = 0$, a straightforward computation yields:

$$\lambda = \frac{n}{n_{Rw}\,(1 + \theta_d)}. \tag{3.9}$$

In the expression (3.9), λ is expressed by means of the macroscopic density n at the wall and the ratio θ_d of the mean values of the incident and the reflected molecule fluxes. The quantities n and θ_d are unknowns and should normally be determined after the resolution of the kinetic equations. However, the geometry of the models can simplify the determination of θ_d.

For instance θ_d is equal to one for models whose velocities have the same absolute value of their normal components with respect to the wall (i.e., $|\,\mathbf{u}_i.\mathbf{n}\,| = |\,\mathbf{u}_j.\mathbf{n}\,|, \forall i, j)$. Furthermore if the velocities of the model are symmetrical with respect to the wall and $\mathbf{u}_w.\mathbf{n} = 0$, then

$$n_{Rw} = \frac{1}{2} \quad \text{and then} \quad \lambda = n. \tag{3.10}$$

We also notice that a change n in βn $(\beta > 0)$ leads to a change λ in $\beta\lambda$. This allows us to adimensionalise λ with respect to n.

In work in progress, boundary conditions have been written for evaporation and condensation problems. We assume that the molecules leaving the condensed phase are in Maxwellian equilibrium with it and that for given macroscopic variables of the condensed phase, the number of evaporating molecules is the same when the vapor is either in equilibrium or in nonequilibrium with the condensed phase. Then using (3.8) and the fact that the fluxes of evaporating and condensing molecules counterbalance each other when the vapor is in equilibrium with its condensed phase, we find :

$$\lambda = \frac{n_{sat}}{n_{Rw}\,(1+\theta_e)} \tag{3.11}$$

where n_{sat} is the density of saturation at the interphase. We notice that n_{sat} is known and that θ_e depends only on the geometry of the model and the macroscopic variables of the condensed phase [9].

4. Couette Flows

The Couette problem of shear flow and heat transfer between parallel infinite and moving plates is one of the simplest problems of gas dynamics but no exact solution of the Boltzmann equation has yet been found for it. This problem is interesting since it helps us clarify the nature of gas flow near a solid surface. The main purpose of this section is to find an exact solution of the Couette problem using the discrete theory of gases in order to explain some qualitative peculiarities of rarefied flows. The plates have arbitrary velocities and temperatures.

We choose the origin O of an orthonormal system for the coordinates (xyz), so that the plates are located at $y = \pm\frac{L}{2}$. The velocities and the temperatures of the plates which are respectively, $\mathbf{u}_w^\pm = c\,(u_w^\pm,0,0)$ and $T_w^\pm$ are constant. The velocities $\mathbf{u}_w^\pm$ are thus taken parallel to the plates (Fig. 4.1). Now we briefly describe the used model which is called C1 model (Fig. 4.2). The velocities of this model are:

$$\mathbf{u}_1 = c(-1,1,1); \quad \mathbf{u}_2 = c(1,1,1); \quad \mathbf{u}_3 = c(-1,-1,1); \quad \mathbf{u}_4 = c(1,-1,1);$$
$$\mathbf{u}_{9-i} = -\mathbf{u}_i, \quad i = 1,2,3,4; \quad \mathbf{u}_9 = c(0,1,0); \quad \mathbf{u}_{10} = -\mathbf{u}_9.$$

It is easy to prove that the C1 model has only physical linearly independent summational invariants for binary collisions [9].

All binary and ternary collisions are taken into account. We denote by S the cross section of binary collisions. The transition probabilities of binary collisions are proportional to $S\,|\mathbf{u}_i - \mathbf{u}_j|$ and such that the resulting pairs

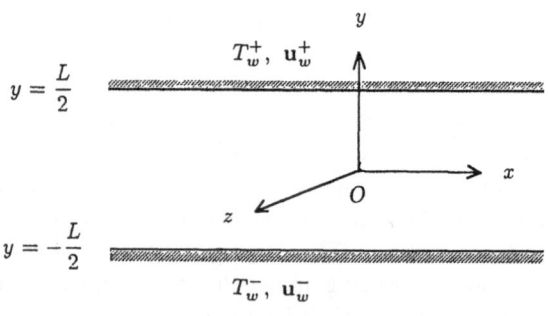

Fig. 4.1.

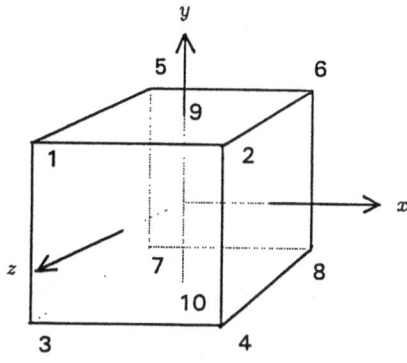

Fig. 4.2.

after the collision are equiprobable. There are two kinds of ternary collisions: the simple ternary collisions between particles having velocities of the same modulus and the mixed ternary collisions between particles having velocities of different moduli. We denote by β_1 and β_2 respectively, the transition probabilities of simple and mixed ternary collisions.

When we assume (as we shall do) that the distribution of velocities is symmetrical with respect to the Oxy plane, the number of unknown densities is reduced to six. They are the microscopic densities N_i, $i = 1, 2, 3, 4, 9, 10$.

By analogy with the classical steady Couette flow problem, we shall look for a solution of the problem under the assumption that the microscopic densities depend only on the variable y. The kinetic equations then take the form:

$$c\frac{dN_1}{dy} = -\left(-c\frac{dN_3}{dy}\right) = Q_1$$

$$c\frac{dN_2}{dy} = -\left(-c\frac{dN_4}{dy}\right) = Q_2 \qquad (4.1)$$

$$c\frac{dN_9}{dy} = -\left(-c\frac{dN_{10}}{dy}\right) = Q_3$$

where

$$Q_1 = P_1 + \left[cS\frac{\sqrt{6}}{2} + \beta_2\left(N_2 + N_4\right)\right]\left(N_3 N_9 - N_1 N_{10}\right)$$

$$Q_2 = -P_1 + \left[cS\frac{\sqrt{6}}{2} + \beta_2\left(N_1 + N_3\right)\right]\left(N_4 N_9 - N_2 N_{10}\right)$$

$$Q_3 = P_2 + 2\beta_2\left[\left(N_1 + N_3\right)\left(N_2 N_{10} - N_4 N_9\right) + \left(N_2 + N_4\right)\left(N_1 N_{10} - N_3 N_9\right)\right]$$

with

$$P_1 = \left[cS \left(\sqrt{2} + \sqrt{3} \right) + \beta_1 \left(N_1 + N_2 + N_3 + N_4 \right) \right] \left(N_2 N_3 - N_1 N_4 \right)$$

$$P_2 = cS\sqrt{6} \left[\left(N_1 + N_2 \right) N_{10} - \left(N_3 + N_4 \right) N_9 \right].$$

The macroscopic variables are the total density n, the tangential component U and the normal component V of the macroscopic velocity $\mathbf{u}$ and the total energy E defined by:

$$n = 2 \left(N_1 + N_2 + N_3 + N_4 \right) + N_9 + N_{10}$$
$$nU = 2c \left(-N_1 + N_2 - N_3 + N_4 \right) \tag{4.2}$$
$$nV = c \left[2 \left(N_1 + N_2 - N_3 - N_4 \right) + N_9 - N_{10} \right]$$
$$\rho E = \frac{mc^2}{2} \left[6 \left(N_1 + N_2 + N_3 + N_4 \right) + N_9 + N_{10} \right]$$

We adopt the diffuse reflection boundary conditions on the two plates. Due to the geometrical properties of the Couette flow and of the model, we have the property (3.10). So the boundary conditions (3.8) are written:

$$N_1(-L/2) = \frac{n^-}{16} \left(2e_w^- - u_w^- - 1 \right), \quad N_2(-L/2) = \frac{n^-}{16} \left(2e_w^- + u_w^- - 1 \right),$$

$$N_9(-L/2) = \frac{n^-}{4} \left(3 - 2e_w^- \right), \tag{4.3}$$

and

$$N_3(L/2) = \frac{n^+}{16} \left(2e_w^+ - u_w^+ - 1 \right), \quad N_4(L/2) = \frac{n^+}{16} \left(2e_w^+ + u_w^+ - 1 \right),$$

$$N_{10}(L/2) = \frac{n^+}{4} \left(3 - 2e_w^+ \right), \tag{4.4}$$

where

$$e_w^\pm = \frac{3kT_w^\pm}{2mc^2} + \frac{u_w^{\pm 2}}{2}; \quad n^\pm = n(\pm L/2).$$

From the kinetic equations and the boundary conditions we deduce that the component V of the macroscopic velocity is zero everywhere and that the density n is constant. So $n^- = n^+ = n$. Now we introduce the nondimensional variables $n_i = \frac{N_i}{n}$, $\eta = \frac{y}{L}$, $u = \frac{U}{c}$, $\tau = \frac{3kT}{mc^2}$, the Knudsen number $\epsilon = (LSn)^{-1}$ and the coefficients $\alpha = \frac{2(\sqrt{2}+\sqrt{3})}{\sqrt{6}}$, $\theta_1 = \frac{2\beta_1 n}{cS\sqrt{6}}$ and $\theta_2 = \frac{2\beta_2 n}{cS\sqrt{6}}$. These last two coefficients match the importance of the other kinds of collisions in relation to the mixed binary collisions.

From the kinetic Eqs. (4.1) we deduce four conservation laws, and consequently the following integrals

$$2 \left(n_1 + n_2 + n_3 + n_4 \right) + n_9 + n_{10} = 1$$
$$2 \left(n_1 + n_2 - n_3 - n_4 \right) + n_9 - n_{10} = 0 \tag{4.5}$$
$$-n_1 + n_2 + n_3 - n_4 = k_1$$
$$6 \left(n_1 + n_2 - n_3 - n_4 \right) + n_9 - n_{10} = k_2$$

where k_1 and k_2 are two constants which will be determined afterwards.

Using the relations (4.5) we show that that the kinetic Eqs. (4.1) can be reduced to two differential equations for n_1 and n_9. The resolution of these equations yields in addition to k_1 and k_2 two new constants: A and F. These four constants are determined by the four independent boundary conditions which are still available. In the case where θ_2 is zero (i.e the mixed ternary collisions are negligible with respect to the binary ones) we can solve analytically the differential equations. For non zero θ_2 only a numerical resolution can be performed. We present here the exact analytical solutions obtained when all binary collisions and simple ternary collisions are taken into account. The solutions for the tangential component of the macroscopic velocity and for the "kinetic" temperature are:

$$u(\eta) = C\left[D exp\left(-F\eta^2 + A\eta + \frac{F}{4}\right) + \frac{\sqrt{6}G\left(e_w^+ - e_w^-\right)}{2\left(4\epsilon + \sqrt{6}\right)}\eta + \frac{G\left(e_w^+ + e_w^- - 1\right)}{4}\right]$$

$$\tau(\eta) = e_w^+ + e_w^- - u^2(\eta) + \frac{2\sqrt{6}\left(e_w^+ - e_w^-\right)}{4\epsilon + \sqrt{6}}\eta. \tag{4.6}$$

The constants are given by:

$$k_2 = \frac{4\epsilon\left(e_w^- - e_w^+\right)}{4\epsilon + \sqrt{6}}, \quad A = \frac{\sqrt{6}k_2}{4\epsilon}\left[\frac{\alpha - 2}{2} + \frac{\theta_1}{8}\left(e_w^+ + e_w^- - 1\right)\right], \quad F = \frac{3k_2^2\theta_1}{64\epsilon^2},$$

$$C = \frac{2}{\left(\frac{1}{2} - e_w^+\right)exp\left(-\frac{A}{2}\right) - \left(\frac{1}{2} - e_w^-\right)exp\left(\frac{A}{2}\right)},$$

$$D = \frac{u_w^- - u_w^+ + 2\left(u_w^+ e_w^- - u_w^- e_w^+\right)}{4},$$

$$G = u_w^- exp\left(\frac{A}{2}\right) - u_w^+ exp\left(-\frac{A}{2}\right), \quad k_1 = \frac{\epsilon CG\left(e_w^- - e_w^+\right)}{2\left(4\epsilon + \sqrt{6}\right)}. \tag{4.7}$$

We remark that the constants k_1 and k_2 are zero for $\epsilon = 0$ and then A and F have finite values. This borderline case will be analysed in the following section.

5. Results

A particular point of interest in the solution of Couette flow is the study of the velocity slip and the temperature jump.

5.1 Study of the Tangential Velocity

The slip velocities on the plates are:

$$u\left(\pm\frac{1}{2}\right) - u_w^\pm = \pm 2k_1. \tag{5.1}$$

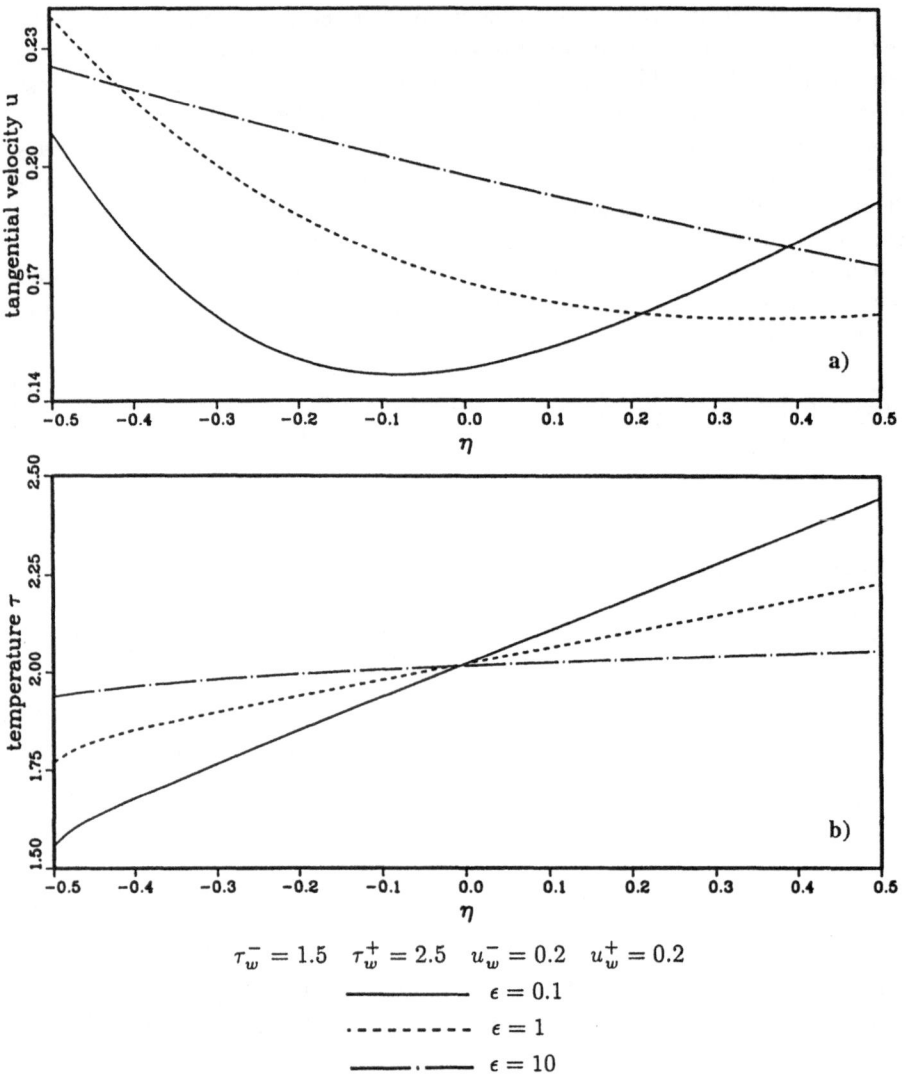

$$\tau_w^- = 1.5 \quad \tau_w^+ = 2.5 \quad u_w^- = 0.2 \quad u_w^+ = 0.2$$

——————— $\epsilon = 0.1$

- - - - - - $\epsilon = 1$

———·——— $\epsilon = 10$

Fig. 5.1. Couette flow with thermophoresis

The slip velocity is zero when both the plates are at rest. The modulus of the slip velocity is a function of ϵ if at least one of the plates is moving (Fig. 5.1(a)), (Fig. 5.2(a)). When ϵ tends towards zero, k_1 tends towards zero and we have the non slip condition in the continuum limit. It is easy to see using (4.7) that the slip velocities tend towards $\pm \frac{(u_w^- - u_w^+)}{2}$ when ϵ tends towards infinity.

For Knudsen numbers of the order of unity, the solution shows Knudsen layers whose thickness decreases when we introduce more ternary collisions.

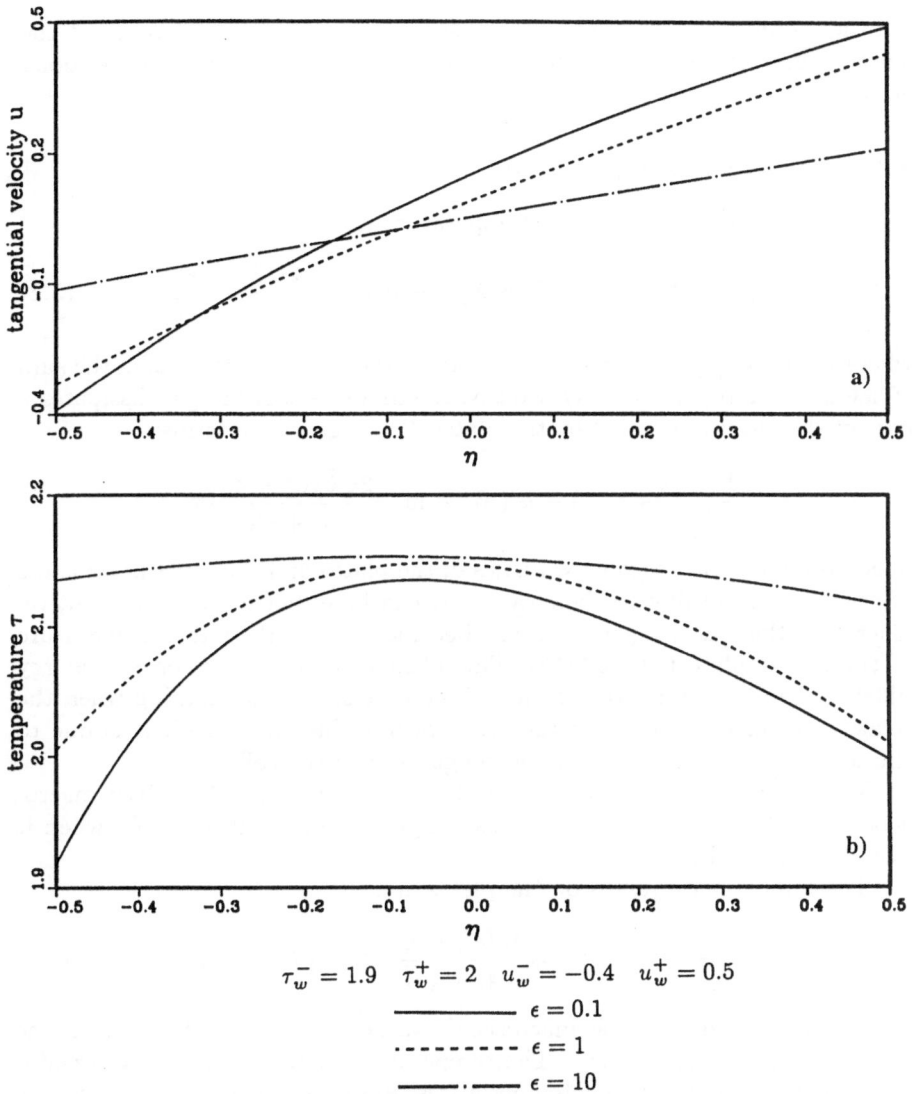

$$\tau_w^- = 1.9 \quad \tau_w^+ = 2 \quad u_w^- = -0.4 \quad u_w^+ = 0.5$$

———————— $\epsilon = 0.1$

- - - - - - - - $\epsilon = 1$

——— · ——— $\epsilon = 10$

Fig. 5.2. Couette flow with temperature rise

Indeed the Knudsen layer is due to the exponential term in the expression of $u(\eta)$ and we remark that F is positive and proportional to θ_1.

We obtain when the plate temperatures are different and the plates are moving in the same direction, a tangential velocity near the plate of the lowest temperature greater than the velocity of the plate. The increasing of this tangential velocity also depends on the Knudsen number and for given velocities and temperatures of the plates, it is more remarkable for Knudsen numbers of the order of unity (Fig. 5.1(a)). This phenomenon is due to the thermophoresis. The thermophoresis is caused by the difference of the temperatures of the plates which accelerate the molecules towards the cooler plate.

5.2 Study of the Temperature

The analytical expression of the temperature is:

$$\tau(\eta) = \frac{1}{2}\left(\tau_w^+ + \tau_w^-\right) + \frac{1}{2}\left(u_w^{+2} + u_w^{-2}\right) - u^2(\eta) + \frac{2\sqrt{6}\left(e_w^+ - e_w^-\right)}{4\epsilon + \sqrt{6}}\eta \qquad (5.2)$$

where $\tau_w^\pm = 3kT_w^\pm/mc^2$. So the temperature varies around the half of the sum of the plates temperatures. Obviously, the movement of the gas dissipates a part of the energy supplied by the plates. The sign of the terms:

$$\frac{1}{2}\left(u_w^{+2} + u_w^{-2}\right) - u^2(\eta) \quad \text{and} \quad \frac{2\sqrt{6}\left(e_w^+ - e_w^-\right)}{4\epsilon + \sqrt{6}}\eta$$

depends on the macroscopic variables of the plates and the other parameters of the flow. According to their values, it can be either negative or positive. Therefore the gas temperature can become greater than the temperature of the hotter plate (Fig. 5.2(b)). This phenomenon depends on the energy difference and the relative velocity of the plates. It is dominating when the relative velocity is large and the temperature difference small. It is due to the work done by the plates upon the gas and to the collisions.

This phenomenon exists even in the continuum limit. For given macroscopic variables of the plates the gas temperature is a function of ϵ when it occurs (Fig. 5.2(b)).

The temperature jumps on the plates are:

$$\tau\left(\pm\frac{1}{2}\right) - \tau_w^\pm = \mp\frac{4\epsilon\left(e_w^+ - e_w^-\right)}{4\epsilon + \sqrt{6}} \mp 4k_1\left(u_w^\pm \pm k_1\right). \qquad (5.3)$$

Their signs depend on the macroscopic variables of the plates and on the other parameters of the flow. The temperature of the gas is always greater than the lowest of the plates temperatures. However the gas temperature can be lower or greater than that of the hottest plate (Fig. 5.1(b)), (Fig. 5.2(b)).

The temperature jumps tend towards zero when ϵ tends towards zero. The temperature of the gas near the plates is thus equal to those of the plates in the continuum limit. When ϵ tend towards infinity, they tend respectively towards $(u_w^\pm - u_w^\mp)\frac{(3u_w^\pm + u_w^\mp)}{4} + e_w^\mp - e_w^\pm$.

We study now the particular case of the transfer of heat between two parallel plates at rest. We deduce from the discussions made in the previous sections that the macroscopic velocity is zero everywhere in the gas. The temperature is:

$$\tau(\eta) = \frac{1}{2}\left(\tau_w^+ + \tau_w^-\right) + \frac{\sqrt{6}\left(\tau_w^+ - \tau_w^-\right)}{4\epsilon + \sqrt{6}}\eta. \qquad (5.4)$$

When ϵ tends towards zero, we find for the temperature the usual expression of the temperature of a fluid between two parallel plates maintained at different temperatures [15]:

$$\tau(\eta) = \frac{1}{2}\left(\tau_w^+ + \tau_w^-\right) + \left(\tau_w^+ - \tau_w^-\right)\eta. \qquad (5.5)$$

The temperatures of the gas near the plates are equal to the temperatures of the plates.

6. Conclusion

The rarefaction effects we pointed out in our analysis are the existence of velocity slip, of temperature jump, of Knudsen layer and of the thermophoresis phenomenon. The increase of the internal energy of the gas occurs even in the continuum limit. However its study for a rarefied flow shows that it is in competition with the thermophoresis phenomenon (Fig. 5.1(b)), (Fig. 5.2(b)).

The thermophoresis is the ability of temperature difference to induce or accelerate the movement of particles of a rarefied gas. For small Knudsen numbers the effect of this phenomenon is remarkable near the plate of the lowest temperature and preponderant when the relative velocity of the plates is small and the temperature difference large. Hence for a given temperature difference the effects of thermophoresis are maximum when the plates have the same velocity. The increasing of the tangential velocity occurs only when at least one of the plates is moving. The reason is that the movement of molecules induced by temperature difference is symmetrical in relation to the direction of the mean flow when the plates are at rest. But the plates movement favours the shear flow which increases the tangential velocity.

This phenomenon is due to the transformation of thermic energy into kinetic energy; while the increasing of the internal energy results to the inverse transformation. It becomes natural that when the gas density increases, the thermophoresis vanishes; since the frictions between the gas layers and between the gas and the plates favour the increasing of the internal energy at the expense of the kinetic energy.

The Knudsen layers brought to the fore for Knudsen numbers of the order of unity, when the plate energies are different vanishes when the plates have the same energy whatever ϵ. This result found using the C1 model is confirmed by other types of discrete models. It is in accordance to the fact that Knudsen layers are permanent energetic unbalanced zones [15].

The same results have been obtained with some quantitative improvements numerically with non zero θ_2 [9].

References

1. C. BARDOS, F. GOLSE and D. LEVERMORE, "Fluid dynamic limits of discrete velocity kinetic equations.", in *"Advances in Kinetic Theory and Continuum Mechanics"*, Ed. by R. Gatignol and Soubbaramayer, Springer- Verlag, 1991.

2. J. E. BROADWELL, "Study of rarefied shear flow by the discrete velocity method.", J. Fluid Mech., 19, (1964), p. 401-414.

3. H. CABANNES, "Couette flow for a gas with a discrete velocity distribution.", J. Fluid Mech., 76, (1976), p. 273-286.

4. C. CERCIGNANI, R. ILLNER and M. SHINBROT, "A boundary value problem for the two dimensional Broadwell model.", Commun. Math. Phys., 114, (1988), p. 687-698.

5. C. CERCIGNANI, "On the thermodynamics of a discrete velocity gas.", private communication.

6. P. CHAUVAT, "Summational invariants in discrete kinetic theory with multiple collisions.", Mechanics Research Communications, 18, (1991), p. 11-16.

7. P. CHAUVAT, F. COULOUVRAT and R. GATIGNOL, "The Euler Description for a Class of Discrete Models of Gases with Multiple Collisions.", in *"Advances in Kinetic Theory and continuum mechanics"*, Ed. by R. Gatignol and Soubbaramayer, Springer- Verlag, 1991.

8. H. CORNILLE, "Temperature and local entropy overshoots for the second fourteen-velocity Cabannes model.", in *"Advances in kinetic Theory and Continuum Mechanics"*, Ed. by R. Gatignol and Soubbaramayer, Springer- Verlag, 1991.

9. A. D'ALMEIDA, Thèse de l'Université Pierre et Marie Curie, Paris, (en préparation).

10. R. GATIGNOL, "Théorie cinétique d'un gaz à répartiton discrète de vitesses.", Z. Flugwissenschaften, 18, (1970), p. 93-97.

11. R. GATIGNOL, *"Théorie cinétique d'un gaz à répartition discrète de vitesses."*, Lecture Notes in Physics, 36, (1975), Springer- Verlag.

12. R. GATIGNOL, "Kinetic theory boundary conditions for discrete velocity gases", The Phys. of Fluids, 20, (1977), p. 2022-2030.

13. T. INAMURO and B. STURTEVANT, "Heat transfert in a discrete velocity gas.", Proc. *17th International Symposium on Rarefied Gas Dynamics*, Aachen, Ed. by A. E. Beylich, (1989), p. 854-861.

14. S. KAWASHIMA and H. CABANNES, "Initial value problem in discrete kinetic theory.", *16th International Symposium on Rarefied Gas Dynamics*, Pasedena, Vol II, (1988), p. 148-154.

15. M. KOGAN, *"Rarefied gas dynamics."*, (1969), Plenum Press.

16. T. PLATKOWSKI and R. ILLNER, "Discrete velocity models of the Boltzmann equation: a survey on the mathematical aspect of the theory.", SIAM Review, 30, (1988), p. 213-255.

Computation of Viscous Transonic Flow Around the F5 Wing

Arthur Rizzi

KTH, The Royal Institute of Technology, S-100 44 STOCKHOLM, Sweden

Summary. Turbulent flow around wings are simulated by solving the three-dimensional Navier-Stokes equations in hexahedronal computational cells. Turbulent flow is described by an algebraic eddy viscosity model. Boundary-conforming O-O type meshes are generated by the transfinite interpolation method. The finite-volume technique is employed for spatial discretization using two cells in each computational coordinate direction, one more than the conventional approach to evaluate the viscous fluxes. An explicit three-stage Runge-Kutta scheme is used for the time integration. The method is applied to compute turbulent transonic flow around the DLR F5 wing, which has been studied extensively both experimentally and numerically in a recent workshop. Our results are compared with those workshop data.

1. Introduction

The computation of viscous transonic flow around airfoils and wings is of current interest to design advanced aircraft components. The solution of the Navier-Stokes equations meets the need of the design engineer for more accurate flow predictions. Faster supercomputers and better numerical methods render calculations more efficient.

Explicit predictor-corrector and implicit approximate factorization time stepping schemes have frequently been applied using finite-difference and finite-volume space discretizations [1], [2] . After successful application to the Euler equations [3], [4], explicit Runge-Kutta finite-volume methods have been developed for the numerical solution of the Navier-Stokes equations [5], [6], [7], [8] and [9].

Apart from different boundary conditions for viscous flow, the stress tensor and the heat flux on all cell interfaces have to be determined (Section 1). The gradients of the velocity components and temperature and the divergence of velocity are calculated either with the help of local coordinate transformations or line integrals along the boundary of another cell straddling the cell interface [10]. The present approach is similar to the latter one, but avoids the definition of straddling cells [11] to [13].

Instead, the spatial derivatives of the velocity components and temperature are approximated first in the cells themselves with the help of line integrals along the cell boundaries. Thus, not only the inviscid flux tensor but also the viscous one can be determined in each cell. The flux tensor on a cell interface is then obtained by the arithmetic average of its values in adjacent cells. Nonlinear second-order differences sensed by the discretized second

131

derivative of the pressure and linear fourth-order differences of the conservative variables are added to the physical flux difference operator to damp short wavelength oscillations. The explicit three-stage Runge-Kutta time integration method is second-order accurate, if constant time stepping is used [14]. Here, local time steps are determined from a linear stability condition derived for discretizations of the Navier-Stokes equations. On a Cartesian mesh, the spatial discretization is second-order accurate (Section 2).

Results are presented for turbulent transonic flow at $M_\infty = 0.82$ and $Re_\infty = 10^7$ (per meter) over the DLR F5 wing in free air at two angles of attack $\alpha = 0°$ and $\alpha = 2°$. Transition is imposed at a prescribed location. The solutions have been obtained on a boundary conforming O-O-type mesh of $193 \times 41 \times 41$ points generated by the transfinite interpolation method [15]. The present results compare reasonably with experimental and other computational data (Section 3).

2. Governing Equations and Boundary Conditions

2.1 Navier-Stokes Equations

Fluid motion is governed by the conservation laws for mass, momentum and energy. The investigated fluid here is a perfect gas obeying Newton's and Fourier's laws. External forces and heat sources are not taken into account. Considering an arbitrary stationary cell $\mathcal{V}$ with the boundary $\partial\mathcal{V}$ and the outer normal unit vector $\mathbf{n}$ in an absolute frame of reference, the Navier-Stokes equations read [10]:

$$\int_\mathcal{V} \frac{\partial \mathbf{u}}{\partial t} d\mathcal{V} + \int_{\partial\mathcal{V}} \mathcal{F} \cdot \mathbf{n} dA = 0 \tag{1}$$

where

$$\mathbf{u} = \begin{pmatrix} \rho \\ \rho\mathbf{v} \\ e \end{pmatrix} \qquad \mathcal{F} = \begin{pmatrix} \rho\mathbf{v} \\ \rho\mathbf{v}\mathbf{v} + p\mathbf{I} - \tau \\ (e+p)\mathbf{v} - \tau \cdot \mathbf{v} + \mathbf{q} \end{pmatrix}$$

$\mathbf{u}$ is the vector of the conservative variables, i.e. density, momentum density and total energy density. $\mathcal{F}$ represents the flux tensor, $\mathbf{I}$ the unit tensor. Pressure and temperature are related to the conservative variables by the equations of state for perfect gas:

$$p = \rho T \tag{2}$$

$$T = (\gamma - 1)\left(\frac{e}{\rho} - \frac{1}{2}|\mathbf{v}|^2\right) \tag{3}$$

with γ the ratio of the constant specific heats. The stress tensor is given by Newton's law:

$$\tau = \mu \, [\, \text{grad } \mathbf{v} + (\text{grad } \mathbf{v})^T] + \lambda \, \text{div}(\mathbf{v}) \, \mathbf{I} \tag{4}$$

with μ and λ the viscosity coefficients. Fourier's law states for the heat flux:

$$\mathbf{q} = -\kappa \, \text{grad} T \tag{5}$$

with κ the thermal conductivity coefficient. The viscosity coefficients are related by Stokes's hypothesis also for polyatomic gas:

$$\lambda = -\frac{2}{3}\mu \tag{6}$$

Since the Prandtl number is assumed to be constant, the following relation between the dimensional thermal conductivity and viscosity coefficients holds:

$$\frac{\bar{\kappa}}{\bar{\kappa}_\infty} = \frac{\bar{\mu}}{\bar{\mu}_\infty} \tag{7}$$

The viscosity coefficient is obtained from the Sutherland formula

$$\frac{\bar{\mu}}{\bar{\mu}_\infty} = \frac{1+S}{T+S} T^{\frac{3}{2}} \tag{8}$$

where $S = S/T_\infty$ with the Sutherland constant $S = 110$ K for air. In three-dimensional Cartesian coordinates, the spatial coordinates are denoted by x, y, and, z and their corresponding velocity components by u, v, and w. These and the other dimensionless variables in Eqs(1) to (8) are defined by the following ratios of their corresponding dimensional quantities (denoted by a bar, which is dropped in the subsequent figures) to freestream values (denoted by index ∞) and a characteristic length L (e.g. chord):

$$
\begin{aligned}
x &= \frac{\bar{x}}{L} \,,\; y = \frac{\bar{y}}{L} \,,\; z = \frac{\bar{z}}{L} \,,\; t = \frac{\bar{t}\sqrt{(\bar{p}_\infty/\bar{\rho}_\infty)}}{L} & (9)\\[2mm]
\rho &= \frac{\bar{\rho}}{\bar{\rho}_\infty} \,,\; p = \frac{\bar{p}}{\bar{p}_\infty} \,,\; T = \frac{\bar{T}}{T_\infty} \,,\; e = \frac{\bar{e}}{\bar{p}_\infty} \,,\\[2mm]
u &= \frac{\bar{u}}{\sqrt{(\bar{p}_\infty/\bar{\rho}_\infty)}} \,,\; v = \frac{\bar{v}}{\sqrt{(\bar{p}_\infty/\bar{\rho}_\infty)}} \,,\; w = \frac{\bar{w}}{\sqrt{(\bar{p}_\infty/\bar{\rho}_\infty)}} \,,\\[2mm]
\mu &= \frac{\bar{\mu}}{\bar{\mu}_\infty} \frac{\sqrt{\gamma} M_\infty}{Re_\infty} \;\; \lambda = \frac{\bar{\lambda}}{\bar{\mu}_\infty} \frac{\sqrt{\gamma} M_\infty}{Re_\infty} \,,\\[2mm]
\kappa &= \frac{\bar{\kappa}}{\bar{\kappa}_\infty} \frac{\sqrt{\gamma} M_\infty}{Re_\infty} \frac{\gamma}{Pr(\gamma-1)} = \mu \frac{\gamma}{Pr(\gamma-1)}
\end{aligned}
$$

The freestream Mach, Reynolds and Prandtl numbers are defined by the reference quantities as follows:

$$M_\infty = \frac{|\bar{\mathbf{v}}|}{\bar{c}_\infty} \ , \ \mathrm{Re}_\infty = \frac{\bar{\rho}_\infty \, |\bar{\mathbf{v}}| \, \bar{L}}{\bar{\mu}_\infty} \ , \ \mathrm{Pr} = \frac{\bar{\mu}_\infty \bar{c}_p}{\bar{\kappa}_\infty} \qquad (10)$$

with $\bar{c}_\infty$ the dimensional freestream speed of sound, and $\bar{c}_p$ the dimensional specific heat at constant pressure.

2.2 Turbulence Model

The algebraic turbulence model of Baldwin and Lomax [16] have been adopted for the Reynolds-averaged Navier-Stokes equations. It is patterned after the two-layer eddy viscosity model of Cebeci-Smith. The eddy viscosity is determined from

$$\mu_t = \min\{\mu_{t\text{ inner}}, \ \mu_{t\text{ outer}}\} \qquad (11)$$

The definition of μ_t in the inner and outer layers is recalled for the application to O-O-meshes. The Prandtl-van Driest formulation is used in the inner region.

$$\mu_{t\text{ inner}} = \rho \, u_{\text{inner}} \, \ell_{\text{inner}} \qquad (12)$$

with the mixing length

$$\ell_{\text{inner}} = 0.4Y \left[1 - \exp\left(-\frac{Y^+}{26} \right) \right] \qquad (13)$$

and the velocity scale

$$u_{\text{inner}} = |\text{ curl } \mathbf{v}| \, \ell_{\text{inner}} \qquad (14)$$

Y denotes the normal distance of the cell midpoint from the wall and Y^+ is defined by

$$Y^+ = Y \left[\frac{\rho \, | \text{ curl } \mathbf{v} |}{\mu} \right]_{\text{wall}}^{1/2} \qquad (15)$$

The wall quantities are approximated by the values in the first cell above the wing surface. For separated flow, $| \text{ curl } \mathbf{v} |_{\text{wall}}$ should be replaced by $| \text{ curl } \mathbf{v} |_{\text{max}}$ to avoid zero eddy viscosity at separation points. The Clauser formulation is modified in the outer region according to

$$\mu_{t\text{outer}} = 0.0168 \rho u_{\text{outer}} \, \ell_{\text{outer}} \, \gamma_K(Y) \qquad (16)$$

with the length scale

$$\ell_{\text{outer}} = C_{CP} Y_{\text{max}} \qquad (17)$$

and the velocity scale

134

$$u_{\text{outer}} = \min\{F_{\max}, \, u_{\text{dif}}^2/F_{\max}\} \tag{18}$$

where $Y_{\max}$ and $F_{\max}$ are defined by

$$F_{\max} = \max_{0<Y}\{F(Y)\} = F(Y_{\max}) \tag{19}$$

and

$$F(Y) = Y \, | \, \text{curl } \mathbf{v} \, | \left[1 - \exp\left(-\frac{Y^+}{26}\right)\right] \tag{20}$$

$Y_{\max}$ and $F_{\max}$ are determined more accurately by quadratic interpolation. u_{dif} is the difference between maximum and minimum total velocity, i.e. for O-O meshes

$$u_{\text{dif}} = \max_{0<Y} | \, \mathbf{v} \, | \tag{21}$$

Note that the present results were obtained with $u_{\text{outer}} = F_{\max}$ in Eq.(18). The intermittancy factor γ_K is given by

$$\gamma_K(Y) = \left[1 + 5.5\frac{(C_K Y)^6}{(Y_{\max})^6}\right]^{-1} \tag{22}$$

The constants $C_{CP} = 1.6$ and $C_K = 0.3$ are adjusted for transonic flow. Thus, in the Navier-Stokes equations the molecular viscosity coefficient μ is replaced by $\mu + \mu_t$, and in the heat flux terms μ/Pr is replaced by $\mu/\text{Pr} + \mu_t/\text{Pr}_t$ with the turbulent Prandtl number $\text{Pr}_t = 0.9$.

2.3 Boundary Conditions

O-O-type meshes introduce periodic boundaries, when "upper" and "lower" lines in 2D and surfaces in 3D are mapped on each other. For the conservative variables to be uniquely defined there, the periodic boundary condition requires:

$$\mathbf{u}_{\text{lower}} = \mathbf{u}_{\text{upper}} \tag{23}$$

For the symmetric boundary $y = 0$ in 3D, ρ, u, w, and e are even functions with respect to y, and v is an odd function:

$$[\rho, u, v, w, e]^T \, (x, y, z) = [\rho, u, -v, w, e]^T (x, -y, z) \tag{24}$$

The no-slip condition holds on the solid wall of an airfoil or a wing:

$$\mathbf{v}_W = 0 \tag{25}$$

Isothermal or adiabatic walls are considered:

$$T_w = \text{constant} \quad \text{or} \quad \frac{\partial T_w}{\partial \mathbf{n}} = 0 \tag{26}$$

The wall pressure is obtained by neglecting the viscous terms in the wall normal momentum equation:

$$\frac{\partial p_w}{\partial n} = 0 \tag{27}$$

The boundary conditions at the farfield boundary are based on the theory of characteristics for locally one-dimensional inviscid flow [17]. For supersonic inflow or outflow, the locally one-dimensional Riemann invariants, entropy and tangential velocity component, i.e. (with the speed of sound c):

$$R_1 = \mathbf{v} \cdot \mathbf{n} - \frac{2}{\gamma - 1}c \tag{28}$$

$$R_2 = \mathbf{v} \cdot \mathbf{n} - \frac{2}{\gamma - 1}c \tag{29}$$

$$R_3 = \ln\left(\frac{p}{\rho^\gamma}\right) \tag{30}$$

$$\mathbf{R}_4 = \mathbf{v} - (\mathbf{v} \cdot \mathbf{n})\mathbf{n} \tag{31}$$

are given from outside or inside the region of interest, respectively:

$$R_m = R_{m\infty} \quad \text{or} \quad \frac{\partial \mathbf{R}_m}{\partial \mathbf{n}} = 0 \tag{32}$$

For subsonic inflow, R_1, R_3 and $\mathbf{R}_4$ are given from outside and R_2 from inside. R_2, R_3 and $\mathbf{R}_4$ are determined from inside and R_1 from outside for subsonic outflow.

3. Numerical Method

3.1 Spatial Discretization

The Navier-Stokes equations (1) are discretized in quadrilaterals in 2D and in hexahedrons in 3D (Fig. 3.1). Since the conservative variables are assumed to be defined by their cell averages, the volume integral in Eq. (1) over a cell P is expressed by:

$$\int_{\mathcal{V}_p} \frac{\partial \mathbf{u}}{\partial t} d\mathcal{V} = \frac{\partial \mathbf{u}_p}{\partial t} \int_{\mathcal{V}_p} d\mathcal{V} \tag{1}$$

The surface integral in Eq. (1) over the boundary of cell P is approximated by assuming the mean-value of the flux tensor on each side to be equal to the arithmetic average of the flux tensor in the adjacent cells:

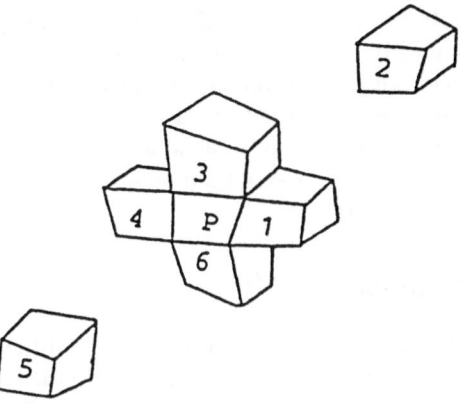

Fig. 3.1. Basic hexahedron P and neighbouring cells 1 to 6 (2 and 5 set apart)

$$\int_{\partial V_p} \mathcal{F} \cdot \mathbf{n} dA \cong \sum_{k=1}^{2d} \mathcal{F}_{Pk} \cdot \int_{\partial V_{pk}} \mathbf{n}\, dA \qquad (2)$$

where

$$\mathcal{F}_{Pk} = \frac{1}{2}(\mathcal{F}_P + \mathcal{F}_k)$$

d is the spatial dimension, i.e. 2 for 2D and 3 for 3D. $\partial V Pk$ denotes the common part of the boundaries of P and its neighbouring cell k. The cell volume in Eq. (1) is computed as the sum of five tetrahedra [4]. The surface normal in Eq. (2) is determined as the sum of the surface normals of two triangles [18]. With the conservative variables given, all terms of the flux tensor are readily available in cell P, except for the gradients of the velocity components and temperature as well as $div\mathbf{v}$. Following the definition of the conservative variables as cell averages, the gradients in cell P are defined by:

$$\mathrm{grad}\ \phi_P = \frac{\int_{V_p} \mathrm{grad}\ \phi\ dV}{\int_{V_p} dV} \qquad (3)$$

where $\phi = u, v, w$, or T. Using the gradient theorem, the volume integral in (3) can be expressed by a surface integral, which is approximated similarly to (2) [13]:

$$\mathrm{grad}\ \phi_P = \frac{\int_{\partial V_p} \phi\ \mathbf{n}\ dS}{\int_{V_p} dV} \cong \frac{1}{V_P} \sum_{k=1}^{2d} \phi_{Pk} \int_{\partial V_{pk}} \mathbf{n}\ dS \qquad (4)$$

where

$$\phi_{Pk} = \frac{1}{2}(\phi_P + \phi_k)$$

and div $\mathbf{v}_P$ is evaluated similarly to $\mathbf{grad}\ \phi$. The present spatial discretization involves two cells more in each computational coordinate direction than the compact discretization of $\mathbf{grad}\ \phi$ on the cell interface either by local coordinate transformation or with the help of a straddling cell. The disadvantage of the gradient discretization (4) is that its second-order truncation error on a smooth mesh is four times greater than that of the compact discretization of $\mathbf{grad}\ \phi$ (Refs [11] to [13]).

3.2 Numerical Damping

The spatial discretization constitutes the physical difference operator $\mathbf{F}_{PH}$ defined by the negative right hand side of (2) divided by the cell volume. These central differences do not damp unphysical oscillations caused by flow discontinuities and waves with short wavelengths. The numerical damping terms $\mathbf{F}_N$, which are therefore added to $\mathbf{F}_{PH}(\mathbf{u})$, comprise nonlinear second-order differences sensed by the discretized second derivative of the pressure, and linear fourth-order differences of the conservative variables [19]:

$$\mathbf{F}_N(\mathbf{u}) = (\mathrm{CFL}/\Delta t)\{ \quad \chi \quad (\delta_I[s_I(p)\delta_I] + \delta_J[s_J(p)\delta_J] \tag{5}$$
$$+ \quad \delta_K[s_K(p)\delta_K]) - \Lambda(\delta_I^4 + \delta_J^4 + \delta_K^4)\}\mathbf{u}$$

with CFL the maximum CFL number used (cf. Section 2.5 on stability below) and Δt the time step. The constants χ and Λ used lie in the ranges: $0 \leq \chi \leq 0.2$ and $0.005 \leq \Lambda \leq 0.02$ (cf. section on stability below). The sensors s_I, s_J and s_K are of similar form, e.g. s_I for a cell indexed by I, J, K:

$$s_I(p_{I\pm\frac{1}{2},J,K}) = \mu_I \mid \delta_I^2 p_{I\pm\frac{1}{2},J,K} \mid / \max_{I',J',K'} \mid \delta_I^2 p_{I',J',K'} \mid \tag{6}$$

The classical finite-difference operators are defined by:

$$\delta_I a_{I,J,K} = a_{I+\frac{1}{2},J,K} - a_{I-\frac{1}{2},J,K} \tag{7}$$
$$\mu_I a_{I,J,K} = \frac{1}{2}(a_{I+\frac{1}{2},J,K} + a_{I-\frac{1}{2},J,K})$$

and similarly for J and K. The numerical damping operator $\mathbf{F}_N$ is modified near solid wall and farfield boundaries to ensure its dissipative property also there [20]. Using the periodic and symmetry boundary conditions (cf. (23) and (24)), $\mathbf{F}_N$ can be determined near these boundaries without modifying (5) or (6).

3.3 Time Integration

Thus, the semi-discrete approximation of the Navier-Stokes equations can be written as:

$$\frac{d\mathbf{u}}{dt} = \mathbf{F}(\mathbf{u}) \tag{8}$$

where

$$\mathbf{F} = \mathbf{F}_{PH} + \mathbf{F}_N$$

Equation (8) represents a large system of first- order ordinary differential equations. It is solved for the steady state by the second-order explicit three-stage Runge-Kutta scheme [14]:

$$
\begin{aligned}
\mathbf{u}^{(0)} &= \mathbf{u}^n \\
\mathbf{u}^{(1)} &= \mathbf{u}^{(0)} + \Delta t \mathbf{F}(\mathbf{u}^{(0)}) \\
\mathbf{u}^{(2)} &= \mathbf{u}^{(0)} + \Delta t[(1-\theta)\mathbf{F}(\mathbf{u}^{(0)}) + \theta\mathbf{F}(\mathbf{u}^{(1)})] \\
\mathbf{u}^{(3)} &= \mathbf{u}^{(0)} + \Delta t[(1-\theta)\mathbf{F}(\mathbf{u}^{(0)}) + \theta\mathbf{F}(\mathbf{u}^{(2)})] \\
\mathbf{u}^{n+1} &= \mathbf{u}^{(3)}
\end{aligned}
\tag{9}
$$

with $\theta = \frac{1}{2}$ and n denotes the time level. Equation (9) is second-order accurate for $\theta = \frac{1}{2}$ and constant Δt. Computing time is saved by evaluating the viscosity coefficient only once per time step, i.e. by approximating $\mu(\mathbf{u}^{(1)})$ and $\mu(\mathbf{u}^{(2)})$ by $\mu(\mathbf{u}^{(3)})$. There was no negative effect of this approximation noticed on the convergence to the steady-state.

3.4 Initialization and Boundary Treatment

The calculation is started from freestream on a mesh obtained by cancelling every other point of the next finer mesh. The converged result is interpolated to the finest mesh. In general three grids, i.e. a coarse, medium and fine one, are used. Suppose that the indices I, J, K denote the chordwise, near normal and spanwise directions. Then, the periodic boundary condition (23) implies:

$$\mathbf{u}_{IMAX,J,K} = \mathbf{u}_{1,J,K} \quad \mathbf{u}_{I,J,KMAX} = \mathbf{u}_{IMAX-I,J,KMAX-1} \tag{10}$$

The pressure and the stress tensor on the solid wall interface of the first cell above the wall are approximated by their values in that cell. The conditions (32) are used to determine R_m in a fictitious cell outside the domain of integration either by freestream or by R_m in the cell adjacent to the farfield.

3.5 Stability

The stability of explicit Runge-Kutta schemes applied to the semi–discretization (8) of the 3D Navier–Stokes equations (1) is studied for the scalar model equation [12], [13] and [21]:

$$
\frac{\partial a}{\partial t} + \lambda_\xi \frac{\partial a}{\partial \xi} + \lambda_\eta \frac{\partial a}{\partial \eta} + \lambda_\zeta \frac{\partial a}{\partial \zeta} = \nu_\xi \frac{\partial^2 a}{\partial \xi^2} + \nu_\eta \frac{\partial^2 a}{\partial \eta^2} + \nu_\zeta \frac{\partial^2 a}{\partial \zeta^2} \tag{11}
$$

$$
+ \nu_{\xi\eta} \frac{\partial^2 a}{\partial \xi \partial \eta} + \nu_{\xi\zeta} \frac{\partial^2 a}{\partial \xi \partial \zeta} + \nu_{\eta\zeta} \frac{\partial^2 a}{\partial \eta \partial \zeta}
$$

$$
+ \varepsilon_\xi \frac{\partial^4 a}{\partial \xi^4} + \varepsilon_\eta \frac{\partial^4 a}{\partial \eta^4} + \varepsilon_\zeta \frac{\partial^4 a}{\partial \zeta^4}
$$

where

$$
\begin{aligned}
\lambda_\eta &= |\mathbf{v} \cdot \mathrm{grad} h| + c |\, \mathrm{grad}\, h\,| \\
\nu_h &= \nu \mathrm{grad}\, h \cdot \mathrm{grad}\, h + \Delta h^2 \chi_h \mathrm{CFL}/\Delta t \\
\nu_{hg} &= 2\nu |\, \mathrm{grad}\, h \cdot \mathrm{grad}\, g\,| + ((\lambda + \mu)/\rho) |\, \mathrm{grad}\, h\,|\,|\, \mathrm{grad}\, g\,| \\
\varepsilon_h &= -\Delta h^4 \Lambda \mathrm{CFL}/\Delta t
\end{aligned}
$$

with c the speed of sound, and

$$
\begin{aligned}
\nu &= \max\{\mu, \lambda + 2\mu, \mu\gamma/\mathrm{Pr}\}/\rho \\
\chi_h &= \chi s_h \text{ and } \mathrm{h, g} \,\epsilon\, \{\xi, \eta, \zeta\}
\end{aligned}
$$

Equation (11) is derived from the differential form of the Navier-Stokes equations in transformed coordinates ξ, η, ζ [10] and from the differential expressions of the damping terms in (5). The coefficients of the model equation are obtained from the spectral radii of the coefficient matrices of the linearized Navier-Stokes equations [22] and from the coefficients of the linearized numerical damping terms. Thus, the scalar ansatz (11) models the full Navier-Stokes equations including the numerical damping terms, whereas inviscid, diffusion and mixed-derivative vector parts of the linearized Navier-Stokes equations are considered separately by [23] to [25] to determine the respective time steps of time splitting schemes. The first and second spatial derivatives in (11) are discretized by second-order central finite-differences corresponding to the finite-volume approximation (2):

$$
\frac{\partial a}{\partial h} |_{I,J,K} = \frac{1}{\Delta h} \mu_h \delta_h a_{I,J,K} + 0(\Delta h^2) \tag{12}
$$

$$
\frac{\partial a^2}{\partial h^2} |_{I,J,K} = \frac{1}{\Delta h^2} (\mu_h \delta_h) a_{I,J,K} + 0(\Delta h^2) \tag{13}
$$

$$\frac{\partial^2 a}{\partial h \partial g}\Big|_{I,J,K} = \frac{1}{\Delta h \Delta g}\mu_h \delta_h \mu_g \delta_g\, a_{I,J,K} + 0(\Delta h^2, \Delta g^2) \qquad (14)$$

The finite-difference operators in (12), (13) and (14) are defined by (7) with ξ corresponding to I, and similarly η to J and ζ to K. The fourth derivatives in (11) are discretized according to (5). Note that $a_{I\pm1,J,K}, a_{I,J\pm1,K}$, and $a_{I,J,K\pm1}$ do not appear in (13). The linear stability of an explicit Runge-Kutta scheme is investigated for the application to the model equation (11) with frozen coefficients and spatially discretized by (12), (13) and (14). The analysis shows that the shortest resolvable waves are undamped by (13) in contrast to the compact differencing δ_h^2. The stability condition reads:

$$\Delta t \leq \min \ \{ \text{CFL} \ \ [|\,\lambda_\xi\,|/\Delta\xi + |\,\lambda_\eta\,|/\Delta\eta + |\,\lambda_\zeta\,|/\Delta\zeta]^{-1} \qquad (15)$$

$$|\,\text{RK}\,| \ \ [\frac{\tilde{\nu}_\xi}{\Delta\xi^2} + \frac{\tilde{\nu}_\xi}{\Delta\xi^2} + \frac{\tilde{\nu}_\eta}{\Delta\eta^2} + \frac{\tilde{\nu}_\zeta}{\Delta\zeta^2} + \frac{\nu_{\xi\eta}}{\Delta\xi\Delta\eta} + \frac{\nu_{\xi\zeta}}{\Delta\xi\Delta\zeta} + \frac{\nu_{\eta\zeta}}{\Delta\eta\Delta\zeta}$$

$$- \ \ 16\left(\frac{\varepsilon_\xi}{\Delta\xi^4} + \frac{\varepsilon_\eta}{\Delta\eta^4} + \frac{\varepsilon_\zeta}{\Delta\zeta^4}\right)\Big]^{-1} \}$$

where

$$\tilde{\nu}_h = \alpha\nu\ \text{grad}\ h \cdot \text{grad}\ h + 4\Delta h^2 \chi_h \text{CFL}/\Delta t$$

with $\alpha = 1$ for (13), but $\alpha = 4$ for the compact differencing δ_h^2. The negative and positive stability bounds RK and CFL, respectively, are chosen such that all complex numbers z with $\text{RK} \leq \text{Re}\,(z) \leq 0$ and $|\,\text{Im}\,(z)\,| \leq \text{CFL}$ lie inside the stability region of the Runge-Kutta method. For the three-stage Runge-Kutta method (9) the following choice is taken: CFL=1.5, RK=-1 (Fig. 3.2). The viscous contributions in (15) (denoted here by S) involving coefficients ν and $(\lambda + \mu)/\rho$ and the second- and fourth-order damping coefficients in (15) (here referred to as D_2 and D_4) share the negative real axis of the stability region. Their distribution may be prescribed by

$$S\Delta t \leq \beta_1\,|\,\text{RK}\,| \qquad (16)$$

$$D_2 \Delta t \leq \beta_2\,|\,\text{RK}\,| \qquad (17)$$

$$D_4 \Delta t \leq \beta_3\,|\,\text{RK}\,| \qquad (18)$$

where $0 \leq \beta_\ell$ and $\sum_{\ell=1}^{3}\beta_\ell = 1$. Equations (17) and (18) yield the following conditions for the damping coefficients:

$$\chi \leq (\beta_2/4d)\,|\,\text{RK}\,|/\text{CFL} \qquad (19)$$

$$\Lambda \leq (\beta_3/16d)\,|\,\text{RK}\,|/\text{CFL} \qquad (20)$$

with d the dimension, i.e. $d = 3$ for 3D.

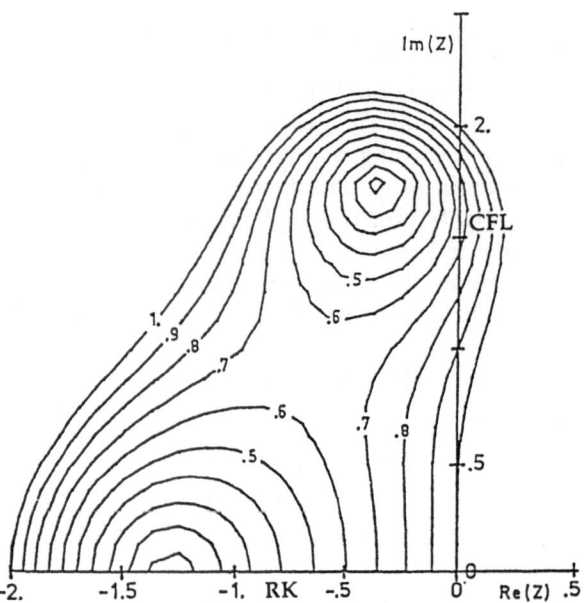

Fig. 3.2. Contours of constant modulus of growth factors of (26) and stability bounds RK and CFL

Using then (16) in (15), the stability condition is obtained:

$$\Delta t \leq \min\{ \quad \text{CFL} \quad \left[\frac{|\lambda_\xi|}{\Delta\xi} + \frac{|\lambda_\eta|}{\Delta\eta} + \frac{|\lambda_\zeta|}{\Delta\zeta}\right]^{-1} \tag{21}$$

$$\beta_1 \quad |\,\text{RK}\,| \quad \left[\frac{\nu_{\xi\xi}}{\Delta\xi^2} + \frac{\nu_{\eta\eta}}{\Delta\eta^2} + \frac{\nu_{\zeta\zeta}}{\Delta\zeta^2} + \frac{\nu_{\xi\eta}}{\Delta\xi\Delta\eta} + \frac{\nu_{\xi\zeta}}{\Delta\xi\Delta\zeta} + \frac{\nu_{\eta\zeta}}{\Delta\eta\Delta\zeta}\right]^{-1}\}$$

where $\nu_{hh} = \alpha\nu\,\mathbf{grad}\,h\cdot\,\mathbf{grad}\,h$ with $\alpha = 1$ for (13), but $\alpha = 4$ for the compact differencing δ_h^2.

In order to apply the stability condition (21) to the finite- volume discretization, the metric terms are related to geometrical quantities (9): cell volume

$$\Delta\xi\Delta\eta\Delta\zeta\mathcal{J}^{-1} \simeq \mathcal{V} = \int_{\mathcal{V}} d\mathcal{V} \tag{22}$$

surface normal in G-direction:

$$\Delta\xi\Delta\eta\Delta\zeta\mathcal{J}^{-1}\mathbf{grad}\,g/\Delta g \simeq \mathbf{S}_G = \int_{\partial\mathcal{V}G} \mathbf{n}\,dS \tag{23}$$

where $g = \xi,\eta,\zeta$ corresponds to $G = I,J,K$ and $\partial_\mathcal{V}$ denotes a cell boundary with constant G, and $\mathcal{J}$ represents the Jacobian determinant of the tranformation $[\xi,\eta,\zeta]^T(x,y,z)$.

Thus, the stability condition (21) with (19) and (20) for an explicit Runge-Kutta method like (9) applied to the finite-volume discretization of the Navier-Stokes equations (8) reads:

$$\Delta t \le \min\{\text{CFL}\ \mathcal{V}\ [|\mathbf{v}\cdot\mathbf{S_I}| + c|\mathbf{S_I}| + |\mathbf{v}\cdot\mathbf{S_J}| + c|\mathbf{S_J}| + |\mathbf{v}\cdot\mathbf{S_K}| + c|\mathbf{S_K}|]^{-1}$$
$$\beta_1\ |\text{RK}|\ \mathcal{V}^2\ [\nu\ \{\ \alpha(\mathbf{S_I}^2 + \mathbf{S_J}^2 + \mathbf{S_J}^2) + 2(|\mathbf{S_I}\cdot\mathbf{S_J}| + |\mathbf{S_I}\cdot\mathbf{S_K}| + |\mathbf{S_J}\cdot\mathbf{S_K}|)\}$$
$$+((\lambda+\mu)/\rho)(|\mathbf{S_I}||\mathbf{S_J}| + |\mathbf{S_I}||\mathbf{S_K}| + |\mathbf{S_J}||\mathbf{S_K}|)]^{-1}\}\ (24)$$

with $\alpha = 1$. If the compact diffrencing δ_h^2 were used instead of (13), the stability condition would be more restictive, since $\mathbf{S_I}\cdot\mathbf{S_I} + \mathbf{S_J}\cdot\mathbf{S_J} + \mathbf{S_K}\cdot\mathbf{S_K}$ would have to be multiplied by $\alpha = 4$. In general, $\beta_1 = \frac{1}{2}, \beta_2 = \frac{1}{10}$, and $\beta_3 = \frac{3}{2}$ are used in (19), (20) and (24) with CFL$= \frac{3}{2}$ and RK$= 1$. Although

$$\sum_{\ell=1}^{3} \beta_\ell > 1$$

stability was maintained indicating that the stability bound RK could have been chosen lower than -1 for the Runge-Kutta method (9). However, for the $129 \times 49 \times 65$ mesh of the delta wing calculation in [13], the CFL number had to be reduced to $\frac{1}{2}$ and RK$= -1, \beta_1 = \frac{1}{2}, \beta_2 = \frac{1}{30}$, and $\beta_3 = \frac{1}{2}$ were used. Computational work is saved by computing the local time steps according to (24) only at those time levels which are powers of 2.

The analysis outlined above can easily be simplified to the 2D Navier-Stokes equation [11]. In general, $\beta_1 = 1$, $\beta_2 = 0$, $0.24 \le \beta_3 \le 0.48$, and RK$= -1$ are used in 2D.

4. Results

Computer programs have been developed based on the Runge-Kutta finite-volume method for the two- and three- dimensional Navier-Stokes equations [11] and [12]. Here we apply the 3D code *NSWINGA* to the large aspect ratio DLR F5 wing [27] proposed by the International Workshop [28] as a means of validating a computer program by comparison with other computed results and experimental measurements [29], [30]. The flow over the wing in free air has been computed at $M_\infty = 0.82$ and $Re_\infty = 10^7$ (based on one meter) and two angles of attack $\alpha = 0°$ and $2°$ with transition imposed at the prescribed location.

4.1 O-O Mesh

The wing is a symmetric trapezoid wing mounted with smooth fairing onto a splitter plate. As it is defined, the wing trailing edge and the wing tip are not closed. This meant that the surface had to be closed (Fig. 4.1), and the

Fig. 4.1. Wing profiles closed in the chordwise and spanwise directions

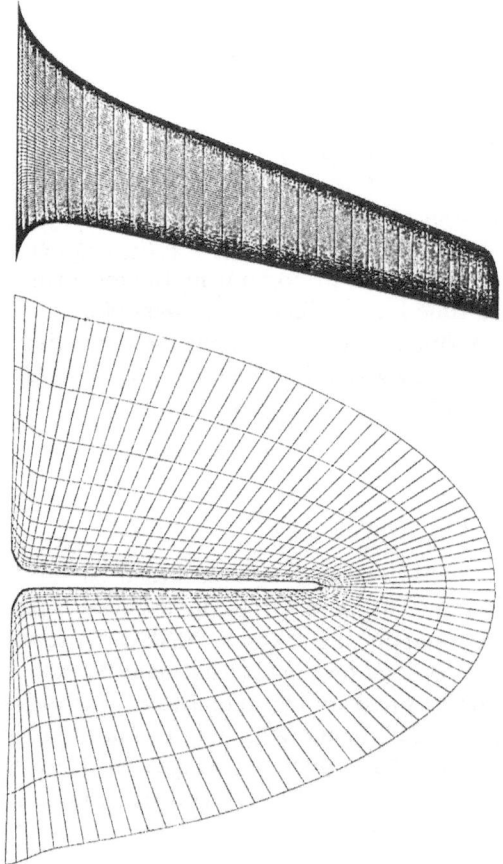

Fig. 4.2. Dense O-O mesh around the DLR F5 wing. $193 \times 49 \times 41$ points

resulting mesh had to be smoothed so that no unnatural sharp edges were left to shed undesired vortices. The final surface mesh is shown in Fig. 4.2, and the complete mesh contains $193 \times 49 \times 41$ points.

4.2 Case 1: $\alpha = 0°$

The lift, drag, and quarter-chord moment coefficients are predicted to $C_L = 0.0009, C_D = 0.0157$, and $C_M = 0.0002$ by the present Navier-Stokes solution. The L_2 norm of the residual $\rho^n - \rho^{n-1}$ had the value 1.2×10^{-5} and the maximum modulus of the residual was 1.4×10^{-4}. The pressure contours on the upper side (Fig. 4.3) show a sharp recompression at about 70% local

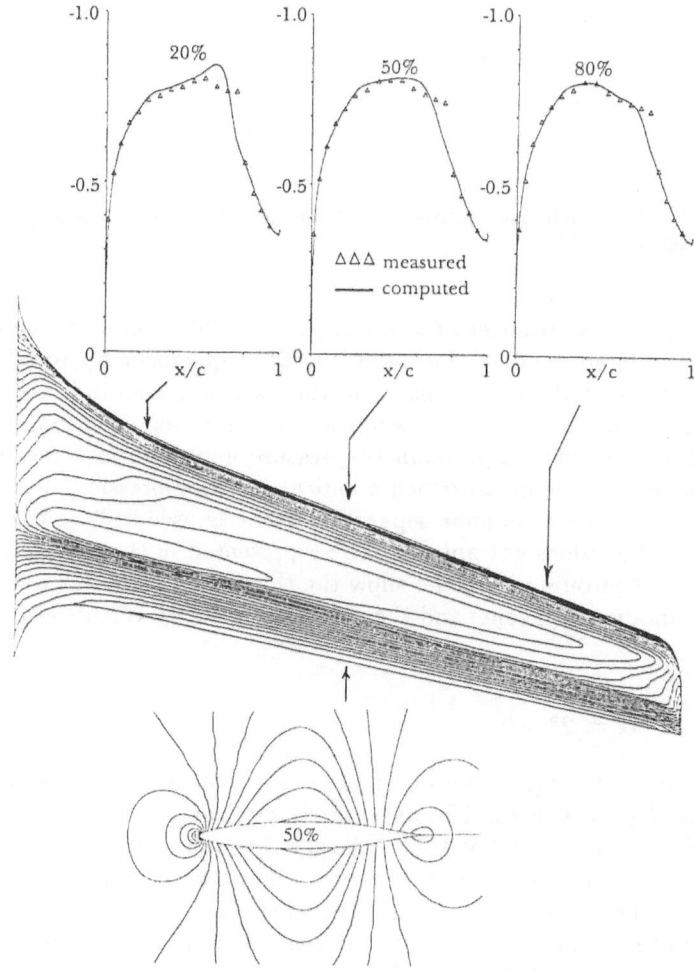

Fig. 4.3. Isobars over the upper wing surface, in the 50% spanwise section, and surface C_p distributions in the 20%, 50%, and 80% sections. Incidence angle 0°

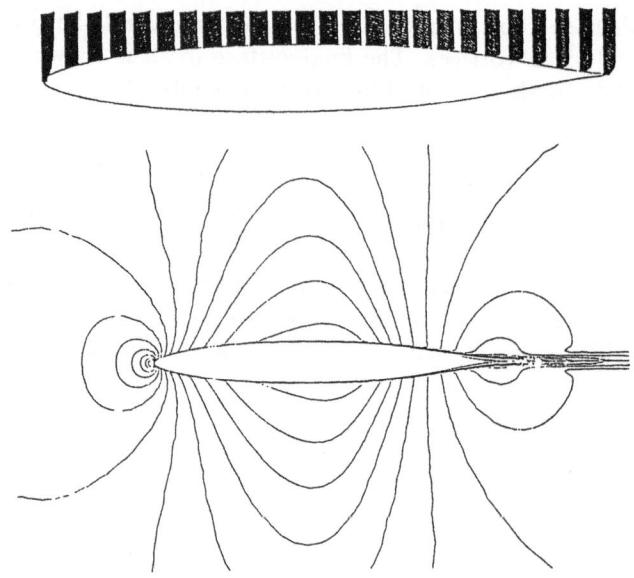

Fig. 4.4. Chordwise velocity vectors and Mach contours in the 50% spanwise section. Incidence angle 0°

chord. The pressure coefficient C_p is compared at 20%, 50% and 80% semi-span with the wind tunnel measurements. The agreement up to 50% local chord is excellent, but then downstream the measurements indicate a shock instead of the continuous recompression that is computed. The computations in the 20% section show a pronounced pressure minimum just ahead of the recompression, but the measurements indicate a recompression ahead of the shock. This suggests a laminar separation (that is assumed by the experimentalists), which does not appear to be represented in the simulation. The Mach number contours in Fig. 4.4 show the thickening of the boundary layer after the recompression zone, and the chordwise velocity vectors also confirm this.

4.3 Case 2: $\alpha = 2°$

The lift, drag, and quarter-chord moment coefficients are $C_L = 0.0747, C_D = 0.0150$, and $C_M = 0.0075$. The average residual had the value 2.1×10^{-5} and the maximum was 1.5×10^{-4}. At this incidence the computed Mach number contours in Fig. 4.5 now indicate a shock wave and an even greater thickening of the boundary layer after the shock. The chordwise velocity vectors also confirm that the velocity profiles are thicker. The skin friction lines on the upper surface of the wing show a complex separation pattern of two separations and one reattachment. Evidently the shock causes a separation

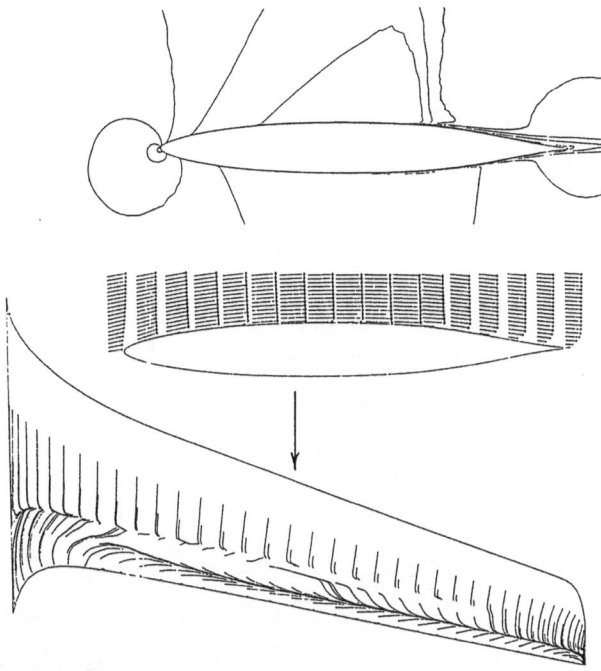

Fig. 4.5. Upper surface skin friction lines, chordwise velocity vectors and Mach contours in the 50% spanwise section. Incidence angle 2°

which may reattach and then separate again. The shock as indicated by the isobars (Fig. 4.6) is at about 70% chord. The comparison of C_p distribution shows that the shock location in the computation is further upstream than in the experiment, especially near the tip. As in the lower incidence case, the measurements indicate a recompression just ahead of the shock wave that is not captured in the simulation. It may be that a laminar separation occurs in the experiment. Why this is not picked up in the computation is a matter for speculation, but one guess is that the results may be sensitive to the location of transition. Where it occurs in the experiment is not known with great certainty.

5. Conclusions

A Navier-Stokes analysis code for compressible flow with turbulent boundary layers over quadrilateral wings has been developed. The finite-volume technique is employed with two more cells in each computational coordinate direction than the conventional compact approach to evaluate the viscous fluxes. A linear stability condition is derived to determine the local time steps and the allowable numerical damping coefficients of the explicit Runge-Kutta

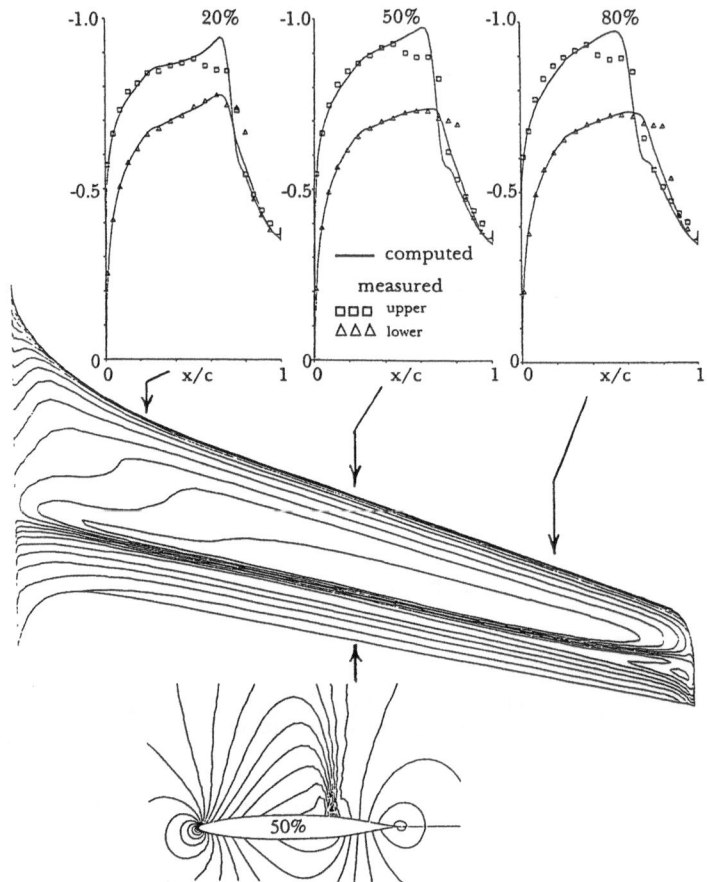

Fig. 4.6. Isobars over the upper wing surface, in the 50% spanwise section, and surface C_p distribution in the 20%, 50%, and 80% sections. Incidence angle 2°

scheme used for time integration. The method has been applied to simulate transonic flow of free air over the DLR F5 wing for the conditions $M_\infty = 0.82$ at incidence angles 0° and 2°. Using an O-O mesh of $193 \times 49 \times 41$ points and the Baldwin-Lomax turbulence model, the computations produce a surface pressure distribution that compares well with experimental measurements over the first half of the wing, but differs just ahead of the shock wave. Of the two cases, the lifting case shows the worse agreement. There are several possible explanations for this disparity, but a likely one is that we do not know the precise location of transition. A better explanation requires further study.

References

1. MacCormack, R.W., 'The Effect of Viscosity in Hypervelocity Impact Cratering', AIAA Paper 69-354, 1969.
2. Beam, R.M., and Warming, R.F., 'An Implicit Factored Scheme for the Compressible Navier-Stokes Equations', AIAA J., Vol. 16, 1978, pp. 393-402.
3. Jameson, A., Schmidt, W., and Turkel, E., 'Numerical Solutions of the Euler Equations by Finite Volume Methods Using Runge-Kutta Time-Stepping Schemes', AIAA Paper 81-1259, 1981.
4. Rizzi, A., 'Damped Euler-Equation Method to Compute Transonic Flow Around Wing-Body Combinations', AIAA J., Vol. 20, 1982, pp. 1321-1328.
5. Haase, W., Wagner, B., and Jameson, A., 'Development of a Navier-Stokes Method Based on a Finite Volume Technique for the Unsteady Euler Equations', in 'Proceedings of the 5th GAMM Conference on Numerical Methods in Fluid Mechanics', (M. Pandolfi, R. Piva, Eds.), Notes on Num. Fluid Mech., Vol. 7, Vieweg, Braunschweig, 1984, pp. 99-107.
6. Agarwal, R.K., and Deese, J.E., 'Computation of Transonic Viscous Airfoil, Inlet, and Wing Flowfields', AIAA Paper 84-1551, 1984.
7. Swanson, R.C., and Turkel, E., 'A Multistage Time- Stepping Scheme for the Navier-Stokes Equations', AIAA Paper 85-0035, 1985.
8. Martinelli, L., Jameson, A., and Grasso, F., 'A Multigrid Method for the Navier-Stokes Equations', AIAA Paper 86-0208, 1986.
9. Reister, H., and Schwamborn, D., 'Viscous Pressure Wave Boundary Layer Interaction', in 'Proceedings of 10th International Conference on Numerical Methods in Fluid Dynamics', (F.G. Zhuang, Y.L. Zhu, Eds.), Springer-Verlag, Berlin, 1986, pp. 533-537.
10. Peyret, R., and Taylor, T.D., 'Computational Methods for Fluid Flow', Springer-Verlag, N.Y., 1983.
11. Mueller, B., and Rizzi, A., 'Runge-Kutta Finite-Volume Simulation of Laminar Transonic Flow over a NACA 0012 Airfoil Using the Navier-Stokes Equations', FFA Tech. Note TN 1986-60, Stockholm 1986.
12. Mueller, B., and Rizzi, A., 'Modeling of Turbulent Transonic Flow Around Aerofoils and Wings', Comm Appl Num Meth, Vol 6, 1990, pp. 603-613.
13. Mueller, B., and Rizzi, A., 'Navier-Stokes Computation of Transonic Vortices Over a Round Leading Edge Delta Wing', Int. J. for Num. Methods in Fluids, Vol. 9, 1989, pp. 943-962.
14. Gary, J., 'On Certain Finite-Difference Schemes for Hyperbolic Systems', Math. Comp. 18, 1964, pp. 1-18.
15. Lindeberg, T., 'The Construction of a Three-Dimensional Finite-Volume Grid Generator for a Wing in a Wind Tunnel with Application to Navier-Stokes Flow Solvers', FFA TN 87-62, Stockholm, 1987.
16. Baldwin, B.S., and Lomax, H., 'Thin Layer Approximation and Algebraic Model for Separated Turbulent Flows', AIAA Paper 78-257, Jan 1978.
17. Jameson, A., and Baker, T.J., 'Solution of the Euler Equations for Complex Configurations', AIAA Paper 83-1929, 1983.
18. Kordulla, W., and Vinokur, M., 'Efficient Computation of Volume in Flow Predictions', AIAA J., Vol. 21, 1983, pp. 917-918.
19. Rizzi, A., and Eriksson, L.-E., 'Computation of Flow Around Wings Based on the Euler Equations', J. Fluid Mech., Vol. 148, 1984, pp. 45-71.
20. Eriksson, L.-E., and Rizzi, A., 'Computer-Aided Analysis of the Convergence to Steady State of Discrete Approximations to the Euler Equations', J. Comp. Physics, Vol. 57, 1985, pp. 90-128.

21. Mueller, B., and Rizzi, A., in Proceedings of the Seventh GAMM Conference on Numerical Methods in Fluid Mechanics", (M. Deville, Ed.) Notes on Num. Fluid. Mech. Vol. 20, Vieweg, Braunschweig, 1988, pp. 247- 255.

22. Mueller, B., 'Linear Stability Condition for Explicit Runge-Kutta Methods to Solve the Compressible Navier-Stokes Equations', Math. Methods in the Applied Sciences, Vol. 12, 1990, pp. 139-151.

23. MacCormack, R.W., and Baldwin, B.S., 'A Numerical Method for Solving the Navier-Stokes Equations with Application to Shock-Boundary Layer Interactions', AIAA Paper 75-1, Jan 1975.

24. Abarbanel, S., and Gottlieb, D., 'Optimal Time Splitting for Two- and Three-Dimensional Navier-Stokes Equations with Mixed Derivatives', J. Comp. Physics, Vol. 41, 1981, pp. 1-33.

25. Nordström, J., 'Stability Criteria for a Second Order Accurate, Time-Split Finite Volume Scheme to Solve the Navier-Stokes Equation', FFA Tech. Note TN 1985-08, Stockholm, 1985.

26. Mueller, B., 'Navier-Stokes Solution for Hypersonic Flow over an Indented Nosetip', AIAA Paper 85-1504, July 1985.

27. Sobieczky, H., Hefer, G., and Tusche, H., 'DLR F5 Test Wing Experiment for Computational Aerodynamics', AIAA Paper No.87- 2485CP, 1987.

28. Sobieczky, H., 'Geometry Generation for Transonic Design', in Advances in Computational Transsonics, Vol 4, Recent Advances in Numerical Methods in Fluids, (ed) W.G. Habashi, Pineridge Press, 1985.

29. Kordulla, W.(ed), 'Numerical Simulation of the Transonic DFVLR-F5 Wing Experiment', NNFM, Vol 22, Vieweg, Braunschweig, 1989.

30. Hasse, W., Bradsma, F., Elsholz, E., Leschziner, M., and Schwamborn, D. (eds), 'EUROVAL - An European Initiative on Validation of CFD Codes', NNFM, Vol 42, Vieweg, Braunschweig, 1993.

A Time-Dependent Space Marching Algorithm for Three-Dimensional PNS Equations

Wang Ru-quan[1] and Xue Ju-kui[2]

[1] The Computing Center, Academia Sinia, Beijing, China
[2] Northwestern Normal University, Lanzhou

1. Introduction

With development of aircraft and space shuttle, numerical simulation of large-scale three-dimensional viscous flowfields becomes more and more important in Computational Fluid Dynamics. Up to now there exist two different approaches to the simulation of compressible viscous flow. One of them is to integrate directly the full Navier-Stokes equations with the time-dependent method [1], another is to solve the Simplified Navier Stokes equations using the space-marching technique [2]. The former is based on the well-posed initial-boundary value problem, but it consumes a lot of CPU time for reaching a steady state and the later is able to reduce CPU time, but it has an inherent instability in subsonic region of the flow [3]. In our opinion, the time-dependent method is a powerful tool for treating local complicated flowfields, where as the space-marching technique is more applicable to large scale predominant supersonic flowfields. Therefore it is necessary to establish such an algorithm, wich can combine both the above mentioned two approaches together. In fact, this was done for inviscid flow in [4]. In recent years several time-dependent PNS marching algorithms have been developed [5,6,7] and an unified marching/iteration algorithm for three-dimensional PNS/NS equations was reported briefly by present authors [8]. In this paper we will describe in detail our algorithm.

It is well known that the explicit algorithm is not suitable for three-dimensional viscous flow, because of its severe restriction on time step. In this case there has been rapid developement in implicit algorithm for solving three-dimensional PNS/NS equations. In the middle of 1970's, Beam and Warming [9] made a key step in development of implicit approximate factorization (AF) algorithm for the Navier-Stokes equations. They reduced a three-dimensional problem to three one-dimensional problems to be solved. Each one-dimensional problem requires a block tridiagonal inversion, which may use the standard program. Unfortunately the AF algorithm needs a lot of computer storage and CPU time. In the middle of 1980's, the Symmetrical Gauss-Seidel (SGS) iteration was introduced into the implicit algorithm [11]. Numerical computations showed that it is more efficient to combine SGS relaxation in the streamwise direction with two block tridiagonal inversions in the normal to the body and the circumferential directions.

In this paper based on CSCM-S algorithm [12], we will describe in detail a more efficient explicit-implicit algorithm. The present algorithm includes two key improvments for CSCM-S algorithm. The first is a combination of the single-marching technique based on the unsteady PNS equations in the predominant supersonic regions of the flow with SGS relaxation for PNS/NS equations in subsonic and local complicated flowfields instead of the global sweep in the whole computational domain as it was done in [11, 12]. The second is implicit treatment in the normal direction to the body and explicit in the circumferential direction. Numerical experiments showed that the new algorithm is fast convergent and robust for most of external flow problems.

2. Governing Equations

The consevative time-dependent three-dimensional Parabolized Navier-Stokes equations have the following form in generalized curvilinear coordinates

$$\frac{\partial U}{\partial t} + \frac{\partial E}{\partial \xi} + \frac{\partial F}{\partial \eta} + \frac{\partial G}{\partial \zeta} = \frac{1}{Re} \left(\frac{\partial Fv}{\partial \eta} + \frac{\partial Gv}{\partial \zeta} \right) \tag{2.1}$$

where $U = \frac{1}{J} (\rho, \rho u, \rho v, \rho w, \rho e_t)^T$, $e_t = \frac{p}{(\gamma-1)\rho} + \frac{u^2+v^2+w^2}{2}$, $J = \frac{\partial(\xi,\eta,\zeta)}{\partial(x,y,z)}$ and $Re = \frac{\overline{\rho_\infty} \overline{U_\infty} \overline{L}}{\overline{\mu_\infty}}$. E, F, G and Fv, Gv represent inviscid and viscous flux vectors, respectively.

Note that Fv, Gv include only such viscous terms which have orders of magnitude of $O(1)$ and $O(Re^{-\frac{1}{2}})$. The velocity, density, temperature, viscosity coefficient and pressure were nondimensioned by $\overline{U_\infty}$, $\overline{\rho_\infty}$, $\overline{T_\infty}$, $\overline{\mu_\infty}$ and $\overline{\rho_\infty}\,\overline{U_\infty^2}$, respectively. The variables with a bar are dimensionless. In this paper we will construct a time-iteration and space-marching algorithm for the PNS equations (2.1), which may be applied to both steady and unsteady problems.

3. The Flux-Difference Splitting

In recent years there has been a great development in high-resolution upwind difference schemes for conservative system of hyperbolic equations. Generally, upwind difference schemes may be classified into the flux-vector splitting [13] and the flux-difference splitting [14]. Some analysis and numerical experiments [15] showed that the flux-difference splitting is supperior to the flux-vector splitting. Therefore, based on the Roe's flux-difference splitting and flux-vector splitting, Lombard et al.[11] developed further a Conservative Supra-Characteristic Method (CSCM) and then applied successfully it to the Euler and Navier-Stokes equations together with their DDADI Solver. CSCM is quite efficient and robust in computation of practical problems.

Here we choose the Lombard's flux-difference splitting as a basical scheme of our time-dependent space-marching algorithm.

Consider a conservative one-dimensional system of hyperbolic equations

$$\frac{\partial U}{\partial t} + \frac{\partial E}{\partial \xi} = 0 \tag{3.1}$$

where U, $\overline{U}$, $\tilde{U}$ represent vectors of conservative, nonconservative and characteristic variables, respectively. We have the following differential relationships

$$\frac{\partial E}{\partial \xi} = A \frac{\partial U}{\partial \xi} = MT\Lambda T^{-1} M^{-1} \frac{\partial U}{\partial \xi} = MT\Lambda T^{-1} \frac{\partial \overline{U}}{\partial \xi} = MA' \frac{\partial \overline{U}}{\partial \xi} = MT\Lambda \frac{\partial \tilde{U}}{\partial \xi} \tag{3.2}$$

where $A = \frac{\partial E}{\partial U}$ is the Jacobian matrix, A' is the coefficient matrix of the nonconservative system equivalent to (3.1). Λ is a diagonal matrix with eigenvalues of A' as its entries. M represents a transform matrix from the nonconservative variables to conservative ones and T^{-1} is a transform matrix from the nonconservative variables to the characteristic.

From the relation (3.2), we can obtain a flux difference

$$\Delta E = \overline{M} \, \overline{A'} \, \Delta \overline{U} \tag{3.3}$$

where " - " denotes the average value of two adjacent points.
Using the following equality

$$\Delta ab = \overline{a} \, \Delta b + \overline{b} \, \Delta a \tag{3.4}$$

one can lead to another relation

$$\overline{A'} \, \Delta \overline{U} = \widetilde{M}^{-1} \, \Delta U \tag{3.5}$$

Substituting (3.5) into (3.3), we have

$$\Delta E = \overline{M} \, \widetilde{M}^{-1} \, \Delta U \tag{3.6}$$

where the form of the matrices $\overline{M}$ and $\widetilde{M}^{-1}$ can be found in [11]. Note that it is difficult to find all eigenvalues of the matrix $\overline{M} \, \widetilde{M}^{-1}$. In this case an approximate treatment was done as follows:

Let $D = D^+ + D^- = I, \; D^{\pm} = \frac{1}{2} \left(I \pm \frac{\overline{\Lambda}}{|\Lambda|} \right)$

Since

$$\overline{T}^{-1} \overline{A'} \, \Delta \overline{U} = \overline{\Lambda} \, \Delta \tilde{U}$$

Then

$$D^+ \overline{T}^{-1} \overline{A'} \nabla \overline{U} = \overline{\Lambda}^+ \nabla \tilde{U}$$
$$D^- \overline{T}^{-1} \overline{A'} \, \Delta \overline{U} = \overline{\Lambda}^- \, \Delta \tilde{U}$$

Added the above equalities, one obtain

$$D^+ \overline{T}^{-1} \overline{A'} \nabla \overline{U} + D^- \overline{T}^{-1} \overline{A'} \Delta \overline{U} = \overline{A} \Delta \tilde{U} \tag{3.7}$$

where Δ and ∇ represent the forward and backward differences, respectively. Multiplying (3.7) by $\overline{MT}$ from the left side and considering (3.5), we derive

$$\overline{MT} D^+ \overline{T}^{-1} \widetilde{M}^{-1} \nabla U + \overline{MT} D^- \overline{T}^{-1} \widetilde{M}^{-1} \Delta U = \overline{MT} \overline{A} \Delta \tilde{U} \tag{3.8}$$

or

$$\Delta E = \Delta E^+ + \Delta E^- = \tilde{A}^+ \Delta U + \tilde{A}^- \Delta U \tag{3.9}$$

where

$$\tilde{A}^{\pm} = \overline{MT} D^{\pm} \overline{T}^{-1} \widetilde{M}^{-1}$$

Finally, the first-order CSCM difference scheme of the equations (3.1) can be written as below

$$\delta_t U_i + \lambda \left(\tilde{A}^+ \nabla U + \tilde{A}^- \Delta U \right)_i = 0 \tag{3.10}$$

where $\lambda = \frac{\Delta t}{\Delta \xi}$, $\nabla U_i = U_i - U_{i-1}$, $\Delta U_i = U_{i+1} - U_i$.
The scheme (3.10) may be extended to second-order accuracy [7], which is used in our computation.

4. Explicit-Implicit Difference Scheme

Based on the one-dimensional flux-difference splitting described in the previous section, we consider the following explicit-implicit difference equations for system (1.1) at any grid point (ξ_i, η_j, ζ_k)

$$\frac{\delta U^{n+1}}{\delta t} + \frac{\Delta E^{n+1}}{\Delta \xi} + \frac{\Delta F^{n+1}}{\Delta \eta} + \frac{\Delta G^n}{\Delta \zeta} = \frac{1}{Re} \left(\frac{\Delta F v^n}{\Delta \eta} + \frac{\Delta G v^n}{\Delta \zeta} \right) \tag{4.1}$$

in which the inviscid flux differende can be expressed as

$$\Delta E = \Delta E^+ + \Delta E^- = \tilde{A}^+ \nabla_\xi U + \tilde{A}^- \Delta_\xi U$$
$$\Delta F = \Delta F^+ + \Delta F^- = \tilde{B}^+ \nabla_\eta U + \tilde{B}^- \Delta_\eta U \tag{4.2}$$
$$\Delta G = \Delta G^+ + \Delta G^- = \tilde{C}^+ \nabla_\zeta U + \tilde{C}^- \Delta_\zeta U$$

where $A = \frac{\partial E}{\partial U}$, $B = \frac{\partial F}{\partial U}$ and $C = \frac{\partial G}{\partial U}$ are the Jacobian matrices and the matrices $\tilde{A}^{\pm}$, $\tilde{B}^{\pm}$, $\tilde{C}^{\pm}$ were obtained by the characteristic information along each coordinate direction (see ref. [11]). i.e.

$$\tilde{A}^{\pm} = \overline{M}\,\overline{T}_{\xi}\,D^{\pm}\,\overline{T}_{\xi}^{-1}\,\widetilde{M}_{\xi}^{-1}$$

$$\tilde{B}^{\pm} = \overline{M}\,\overline{T}_{\eta}\,D^{\pm}\,\overline{T}_{\eta}^{-1}\,\widetilde{M}_{\eta}^{-1} \tag{4.3}$$

$$\tilde{C}^{\pm} = \overline{M}\,\overline{T}_{\zeta}\,D^{\pm}\,\overline{T}_{\zeta}^{-1}\,\widetilde{M}_{\zeta}^{-1}$$

The viscous terms are approximated by centered differences. Finally, the resulting difference scheme has the following form

$$\delta U_{i,j,k}^{n+1} + \left(\tilde{A}^+\,\nabla_\xi\,U + \tilde{A}^-\,\Delta_\xi\,U\right)^{n+1} + \left(\tilde{B}^+\,\nabla_\eta\,U + \tilde{B}^-\,\Delta_\eta\,U\right)^{n+1} +$$

$$\left(\tilde{C}^+\,\nabla_\zeta\,U + \tilde{C}^-\,\Delta_\zeta\,U\right)^{n} = \frac{\Delta t}{Re}\left(\frac{\Delta Fv}{\Delta \eta} + \frac{\Delta Gv}{\Delta \zeta}\right)^{n} \tag{4.4}$$

The nonlinear terms appearing in (4.4) may be linearized as below

$$\Delta_\xi E^{n+1} = \tilde{A}^{n+1}\Delta_\xi U^{n+1} = \tilde{A}^n\Delta_\xi\left(U^n + \delta U^{n+1}\right) = \tilde{A}^n\Delta_\xi U^n + \tilde{A}^n\Delta_\xi\left(\delta U^{n+1}\right)$$

$$= \left(\tilde{A}^+\right)^n\nabla_\xi U^n + \left(\tilde{A}^-\right)^n\Delta_\xi U^n + \left(\tilde{A}^+\right)^n\nabla_\xi\delta U^{n+1} + \left(\tilde{A}^-\right)^n\Delta_\xi\delta U^{n+1} \tag{4.5}$$

Substituting the linearized expressions into (4.4), we obtain

$$\left(I + \tilde{A}^+\,\nabla_\xi + \tilde{A}^-\,\Delta_\xi + \tilde{B}^+\,\nabla_\eta + \tilde{B}^-\,\Delta_\eta\right)\delta U^{n+1} = -LU^n$$

$$LU^n = \left(\tilde{A}^+\,\nabla_\xi + \tilde{A}^-\,\Delta_\xi + \tilde{B}^+\,\nabla_\eta + \tilde{B}^-\,\Delta_\eta + \tilde{C}^+\,\nabla_\zeta + \tilde{C}^-\,\Delta_\zeta\right)\delta U^n$$

$$- \frac{\Delta t}{Re}\left(\frac{\Delta Fv}{\Delta \eta} + \frac{\Delta Gv}{\Delta \zeta}\right)^{n} \tag{4.6}$$

Following Lombard et al. [12], we use DDADI solver for solving (4.6), in the case the convection terms in the ξ direction may be rewritten as

$$(L_\xi\,\delta U)_{i,j,k}^{n+1} = \left(\tilde{A}^+\right)^n\nabla_\xi\left(\delta U_{i,j,k}^{n+1}\right) + \left(\tilde{A}^-\right)^n\Delta_\xi\left(\delta U_{i,j,k}^{n+1}\right)$$

$$= -\left(\tilde{A}^+\right)_{i-1,j,k}^n\delta U_{i-1j,k}^{n+1} + \left(\tilde{A}_{i-1,j,k}^+ - \tilde{A}_{i,j,k}^-\right)^n\delta U_{i,j,k}^{n+1} + \left(\tilde{A}^-\right)_{i,j,k}^n\delta U_{i+1,j,k}^{n+1} \tag{4.7}$$

Substituting the above formula into (4.6), we can obtain a new system of linear algebraic equations with dominant diagonal entries

$$\left(-\tilde{B}^+, D, \tilde{B}^-\right)\delta U^{n+1} = L^{n,n+1}U$$

$$U^{n+1} = U^n + \delta U^{n+1}$$

$$L^{n,n+1}U = -LU^n + \left(\tilde{A}^+\,\delta U_{i-1,j,k}^{n+1} - \tilde{A}^-\,\delta U_{i+1,j,k}^{n+1}\right) \tag{4.8}$$

$$D = I + \left(\tilde{A}^+ - \tilde{A}^-\right) + \left(\tilde{B}^+ - \tilde{B}^-\right)$$

Up to now we derived the unified form of the explicit-implicit difference scheme different from Lombard's fomulations in [11], where two systems of block-tridiagonal linear equations must be solved at each space-marching station and the global sweep along the ξ -direction is required. Instead only one linear system must be solved and the global sweep is replaced by the combination of the single-marching with the global sweep in our algorithm.

4.1 Multi-Sweep Technique

Following Lombard et al. [12], in subsonic region and local complicated flow-fields the Symmetrical Gauss-Seidel (SGS) iteration is used in the ξ direction. In the case the formulation (4.8) may be decomposed into two step, i.e., A Forward-sweep step

$$\left(-\tilde{B}^+, \, D, \, \tilde{B}^-\right) \delta U^*_{i,j,k} = L^{n,*} U_{i,j,k}$$

$$U^*_{i,j,k} = U^n_{i,j,k} + \delta U^*_{i,j,k} \tag{4.9}$$

$$L^{n,*} U_{i,j,k} = -L U^n_{i,j,k} + \tilde{A}^+ \delta U^*_{i-1,j,k} - \tilde{A}^- \delta U^n_{i+1,j,k}$$

and a backward-sweep step

$$\left(-\tilde{B}^+, \, D, \, \tilde{B}^-\right) \delta U^{n+1}_{i,j,k} = L^{*,n+1} U_{i,j,k}$$

$$U^{n+1}_{i,j,k} = U^n_{i,j,k} + \delta U^{n+1}_{i,j,k} \tag{4.10}$$

$$L^{*,n+1} U_{i,j,k} = -L U^*_{i,j,k} + \tilde{A}^+ \delta U^*_{i-1,j,k} - \tilde{A}^- \delta U^{n+1}_{i+1,j,k}$$

4.2 Single-Sweep Technique

For dominant supersonic flowfields the backward-sweep step (4.10) along the ξ negative direction may be ignored and the numerical solution is found only by using the forward-sweep step (4.9). This can be easily done putting $\tilde{A}^- \equiv 0$ in the system (4.9). As result, the forward-sweep is called the single-marching method and (4.9) becomes the following form

$$\left(-\tilde{B}^+, \, D^*, \, \tilde{B}^-\right) \delta U^{n+1}_{i,j,k} = L^{n,n+1} U_{i,j,k}$$

$$U^{n+1}_{i,j,k} = U^n_{i,j,k} + \delta U^{n+1}_{i,j,k} \tag{4.11}$$

$$L^{n,n+1} U_{i,j,k} = -L U^n_{i,j,k} + \tilde{A}^+ \delta U^{n+1}_{i-1,j,k}$$

$$L U^n_{i,j,k} = \left(\tilde{A}^+ \nabla_\xi + \tilde{B}^+ \nabla_\eta + \tilde{B}^- \Delta_\eta + \tilde{C}^+ \nabla_\zeta + \tilde{C}^- \Delta_\zeta\right) \delta U^n$$

$$- \frac{\Delta t}{Re} \left(\frac{\Delta Fv}{\Delta \eta} + \frac{\Delta Gv}{\Delta \zeta}\right)^n$$

$$D^* = I + \tilde{A}^+ + \tilde{B}^+ - \tilde{B}^-$$

Using the single-marching method, the numerical solution at each marching station is obtained so that the known solution at the i-th station is taken as the inital data of the (i+1)-th station, then the local iteration in the time at the new station is not completed until the steady solution is reached.

5. Numerical Tests

In order to validate the efficency of the new time-dependent space marching method, we solved some large-scale three-dimensional hypersonic problems such as flows past a sphere-cone, a sharppointed cone at high angle of attack and a simplified space-shuttle orbiter at low angle of attack. In all cases the converged solutions were obtained by using the time-dependent single-marching technique except the local nose subsonic flowfields and separated regions, in which the time-dependent multi-sweep technique must be required.

5.1 The Hypersonic Flow Past the Sphere-Cone at High Angle of Attack

Following reference [16], the flow conditions are given as follows

$$M_\infty = 10.6, Re = 1.318 \times 10^5, T_\infty = 47.34K \text{ and } T_w = 300K$$

The test case was carried out for 15° sphere-cone at 20° angle of attack. In this case the explicit-implicit method may reduce one third of CPU time in comparision with the full implicit one (see Table I).

Figs. 5.1 and 5.2 represent axial distribution of surface pressures and heating rates of the different meridian planes. The points indicate experimental data and the solid lines are the numerical results obtained by the present method. Clearly, they are in good agreement to each other. Fig. 5.3 shows distribution of the circumferential velocities at x = 17.5 Rn and the separated vortices on the leeside may be observed. The present results are in good agreement with those offered by reference [17].

5.2 Hypersonic Flow Around the Simplified Space-Shuttle Orbiter

The present time-dependent space marching method was also applied to simulate the hypersonic flowfield around the simplified space-shuttle orbiter shown as in Fig. 5.4. Such a geometry is noncircular and its flowfield feature has significant difference from those around the sharp-pointed cone or sphere-cone. In computation, the flow was assumed to be perfect viscous gas

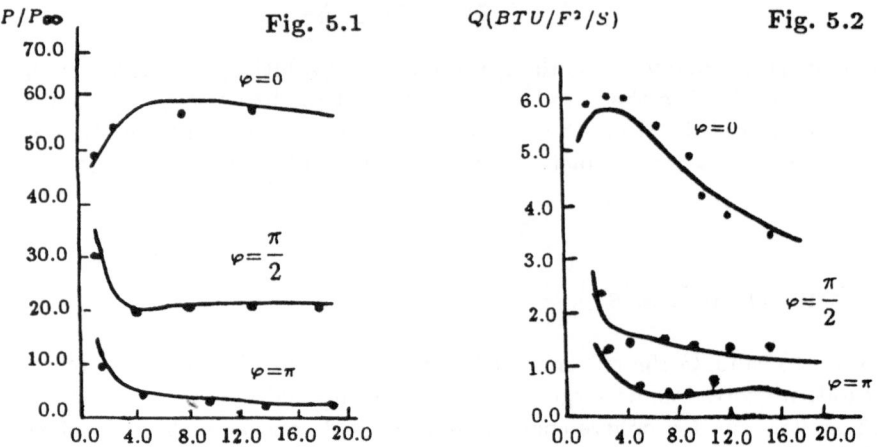

Fig. 5.1. Axial distribution of wall pressures for 15° sphere-cone at 20° angle of attack

Fig. 5.2. Axial distribution of wall heating-rates for 15° sphere-cone at 20° angle of attack

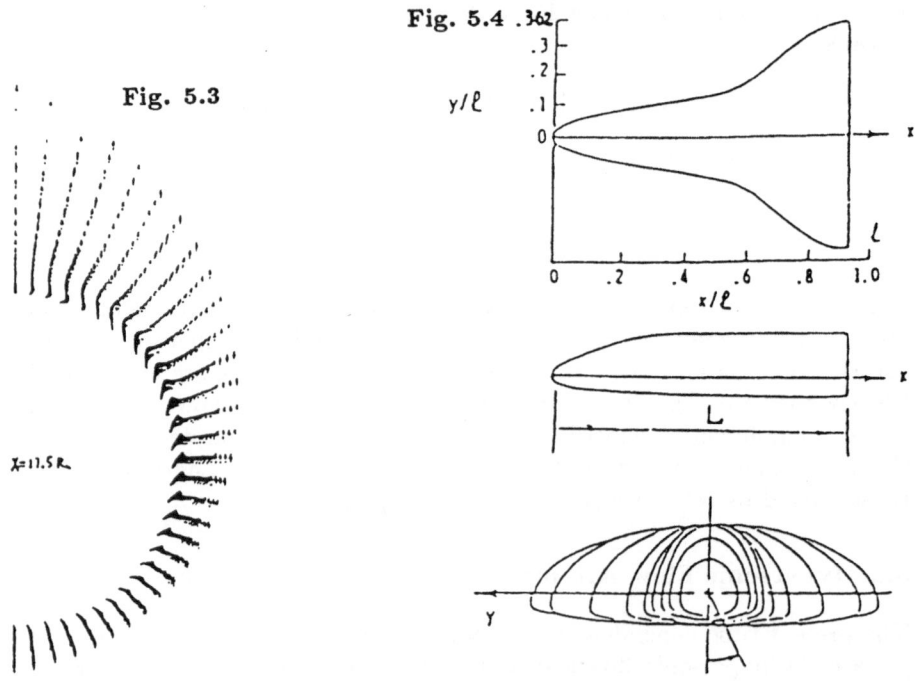

Fig. 5.3. Distribution of crossflow velocities for 15° sphere-cone at 20° angle of attack

Fig. 5.4. Configuration of simplified space-shuttle orbiter

Table 5.1. Comparison of different schemes at one marching station

Scheme Cost. (sec.)	Explicit-Implicit	Implicit
CPU time of a time step	16.36	24.93
CPU time of a grid point	0.021	0.032
Total CPU time	196.32	299.16
Total number of iterations	12	12

and laminar. The freestream conditions were given as below

$$M_\infty = 7, \ Re_L = 5 \times 10^6, \ T_\infty = 67K \text{ and } T_w = 300K$$

The Prandtl number $Pr = 0.72$ and the angles of attack $\alpha = 0^o, 5^o$. At the body surface, no-slip condition and $\frac{\partial P}{\partial n} = 0$ were used. The outer boundary far from the body is a surface, wich surrounds the space-shuttle orbiter and the bow shock so that all discontinuities appearing in the flowfield may be automatically captured. In the present paper a part of computational results are given. Figs. 5.5 and 5.6 present the suface pressures and the heating rates along the circumferential direction at different axial stations for 5° angle of attack. It is clear that the pressures and heating rates reach their peak values near the wing leading edge of the space-shuttle. Fig. 5.7 shows the density contours on the meridianal planes $\phi = 0^o, 90^o$ and 180° respectively. We can see that near the wing-body juncture there exist the interaction phenomena of the bow shock wave with the secondary shock wave reduced by the wing of the space-shuttle.
Numerical results on the flow past the sharp-pointed cone at high angles of the attack will be given in detail in another paper.

6. Conclusions

The time-dependent Navier-Stokes Solver requires a great amount of CPU time to obtain the converged numerical solution, it is more applicable to local complex flowfields. The space marching PNS algorithm can save an order of magnitude CPU time in comparison with the time-dependent method. However, it has the inherent problem with stability. The present time-dependent

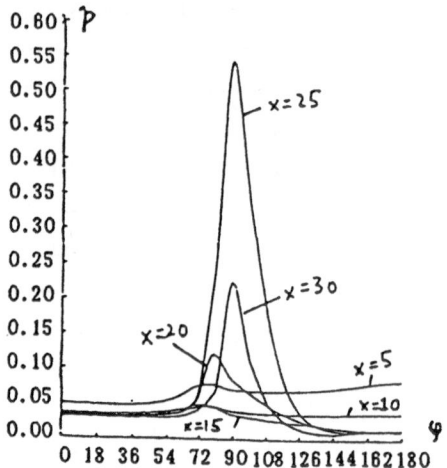

Fig. 5.5. Circumferential distribution of wall pressures for space-shuttle orbiter at 5° angle of attack

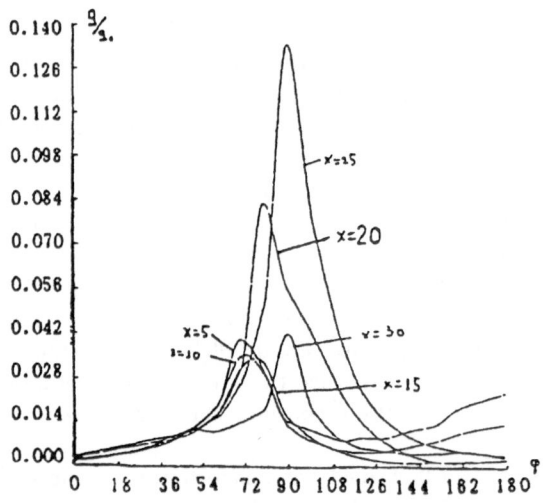

Fig. 5.6. Circumferential distribution of wall heating-rates for space-shuttle orbiter at 5° angle of attack

space marching algorithm consists of the space marching technique and the time iteration and it is free of instability and needs less CPU time than the time-dependent N-S Solver. The present algorithm is more efficient and robust to the large-scale complicated viscous flowfields.

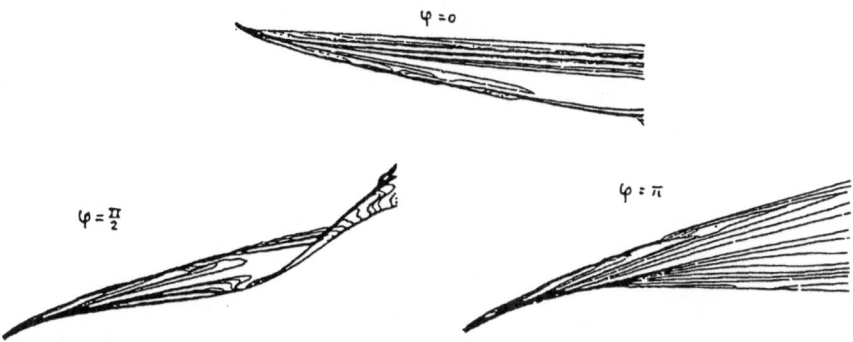

Fig. 5.7. Density contours of different meridianal planes for space-shuttle orbiter at 5° angle of attack

Acknowledgments
This work was supported by National Natural Science Foundation of China and in part by the Laboratory of Numerical Modelling for Atmospheric Sciences and Geophysical Fluid Dynamics, Institute of Atmospheric Physics, Academia Sinica.

References

1. R.W. MacCormack, Annu. Rev. Fluid Mech., Vol. 11, 289-316, 1979.
2. S.G. Rubin et al, Annu. Rev. Fluid Mech., Vol. 24, 117-144, 1992.
3. Ru-Quan Wang, Lecture Notes in Physics, Vol. 141, 1980.
4. S.R. Charavarthy et al., J. of Aircraft, Vol. 24, 73-83, 1987.
5. R.W. Newsome et al., AIAA paper 87-1113 CP, 1987.
6. C.L. Chang et al., J. Computational Physics, Vol., 80, 344-361, 1989.
7. Ju-Kui Xue et al., ACTA Mechanica Sinaca, Vol. 7, No. 3, 1991.
8. Ru-Quan Wang et al., Proceeding of 13th IMACS World Congress, Dublin-Ireland, 22-26 July, 1991.
9. R.M. Beam et al., AIAA J., Vol., 16, 393-401, 1976.
10. R.W. MacCormack, AIAA paper 90-1520, 1990.
11. C.K Lombard, et al., AIAA paper 83-1895, 1983.
12. C.K. Lombard et al., paper 84-1523, 1984.
13. J.L. Steger et al., J. Computational Physics, Vol., 40, 263, 1981.
14. P.L. Roe, Lecture Notes in Physics, Vol. 141, 1980.
15. W.K. Anderson et al., AIAA paper 85-0102, 1985.
16. J.W. Cleary, NASA TN D-5450, 1969.
17. B.A. Bhutta et al., AIAA paper 88-2696, 1988.

New Potential-Field Properties of General Laminar and Turbulent Motions of Newtonian Fluids

H. Bischoff[1] and E. Kaucher[2]

[1] Institut für Wasserbau, Technische Hochschule Darmstadt, D-64287 Darmstadt,
[2] Institut für Angewandte Mathematik, Universität Karlsruhe, D-76128 Karlsruhe,
  Germany

Summary. In this paper we show that a more complete formulation of the motion of Newtonian Fluids leads to very new field properties of laminar and turbulent flows. The Navier Stokes equation is completed by the balance of mechanical energy and by fundamental laws of thermodynamics of non-equilibrium. A new detailed review of some terms of the Navier Stokes Equation on the base of these principles leads to some new surprising field properties. For example the kinetic energy, the pressure and the acceleration terms turn out to be potential fields. These results are surprising because they hold for general laminar as well as for general turbulent motions in temporal mean.

1. Introduction

Hydromechanics has come to a point of stagnation because most theoretical considerations have mostly been restricted to describe the motion of Newtonian Fluids (NF) by a restricted model of the so called Navier-Stokes-equations (NSE), which is the balance equation of linear momentum and the equation of conservation of mass.

In this paper we show that a more complete formulation of the motion of NF leads to very new field properties of laminar and turbulent flows. The Navier Stokes equation is completed

- by the balance of mechanical energy
- by fundamental laws of thermodynamics of non-equilibrium (TDNE).

The latter laws are concerned with the Casimir-Onsager reciprocal relations of the thermodynamics of non-equilibrium, in particular we make essential use of the extremum principle of entropy production (minimum of dissipation of mechanical energy).

A new detailed review of some terms of the NSE on the base of both principles leads to some new surprising field properties. For example the kinetic energy, the pressure and the acceleration terms turn out to be potential fields. These results are surprising because they hold for general laminar as well as for general turbulent (GT) motions in temporal mean.

Properties which have been valid only for inviscid fluids are proven to be valid also for general laminar and even some turbulent flows.

Furthermore we can prove that even the term which describes the dissipation of energy is also a potential function in the temporal mean of the energy equation. Consequences of the extremal principle of TDNE are some criterions of stability of stationary laminar and turbulent flows in the temporal mean.

It seems, based on these ideas, that in future a lot of fundamental and new properties can be discovered by use of the energy– and TDNE–principle.

2. Physical and Mathematical Preliminaries

We consider throughout this paper fluid fields of isothermal incompressible Newtonian fluids of constant density and constant viscosity. The flow may be laminar or turbulent. The following five cases are considered:

$$
\text{"laminar"}\;\begin{cases}
\text{(a)} & \text{stationary laminar} & \text{(SL)} \\
& \text{motion} \\[4pt]
\text{(b)} & \text{weak instationary} & \text{(WI)} \\
& \text{laminar motion} \\[4pt]
\text{(c)} & \text{instationary turbulent} & \text{(IT)} \\
& \text{motion}
\end{cases}\qquad (2.1)
$$

$$
\text{"turbulent"}\;\begin{cases}
\text{(d)} & \text{turbulent} & \text{(TS)} \\
& \text{stationary motion} \\
& \text{(temporal mean)} \\[4pt]
\text{(e)} & \text{weak instationary} & \text{(WS)} \\
& \text{turbulent motion} \\
& \text{(temporal mean)}
\end{cases}\qquad (2.2)
$$

2.1 Mathematical Laws

Here $\mathbf{u}, \mathbf{v}, \mathbf{w}, \delta\mathbf{u}$ etc. are vector functions (velocity, vorticity) which are continous, one, twice or more continous differentiable on a given compact region $G \subset \mathcal{R}^3$. The set of this functions in denoted by $Z(G)$.

$$\delta\mathbf{u} \quad \text{variation of } \mathbf{u} \qquad (2.3)$$

$$\mathbf{u} \times \mathbf{u} = 0 \qquad (2.4)$$

$$\mathbf{u} \cdot (\mathbf{v} \times \mathbf{w}) = (\mathbf{u} \times \mathbf{v}) \cdot \mathbf{w} \qquad (2.5)$$

$$\nabla \cdot (\varphi\mathbf{u}) = \mathbf{u} \cdot \nabla\varphi + \varphi(\nabla \cdot \mathbf{u}) \quad (\varphi \text{ scalar}) \qquad (2.6)$$

$$\nabla \cdot (\nabla \times \mathbf{u}) = 0 = \nabla \cdot \mathbf{w} \qquad (2.7)$$

$$\nabla \times (\nabla\varphi) = 0 \quad (\text{furthermore } \nabla \times \mathbf{v} = 0 \text{ implies } \mathbf{v} = \nabla\varphi) \qquad (2.8)$$

$$\Delta \mathbf{u} = \nabla(\nabla \cdot \mathbf{u}) - \nabla \times \nabla \times \mathbf{u} = \nabla(\nabla \cdot \mathbf{u}) - \nabla \times \mathbf{w} \qquad (2.9)$$

$$(\mathbf{u} \cdot \nabla)\mathbf{u} = \nabla(u^2/2) - \mathbf{u} \times \nabla \times \mathbf{u} = \nabla(u^2/2) - \mathbf{u} \times \mathbf{w} \qquad (2.10)$$

$$\nabla \cdot (\mathbf{u} \cdot \nabla)\mathbf{u} = \nabla \cdot \nabla(u^2/2) - \nabla \cdot (\mathbf{u} \times \nabla \times \mathbf{u}) = \Delta(u^2/2) - \nabla \cdot (\mathbf{u} \times \mathbf{w}) \quad (2.11)$$

$$\nabla \cdot (\mathbf{u} \times \mathbf{v}) = \mathbf{v} \cdot (\nabla \times \mathbf{u}) - \mathbf{u} \cdot (\nabla \times \mathbf{v}) \qquad (2.12)$$

$$\nabla \cdot (\mathbf{u} \times \mathbf{w}) = \mathbf{w}^2 - \mathbf{u} \cdot (\nabla \times \mathbf{w}) \qquad (2.13)$$

$$\nabla \times (\mathbf{u} \times \mathbf{w}) = \mathbf{u}(\nabla \cdot \mathbf{w}) - \mathbf{w}(\nabla \cdot \mathbf{u}) + (\mathbf{w} \cdot \nabla)\mathbf{u} - (\mathbf{u} \cdot \nabla)\mathbf{w} \qquad (2.14)$$

$$\nabla(\mathbf{u} \cdot \mathbf{w}) = \mathbf{u} \times \nabla \times \mathbf{w} + \mathbf{w} \times \nabla \times \mathbf{u} + (\mathbf{w} \cdot \nabla)\mathbf{u} + (\mathbf{u} \cdot \nabla)\mathbf{w} \qquad (2.15)$$

$$(\mathbf{v}^T \cdot \nabla)\mathbf{u} = (\nabla \mathbf{u}) \cdot \mathbf{v}, \quad (\nabla \mathbf{u} \text{ tensor, dyadic of } \nabla \text{ and } \mathbf{u}) \qquad (2.16)$$

In this paper we make essential use of a variational principle applied on the expansion rate (divergence) of vector fields.

Lemma 2.1

Let $G \subset \mathcal{R}^3$ be a given compact region and $a(\mathbf{u})$ be an arbitrary operator $a : Z(G) \to Z(G)$ such that

$$a(\mathbf{u}) \in C^0(G) \qquad (2.17)$$

is a continous scalar field.

a) Then

$$\delta J(\mathbf{u}) := J(\mathbf{u} + \delta \mathbf{u}) - J(\mathbf{u}) = 0 \quad \text{on } G_1 \qquad (2.18)$$

for all variations $\delta \mathbf{u}$ of $\mathbf{u}$ (which may obey some boundary conditions and additional properties) implies that

$$J(\mathbf{u}) = \text{const} \quad \text{on } G_1. \qquad (2.19)$$

b) If (2.19) hold for arbitrary subregions $G_1 \subset G$ then this implies

$$J(\mathbf{u}) = 0 \quad \text{and} \quad a(\mathbf{u}) \equiv 0 \quad \text{on } G. \qquad (2.20)$$

Proof:

a) Suppose $J(\mathbf{u})$ ist not constant on a fixed region G_1, then there exists an $\delta \mathbf{u}$ such that $J(\mathbf{u} + \delta \mathbf{u}) \neq J(\mathbf{u})$, but this contradicts (2.18) which proves (2.19).

b) From (a) and the mean value theorem for integrals we have with a $\xi \in G_1$

$$|\text{constant}| = |J(\mathbf{u})| = \left| \int_{G_1} a(\mathbf{u}) \, dV \right| = V(G_1) \, |a(\mathbf{u})(\xi)| \qquad (2.21)$$

which implies for $V(G_1) \to 0$ that

$$|J(\mathbf{u})| = 0. \qquad (2.22)$$

On the other hand, suppose w. l. o. g. $a(\hat{\mathbf{u}})(\xi) > 0$ for a $\hat{\mathbf{u}} \in Z(G)$ and $\xi \in G$, then the continuity implies for a sufficient small neighbourhood $U(\xi)$ of ξ

$$0 < \int_{U(\xi)} a(\mathbf{u})\,dV = J(\mathbf{u}) \tag{2.23}$$

which contradicts (2.22). Thus (2.20) holds.

Lemma 2.2

Let $\mathbf{u}$ be a twice continous differentiable vectorfield (velocity) on a region G satisfying (2.42) and $\delta\mathbf{u}$ be an arbitrary variation of $\mathbf{u}$ with

$$\delta\mathbf{u} = 0 \quad \text{on the boundary } \partial G_1 \tag{2.24}$$

for each arbitrary subregion $G_1 \subset G$, and

$$\nabla \cdot \delta\mathbf{u} = 0 \quad \text{on } G\,. \tag{2.25}$$

Then a variational study of

$$J(\mathbf{u}) = \int_{G_1} \nabla \cdot ((\mathbf{u} \cdot \nabla)\mathbf{u}\,dV \quad \text{for each } G_1 \subset G \tag{2.26}$$

implies

$$\boxed{\nabla \cdot ((\mathbf{u} \cdot \nabla)\mathbf{u}) = 0 \quad \text{on } G\,.} \tag{2.27}$$

Proof:

We start with (2.26) and find

$$J + \delta J = \int_{G_1} 2\nabla \cdot [(\mathbf{u} + \delta\mathbf{u}) \cdot \nabla(\mathbf{u} + \delta\mathbf{u})]\,dV, \tag{2.28}$$

$$\delta J = \int_{G_1} 2\nabla \cdot [(\delta\mathbf{u} \cdot \nabla)(\mathbf{u} + \delta\mathbf{u}) + (\mathbf{u} \cdot \nabla)\delta\mathbf{u}]\,dV \tag{2.29}$$

and due to Gauss' theorem

$$\delta J = \int_{\partial G_1} 2\mathbf{n} \cdot [(\delta\mathbf{u} \cdot \nabla)(\mathbf{u} + \delta\mathbf{u}) + (\mathbf{u} \cdot \nabla)\delta\mathbf{u}]\,dA\,. \tag{2.30}$$

With (2.24) we get

$$\boxed{\delta J = \int_{\partial G_1} 2\mathbf{n} \cdot [(\mathbf{u} \cdot \nabla)\delta\mathbf{u}]\,dA} \tag{2.31}$$

From (2.14) (setting $\mathbf{w} := \delta\mathbf{u}$) and (2.42) and (2.25) we have

$$\nabla \times (\mathbf{u} \times \delta\mathbf{u}) = (\delta\mathbf{u} \cdot \nabla)\mathbf{u} - (\mathbf{u} \cdot \nabla)\delta\mathbf{u}, \tag{2.32}$$

$$(\mathbf{u} \cdot \nabla)\delta\mathbf{u} = (\delta\mathbf{u} \cdot \nabla)\mathbf{u} - \nabla \times (\mathbf{u} \times \delta\mathbf{u})\,, \tag{2.33}$$

and inserting in (2.31)

$$\delta J = \int_{\partial G_1} 2\mathbf{n} \cdot [(\delta\mathbf{u} \cdot \nabla)\mathbf{u} - \nabla \times (\mathbf{u} \times \delta\mathbf{u})] \, dA \, .$$

With (2.24) and applying Stokes theorem to the second term it follows that

$$\delta J(\mathbf{u}) = 0 \quad \text{for all } \mathbf{u} \in G_1 \, .$$

Thus with (2.26) which is of the form

$$J(\mathbf{u}) = \int_{G_1} a(\mathbf{u}) \, dV$$

with a continous $a(\mathbf{u})$ and with Lemma 2.1 (2.20) we follow that

$$\nabla \cdot ((\mathbf{u} \cdot \nabla)\mathbf{u}) = 0 \quad \text{on } G \, . \tag{2.34}$$

which proves (2.27).

Remark:

This Lemma 2.2 is a special case of a more general result:
Let $\mathbf{u}$ and $\delta\mathbf{u}$ as in Lemma 2.2 and $f(\mathbf{u}), g(\mathbf{u})$ be differentiable vector functions. Then a variational study of

$$J(\mathbf{u}) = \int_{G_1} \nabla \cdot (f(\mathbf{u}) \cdot \nabla)g(\mathbf{u}) \, dV \quad \text{for each } G_1 \subseteq G$$

implies

$$\boxed{\nabla \cdot (f(\mathbf{u}) \cdot \nabla)g(\mathbf{u}) = 0 \quad \text{on } G \, .}$$

Since with (2.24)

$$\delta f(\mathbf{u})|_{\partial G_1} = f(\mathbf{u} + \delta\mathbf{u}) - f(\mathbf{u})|_{\partial G_1} = 0, \quad \delta g(\mathbf{u})|_{\partial G_1} = 0$$

Lemma 2.3

Let L be a linear operator applied on vector–functions $\mathbf{v}$ of $Z(G)$ with $\nabla \cdot \mathbf{v} = 0$ such that $L^k \mathbf{v}$ are continous and bounded functions for some $1 \le k \le m$. Then the minimum condition

$$J(\mathbf{v}) = \int_G (L\mathbf{v})^2 \, dV = \min \tag{2.35}$$

leads (by means of analytical vector–transformations) to a variety of equivalent conditions of the form

$$\delta J(\mathbf{v}) = 2 \int_G M\mathbf{v} N\delta\mathbf{v} \, dV + \int_G (L\delta\mathbf{v})^2 \, dV \ge 0 \tag{2.36}$$

which has to be satisfied for all variations $\delta\mathbf{v}$ on G with $\delta\mathbf{v}|_{\partial G} = 0$ and $\nabla \cdot \delta\mathbf{v} = 0$. The operators M and N are also linear and of the form $M = L^m$ and $N = L^n$ with $m, n \in \mathcal{N}_0$, $m \geq n$.

These conditions in context with (2.36) uniquely define the solution

$$M\mathbf{v} = 0 \quad \text{on } G \tag{2.37}$$

and is thus a characterisation of a minimum of (2.35).

Proof:

The fact that the latter term in (2.36) is always positiv leads to the necessary and sufficient condition that $M\mathbf{v} = 0$. Assume $M\mathbf{v} \neq 0$, then w. l. o. g. $M\mathbf{v} > 0$ for an $x \in G$. The continuity of $M\mathbf{v}$ then implies that $M\mathbf{v} > 0$ even on a neighbourhood $U(x)$. Thus the integral $\displaystyle\int_{U(x)} M\mathbf{v}N\delta\mathbf{v}\, dV$ can be made negative such that

$$\int_G M\mathbf{v}N\delta\mathbf{v}\, dV < -\int_G (L\delta\mathbf{v})^2\, dV \tag{2.38}$$

holds for some variations $\delta\mathbf{v} := \lambda L^{m-n}\mathbf{v}$ with $\lambda < 0$ on $U(x)$ and $\delta\mathbf{v} \equiv 0$ on $G \setminus U(x)$. Inserting $\delta\mathbf{v} = \lambda L^{m-n}\mathbf{v}$ in (2.38) gives

$$\lambda\int_G (L^m\mathbf{v})^2 dV = \lambda\int_{U(x)} (L^m\mathbf{v})^2 dV < -\lambda^2\int_{U(x)} (L^{m-n+1}\mathbf{v})^2 dV =$$

$$= -\lambda^2\int_G (L^{(m-n+1)}\mathbf{v})^2 dV$$

which implies for $\lambda \neq 0$ that $\lambda < 0$ with

$$0 < -\lambda < \frac{\displaystyle\int_G (L^m\mathbf{v})^2\, dV}{\displaystyle\int_G (L^{m-n+1}\mathbf{v})^2\, dV} \quad \text{exists.}$$

Hence (2.38) holds and implies $\delta J(\mathbf{v}) < 0$ which contradicts (2.36). Therefore $M\mathbf{v} = 0$ on G.

The following very important and powerful Theorem comes from Stokes (1849).

Theorem 2.1 (Fundamental Theorem for vector fields, Stokes–Helmholtz decomposition)

Let

$$\mathbf{v} : G \subset \mathcal{R}^3 \rightarrow \mathcal{R}^3 \tag{2.39}$$

be an arbitrary vector field G with the following properties

1) $\mathbf{v}$ is piecewise continuously differentiable
2) $\mathbf{v}$ can be assumed to be zero outside of a sphere containing G (resp. containing all regions, where $\mathbf{v}$ is considered)

Then there exists in $G \setminus S$ a decomposition

$$\mathbf{v} = \operatorname{rot} \mathbf{a} - \operatorname{grad} \varphi , \qquad (2.40)$$

where φ is a scalar potential function and $\mathbf{a}$ a vector function and S is the union of all surfaces of discontinuities of $\mathbf{v}$. Furthermore there exists always a unique decomposition such that

$$\operatorname{div} \mathbf{a} = 0 \qquad (2.41)$$

holds.

Proof:

cf. [Mar], where also an explicit integral formulation is given for $\mathbf{a}$ and φ.
Remark:

Note that each operator M in (2.37) characterises a minimum solution of (2.35). A full study of (2.35) in case of the operator $L = \nabla \times$ which we use throughout in the following theory will deliver the fact, that there exist an infinite set of solutions (at least weak solutions) which we call "eigenstates" of turbulent flows.

2.2 Physical Terminologies and Fundamental Laws

$g = \text{const} = \text{gravity constant}$
$\varrho = \text{const} = \text{constant density of the fluid}$
$\nu = \text{const} = \text{constant kinematic viscosity of the fluid}$
$\eta = \nu \varrho = \text{dynamic viscosity}$
$T = \text{const} = \text{temperature of the quasi–isotherm fluid}$
$\mathbf{Q} = \text{vector field of heath flow}$
$\mathbf{u} = \text{instantaneous velocity–field}$
$\langle \mathbf{u} \rangle = \frac{1}{\Delta t} \int_0^{\Delta t} \mathbf{u}\, d\tau; \quad \Delta t \to \infty$
$\mathbf{u} = \underline{\mathbf{v}} + \mathbf{v}' \qquad$ Reynolds decomposition in mean and fluctuation
$\underline{\mathbf{v}} = \langle \mathbf{u} \rangle \qquad$ time–averaged velocity–field (temporal mean)
$\mathbf{v}' \qquad$ turbulent fluctuations with $\langle \mathbf{v}' \rangle = 0$

$$\operatorname{div} \mathbf{u} = \nabla \cdot \mathbf{u} = 0 \quad \text{source free vector field } \mathbf{u} \qquad (2.42)$$

$$\operatorname{div} \underline{\mathbf{v}} = \nabla \cdot \underline{\mathbf{v}} = 0 \quad \text{source free in temporal mean} \qquad (2.43)$$

$$\operatorname{div} \mathbf{v}' = \nabla \cdot \mathbf{v}' = 0 \quad \text{source free in turbulent fluctuation} \qquad (2.44)$$

$$\operatorname{rot} \mathbf{u} = \nabla \times \mathbf{u} = \mathbf{w} \quad \text{deformation–vorticity} \qquad (2.45)$$

$$\operatorname{rot} \underline{\mathbf{v}} = \nabla \times \underline{\mathbf{v}} = \underline{\mathbf{w}} \quad \text{deformation–vorticity in temporal mean} \qquad (2.46)$$

$$\text{rot } \mathbf{v}' = \nabla \times \mathbf{v}' = \mathbf{w}' \quad \text{deformation–vorticity of turbulent fluctuations} \tag{2.47}$$

$$\text{div } \mathbf{Q} = 0 \quad \text{isothermal fluid} \tag{2.48}$$

$$\mathbf{q} = \mathbf{u} \times \mathbf{w} \quad \text{transversal acceleration (vortex force)} \tag{2.49}$$

$$\mathbf{f} = \nu \nabla \times \nabla \times \mathbf{u} \quad \text{viscous force} \tag{2.50}$$

Balance of linear momentum (NSE)

$$\frac{\partial \mathbf{u}}{\partial t} + (\mathbf{u} \cdot \nabla)\mathbf{u} + \nabla \left(\frac{p}{\varrho} + gz \right) + \nu \nabla \times \nabla \times \mathbf{u} = 0 \tag{2.51}$$

$$\frac{\partial \mathbf{u}}{\partial t} + \nabla \left(\frac{u^2}{2} + \frac{p}{\varrho} + gz \right) - \mathbf{u} \times \mathbf{w} + \nu \nabla \times \mathbf{w} = 0 \tag{2.52}$$

From (2.46) follows the **curl of momentum equation** (vorticity equation) (cf. (2.45))

$$\partial(\mathbf{w})/\partial t + \nu \nabla \times \nabla \times \mathbf{w} - \nabla \times (\mathbf{u} \times \mathbf{w}) = 0 \tag{2.53}$$

The divergence of (2.51) leads to

$$\nabla \cdot (\mathbf{u} \cdot \nabla)\mathbf{u} + \Delta \left(\frac{p}{\varrho} + gz \right) = 0 \tag{2.54}$$

or

$$\Delta \left(\frac{u^2}{2} + \frac{p}{\varrho} + gz \right) - \nabla \cdot (\mathbf{u} \times \mathbf{w}) = 0. \tag{2.55}$$

Scalar product with $\mathbf{u}$ gives us the equation of

Balance of mechanical energy (ME)

$$\mathbf{u} \cdot \frac{\partial \mathbf{u}}{\partial t} + \mathbf{u} \cdot \nabla \left(\frac{u^2}{2} + \frac{p}{\varrho} + gz \right) + \mathbf{u} \cdot (\nu \nabla \times \nabla \times \mathbf{u}) = 0, \tag{2.56}$$

where the term $\mathbf{u} \cdot (\mathbf{u} \times \mathbf{w}) = 0$ due to (2.5).

The first law of thermodynamics (conservation of total energy)

$$\begin{aligned}
0 = & \frac{\partial(\varrho e)}{\partial t} + \nabla \cdot (\varrho e \, \mathbf{u}) + \nabla \cdot \mathbf{Q} + p\nabla \cdot \mathbf{u} - \eta(\nabla \times \mathbf{u})^2 - 2\eta \nabla \cdot (\mathbf{u} \cdot \nabla \mathbf{u}) \\
& + 2\eta \mathbf{u} \cdot \nabla(\nabla \cdot \mathbf{u}) + \frac{\partial(\varrho u^2/2)}{\partial t} + \nabla \cdot (\varrho \, \mathbf{u} \, u^2/2) - \varrho \mathbf{u} \cdot \mathbf{g} + \mathbf{u} \cdot \nabla p \\
& + \eta \mathbf{u} \cdot (\nabla \times \nabla \times \mathbf{u}) - 2\eta \mathbf{u} \cdot \nabla(\nabla \cdot \mathbf{u}) = 0
\end{aligned} \tag{2.57}$$

The results of the theory of non–equilibrium thermodynamics (TDNE) of Gyarmati [Gy] based on de Groot, Masur and Prigogine [GM] [P] lead to the fundamental principle of minimum entropy production. The entropy production (dissipation) is at most produced by the so called *dissipation term*, which even Rayleigh, Wieghardt [Wi] and others have defined as

Dissipation term

$$d_c = \nu(\nabla \times \mathbf{u})^2 + 2\nu\nabla \cdot ((\mathbf{u} \cdot \nabla)\mathbf{u}) - 2\nu\mathbf{u} \cdot \nabla(\nabla \cdot \mathbf{u}), \qquad (2.58)$$

where d_c denotes the historically defined (classical) dissipation function. In case of **stationary flow** the TDNE demands that under the conditions (2.42) – (2.57)

$$D_c := \int_G d_c(\mathbf{u}) \, dV = \min , \qquad (2.59)$$

where the minimum relates to all variations $\delta\mathbf{u}$ of $\mathbf{u}$ on G such that the properties (2.24) and (2.25) prevail. The region G is fixed in the x-coordinate system and is therefore independent of x and t.

First we observe that under the condition (2.42) the last term in (2.58) vanishes. Furthermore the second term of the right side in (2.58)

$$\boxed{\nabla \cdot ((\mathbf{u} \cdot \nabla)\mathbf{u}) = 0} \qquad (2.60)$$

vanishes also due to Lemma 2.2 (2.27). Note that (2.60) holds not only in (2.58) and (2.57) but also generally in all equations where this term occurs throughout in this theory.

The physical interpretation of these phenomena is that the **source of substantial acceleration has no effect on the internal energy production**. Thus the conservation law of total energy (2.57) is reduced by (2.48) and (2.60) to

$$\frac{de}{dt} - \nu(\nabla \times \mathbf{u})^2 + \mathbf{u} \cdot \left[\frac{\partial \mathbf{u}}{\partial t} + \nabla \left(\frac{u^2}{2} + gz + \frac{p}{\varrho} \right) + (\nu\nabla \times \nabla \times \mathbf{u}) \right] = 0 \ (2.61)$$

Thus we are able to simplify (2.58) by the actual dissipation function of TDNE

$$d := \nu(\nabla \times \mathbf{u})^2 \qquad (2.62)$$

and (2.59) becomes

$$\boxed{D := \int_G d(\mathbf{u}) \, dV = \int_G \nu(\nabla \times \mathbf{u})^2 \, dV = \min} . \qquad (2.63)$$

The formulation of this principle for NF is based more precisely on the formulation of viscous **entropy production** (dissipation function) due to [Bff].

$$\nu \, (\text{diss} \, u) =$$

$$\nu \begin{pmatrix} 0 & (v_x - u_y) & (u_z - w_x) \\ (v_x - u_y) & 0 & (w_y - v_z) \\ (u_z - w_x) & (w_y - v_z) & 0 \end{pmatrix} : \begin{pmatrix} 0 & (v_x - u_y) & (u_z - w_x) \\ (v_x - u_y) & 0 & (w_y - v_z) \\ (u_z - w_x) & (w_y - v_z) & 0 \end{pmatrix} \geq 0.$$

$$(2.64)$$

Obviously, the two tensors of deformation vorticity have the necessary property of Gyarmati's symmetric tensor with zero trace.

Thus (2.62) is a short notation of

$$\nu(\nabla \times u)^2 = \nu \operatorname{diss} u = \nu(\nabla \times u)^s : (\nabla \times u)^s \geq 0 . \tag{2.65}$$

The tensor $(\nabla \times u)^s$ is the symmetric counterpart of the asymmetric tensor $(\nabla \times u)^a$ which describes the rigid body rotation of the fluid.

Note, that the minimum principle of dissipation (entropy production by viscous forces) delivers a posteriori so to speak new field properties which may be enclosed to the rules (2.42) to (2.59) and play the role of additional physical fundamental laws.

3. Derivation of Some Field Properties of NSE

In this section we derive some essential new field properties by means of principles of vector analysis and of consequences of the previous minimum principle.

3.1 Relation Between Mechanical and Internal Energy

Rewriting (2.56) in a special form and subtracting this from (2.61) gives the rate of change of internal energy $\frac{de}{dt}$.

$$\nu u \cdot (\nabla \times \nabla \times u) \overset{\leftarrow}{=} - \left(u \cdot \left(\frac{\partial u}{\partial t} \right) + u \cdot \nabla \left(\frac{u^2}{2} + \frac{p}{\varrho} + gz \right) \right) \nu(\nabla \times u)^2 \overset{\rightarrow}{\equiv} \frac{de}{dt}$$

$$\tag{3.1}$$

Due to the second thermodynamical theorem (irreversibility) it is clear that the work of dissipation production is uniquely defined by the term (2.62) (cf. [Wi] p. 142ff). Thus the inner energy (coming from the mechanical work of viscous forces as entropy production) can only be transported corresponding to the arrows in (3.1). Thus we have the following identity

$$\nu \cdot u \cdot (\nabla \times \nabla \times u) = \nu(\nabla \times u)^2$$

or with (2.13)

$$\boxed{0 = \nu u \cdot (\nabla \times w) - \nu w^2 = \nu \nabla \cdot (u \times w)} . \tag{3.2}$$

Note, this property holds for instationary incompressible flows.

3.2 The Dissipation Minimum Principle in the Stationary Case

We now apply the minimum principle (2.63) of TDNE

$$J(\mathbf{u}) = D = \int_G (\nabla \times \mathbf{u})^2 \, dV = \min \tag{3.3}$$

to derive some essential conditions and properties of the fields $\mathbf{u}$, $\mathbf{w} = \nabla \times \mathbf{u}$, $\nabla \times \mathbf{w} = \nabla \times \nabla \times \mathbf{u}$ and so on.

Applying Lemma 2.3 with the operator $M \equiv N \equiv L \equiv \nabla \times$, (i. e. $m = n = 1$) and $\mathbf{v} = \mathbf{u}$ we find at once the condition

$$\nabla \times \mathbf{u} = 0 \quad \text{on } G \tag{3.4}$$

which is obviously the classical minimum of (3.3) of potential theory. Trivially, (3.4) is fulfilled by

$$\mathbf{u} = 0 \quad \text{on } G \tag{3.5}$$

which is the case for fluids in rest (**hydrostatic**).
The more interesting case (3.4) $u \not\equiv 0$ and $\nabla \times \mathbf{u} = 0$ on G describes the **inviscid potential ideal flow**, i. e.

$$\boxed{\mathbf{u} = \nabla \phi \quad \text{with } \Delta \phi = 0} \tag{3.6}$$

and ϕ a scalar function.
We now proceed to a more complex fluid flow with nonvanishing vorticity

$$\mathbf{w} = \nabla \times \mathbf{u} \not\equiv 0. \tag{3.7}$$

Partial integration of the term $\int_G (\nabla \times \mathbf{u})(\nabla \times \delta\mathbf{u}) \, dV$ using (2.12) replacing $\mathbf{u}$ by $\delta\mathbf{u}$ and $\mathbf{v}$ by $\mathbf{w}$ and using Gauss' theorem we find

$$\delta J(\mathbf{u}) = 2 \int_G (\nabla \times \nabla \times \mathbf{u}) \delta\mathbf{u} \, dV + \int_{\partial G} \mathbf{n} \cdot (\delta\mathbf{u} \times \mathbf{w}) \, dA + \int_G (\nabla \times \delta\mathbf{u})^2 \, dV \geq 0 \tag{3.8}$$

where the second term vanishes because $\delta\mathbf{u}|_{\partial G} = 0$. Again applying Lemma 2.3 with

$$M = L^2 \quad \text{and} \quad N = L^0 = I \quad \text{(i. e. } m = 2 \text{ and } n = 0)$$

we come to a second minimum condition for $\mathbf{u}$

$$\nabla \times \nabla \times \mathbf{u} = 0 \quad \text{on } G \tag{3.9}$$

A discussion shows that all previous cases (3.5) and (3.6) are special cases.

Since (3.7) and (3.9) hold we have fields of **flows with potential kinematics and vanishing effects of viscous forces.**

In the next step we discuss the case where

$$\nabla \times \nabla \times \mathbf{u} \neq 0 \quad \text{on } G \tag{3.10}$$

and look for a further minimum condition.
We decompose

$$\nabla \times \nabla \times \mathbf{u} = \nabla \alpha + \nabla \times \mathbf{a} \tag{3.11}$$

by use of the Stokes–Helmholtz decomposition (cf. Theorem 2.1, (2.47)). Inserting in (3.8) we find

$$\delta J(\mathbf{u}) = 2 \int_G (\nabla \times \mathbf{a}) \delta \mathbf{u} \, dV + \int_G \delta \mathbf{u} (\nabla \alpha) \, dV + \int_G (\nabla \times \delta \mathbf{u})^2 \, dV. \tag{3.12}$$

The second term can be performed with the identity (2.6) and (2.42) (replacing $\mathbf{u}$ by $\delta \mathbf{u}$) to

$$\int_G \nabla \cdot (\alpha \, \delta \mathbf{u}) \, dV = \int_{\partial G} \mathbf{n} \cdot \alpha \, \delta \mathbf{u} \, dA = 0$$

because $\delta \mathbf{u}|_{\partial G} = 0$.
Again we can apply Lemma 2.3 on $\delta J(\mathbf{u})$ of (3.12) and come to a third indirect minimum condition for $\mathbf{u}$

$$\nabla \times \mathbf{a} = 0 \quad \text{on } G. \tag{3.13}$$

The trivial case

$$\mathbf{a} = 0 \quad \text{on } G \tag{3.14}$$

as well as the case

$$\nabla \times \mathbf{a} = 0 \text{ and } a \not\equiv 0 \quad \text{on } G \tag{3.15}$$

leads to (cf. (3.14), (3.15))

$$\nabla \times \nabla \times \mathbf{u} = \nabla \alpha, \tag{3.16}$$

i. e. the viscous force is the gradient of a scalar potential field α with the property

$$\boxed{\nabla \cdot (\nabla \alpha) = \Delta \alpha = 0}. \tag{3.17}$$

Equation (3.16) indicates that rot rot rot $\mathbf{u} = 0$. Flows with $\alpha \neq 0$ satisfying (3.16) are **flows with potential kinematics with nonvanishing viscous forces**. Such flows can be characterized in a slightly more general form. Since the operators L are linear, one can decompose

$$\mathbf{u} = \alpha \mathbf{u} + \beta \mathbf{u} + \gamma \mathbf{u}$$

with $\alpha + \beta + \gamma = 1$, (where α, β, γ are functions or even tensors depending on space variable) and we can summarize (3.4), (3.8) and (3.9) in the form

$$\boxed{\alpha \mathbf{u} + \beta \text{rot } \mathbf{u} + \gamma \text{rot rot } \mathbf{u} = 0}$$

with $\alpha + \beta + \gamma = 1$.

Remark:

1. This process can be continued in some sense ad infinitum, which means that for the operator $L = \nabla \times$

$$L^k \mathbf{v} \equiv 0 \quad \text{on } G \qquad (3.18)$$

for a $k \in \mathcal{N}$ and a field $\mathbf{v}$ related to $\mathbf{u}$. This gives us the impression that the theory of turbulent flow may be analysed by recursive taking the curl of viscous forces leading to weak and weaker solutions in some Sobolev Spaces. Note that the set of all minimal solutions is a convex subset of the linear space spanned by the basis $L^k \mathbf{v} \not\equiv 0$, $k \in \mathcal{N}$. $\sum\limits_{i=1}^{n} \alpha_i(x,t) L^i \mathbf{v} = 0$ with $\sum\limits_{i=1}^{n} \alpha_i(x,t) = 1$, where $\alpha_i \in \mathcal{R}$ are arbitrary functions or tensors, $L^{n+1} \mathbf{v} = 0$ on G, and $n \in \mathcal{N}$ fixed.

In this sense we have the feeling that the cascades of the hierarchical minimum conditions $L^k \mathbf{v_k} = 0$ of (3.3) play the role of generalized "eigenstates" of the minimum function and thus seem to be qualifications for different stages of turbulence. This idea will be published in seperate papers.

A physical meaningful and mathematical restriction for this hierarchical process may be the classification of the functions $\mathbf{u} \in C^3(G)$ (laminar) if we restrict us on **classical solutions**. Thus a further decomposition such that rot rot rot $\mathbf{u} \not\equiv 0$ and rot rot rot $\mathbf{u} = 0$ make no sense since rot rot rot $\mathbf{u} \notin C^0(G)$ is no longer continous. The hierarchical minimum solutions ("eigenstates") of (3.3) are thus for $k \geq 4$ **weak solutions** which are able to describe turbulent flows even of extreme character.

2. It might be interesting to observe that the results $\text{rot}^k \mathbf{u} = 0$ are also minimum solutions of the kinematic minimum principle, i. e.

$$\int_G \frac{u^2}{2} \, dV = \min.$$

The continued partial integration step as done in (3.8) lead obviously to the same minimum conditions on weaker solutions $\mathbf{u}$.

3.3 Potential Field Properties in the Stationary Case

We continue our considerations and draw some consequences from the previous results.
From (2.57) and (2.60) we derive at once

$$\boxed{\nabla \cdot \left(\nabla \left(\frac{p}{\varrho} + g \mathbf{z} \right) \right) = 0} \qquad (3.19)$$

which is a new result.

From $\nabla \cdot (\nabla z) = 0$ and $\varrho \cdot g = \mathrm{const}$ it follows

$$\boxed{\nabla \cdot (\nabla \varrho g z) = \Delta (\varrho g z) = 0}\,. \tag{3.20}$$

and

$$\boxed{\nabla \cdot (\nabla p) = \Delta p = 0} \tag{3.21}$$

With (2.60) and the identity

$$0 = \nabla \cdot ((\mathbf{u} \cdot \nabla)\mathbf{u}) = \nabla \cdot (\nabla \frac{u^2}{2} - \mathbf{u} \times \mathbf{w})$$

and (3.2) we derive the very important result

$$\boxed{\Delta \frac{u^2}{2} = 0} \tag{3.22}$$

In particular the properties (3.21) and (3.22) are very remarkable and important characterisations of the pressure and of the field of kinetic energy $\frac{u^2}{2}$, which are consequences of the minimum principle of TDNE. Thus p and u^2 are not only **potential functions** in case of ideal fluids but also **in the rather general case of stationary laminar flows**.

*

For the next step we take the vorticity equation (2.53) in the stationary case

$$\frac{\partial \mathbf{w}}{\partial t} = 0. \tag{3.23}$$

Due to the fact that with (3.16) and (2.8)

$$\nu(\nabla \times \nabla \times \mathbf{w}) = \nu \nabla \times (\nabla \alpha) = 0, \tag{3.24}$$

in particular

$$\Delta \mathbf{w} = 0,$$

we have

$$\nabla \times (\mathbf{u} \times \mathbf{w}) = 0. \tag{3.25}$$

As we see, (3.24) is the second term in the vorticity equation (2.53) where ν occurs, so that we arrive at the same type of flow as we would have with

$$\nu = 0. \tag{3.26}$$

Furthermore, we derive from (3.25) again a new and remarkable result

$$\boxed{(\mathbf{w} \cdot \nabla)\mathbf{u} - (\mathbf{u} \cdot \nabla)\mathbf{w} = 0} \tag{3.27}$$

This condition (3.27) is in some sense an invariance condition (independent of ν) for any laminar stationary flow.

With $q = u \times w$ we have due to (3.25)

$$\nabla \times q = 0 \qquad (3.28)$$

and by (2.8) we know that there exists a scalar function γ such that

$$\boxed{q = \nabla\gamma}. \qquad (3.29)$$

From (3.2) we know

$$\nabla \cdot q = 0 \qquad (3.30)$$

and hence

$$\boxed{\Delta\gamma = 0}. \qquad (3.31)$$

This is also a new and surprising result which holds generally for laminar stationary flows.

Summarising, we remind that due to (3.26) and (3.28) we have additional laws which we repeat here for completeness

$$\boxed{\Delta p = 0}, \quad \boxed{\Delta(\varrho g z) = 0}, \quad \boxed{\Delta\left(\frac{u^2}{2}\right) = 0} \text{ and } \boxed{\Delta\alpha = 0}. \qquad (3.32)$$

Note, that all these potential properties obviously hold even in the case of rotational flows with rot $u \neq 0$ or even rot rot $u \neq 0$, derived under the condition of incompressibility.

4. Thermodynamical Stability Criterion for Instationary Flows

In this section we define some (as we think) new conditions and properties for instationary flows. We assure stability of instationary flows which may be excited by local perturbations.

First we state that the dissipation function $D(u)$ of (2.63) is constant in case of $\frac{\partial u}{\partial t} = 0$ for a fixed field u.
Now, in some analogue to Gyarmati and de Groot, Mazur, we demand for sufficient conditions in order to assure stabilisation in the instationary case $\frac{\partial u}{\partial t} \neq 0$, (e. g. excited by a small perturbation):

$$\dot{D} = \frac{d\,D(u)}{dt} \leq 0. \qquad (4.1)$$

We have to take the total derivation of D in time direction since the region G is fixed (cf. (2.59)). This condition assures that the dissipation production

$$D(u,t) \rightarrow \hat{D}(u) \quad \text{for } t \rightarrow \infty. \qquad (4.2)$$

Let G be an arbitrary but fixed region, then with (2.63) and (4.1) we derive (due to the Leibniz rule),

$$\boxed{\dot{D} = \int_G 2\nu \mathbf{w} \cdot \frac{\partial \mathbf{w}}{\partial t}\, dV + 2\nu \int_{\partial G} \mathbf{w}^2 \mathbf{U} \cdot \mathbf{n}\, dA \le 0}, \tag{4.3}$$

where the latter term vanishes since the transport of the region G $\mathbf{U} = \mathbf{U}(G) = 0$. We see at once that $\dot{D} = 0$ for $\frac{\partial \mathbf{u}}{\partial t} = 0$ or equivalently $\frac{\partial \mathbf{w}}{\partial t} = 0$ which is the stationary (stable) case.

A sufficient condition satisfying (4.3) is the local condition (in case of a fixed region G)

$$\boxed{\mathbf{w} \cdot \frac{\partial \mathbf{w}}{\partial t} \le 0 \quad \text{on } G}. \tag{4.4}$$

From (2.53) we find

$$\mathbf{w} \cdot \nabla \times (\mathbf{u} \times \mathbf{w}) - \nu \mathbf{w} \cdot \nabla \times \nabla \times \mathbf{w} \le 0 \tag{4.5}$$

and with (2.49) and (2.50)

$$\mathbf{w} \cdot \nabla \times (\mathbf{q} - \mathbf{f}) \le 0, \quad \int_G \mathbf{w} \cdot \nabla \times (\mathbf{q} - \mathbf{f})\, dV \le 0, \quad \text{resp.} \tag{4.6}$$

The second formulation says that locally $\mathbf{w} \cdot \nabla \times (\mathbf{q} - \mathbf{f}) > 0$ may be holds but in the spartial mean we have stability.

Note that in general $\mathbf{q}$ and $\mathbf{f}$ are **not** linear dependent so that for example $|\mathbf{f}| > |\mathbf{q}|$ etc. are not sufficient to satisfy (4.6). With (2.14), (2.42) and (2.16) we have

$$\nabla \times \mathbf{u} \times \mathbf{w} = (\nabla \mathbf{u}) \cdot \mathbf{w} - (\mathbf{u} \cdot \nabla)\mathbf{w}, \tag{4.7}$$

where $\nabla \mathbf{u}$ is the Jacobian matrix, and from (2.9) and (2.42) we have

$$\nabla \times \nabla \times \mathbf{w} = -\Delta \mathbf{w}.$$

Thus we rewrite (4.3) and (4.5) in the form

$$\boxed{\int_G \mathbf{w}^T H(\mathbf{u}) \mathbf{w}\, dV \le 0}, \quad \boxed{\mathbf{w}^T \cdot H(\mathbf{u}) \mathbf{w} \le 0 \quad \text{on } G}, \quad \text{resp:} \tag{4.8}$$

where H is the operator

$$H(\mathbf{u}) := (\nabla \mathbf{u}) - \mathbf{u} \cdot \nabla + \nu \Delta. \tag{4.9}$$

A sufficient condition for a self-stabilisating flow is the semi-positivitivness of the operator $-H(\mathbf{u})$ due to the inner product introduced by (4.8) $(\mathbf{w}, H(\mathbf{u})\mathbf{w})$ $:= \int_G \mathbf{w}^T H(\mathbf{u}) \mathbf{w}\, dV$, in particular the spectrum is nonnegativ and thus necessarily all existing eigenvalues $\lambda(\mathbf{u})$ of

$$- H(\mathbf{u}) \cdot \mathbf{v} = \lambda(\mathbf{u}) \cdot \mathbf{v} \tag{4.10}$$

are nonnegativ

$$\lambda(\mathbf{u}) \geq 0. \tag{4.11}$$

This property seems to have inner connections with the Hamiltonprinciple and the energy-operator principles. From (2.53) we have the formal solution $\mathbf{w} = e^{t\,H(\mathbf{u})}\mathbf{w}_0$ which shows also the asymptotic behaviour.

Since

$$-\Delta = \begin{pmatrix} -\Delta & 0 & 0 \\ 0 & -\Delta & 0 \\ 0 & 0 & -\Delta \end{pmatrix} \tag{4.12}$$

is a positiv definite operator matrix we see that $H(\mathbf{u})$ is at least semi-positiv definit when in some Sobolev space norm $\| \cdot \|$ on G

$$\nu \geq \frac{\|\nabla \mathbf{u} - \mathbf{u} \cdot \nabla\|}{\|\Delta\|} \tag{4.13}$$

or with $\mathbf{w}^T H(\mathbf{u})\mathbf{w} = \mathbf{w}^T(\nabla\mathbf{u}) \cdot \mathbf{w} - \mathbf{w}^T(\nabla\mathbf{w}) \cdot \mathbf{u} + \nu\mathbf{w}^T \Delta\mathbf{w}$ and some Hilbertspace norms

$$\nu \geq \left(\frac{\|\nabla \mathbf{u}\|}{\|\mathbf{u}\|} + \frac{\|\nabla \mathbf{w}\|}{\|\mathbf{w}\|} \right) \frac{\|\mathbf{u}\|\,\|\mathbf{w}\|^2}{|\mathbf{w}^T \Delta\mathbf{w}|} . \tag{4.14}$$

To give a better and more constructive insight into the estimation (4.14) let us consider a G which is a rectangular cube with the sizes a, b, c and the volume abc. Since the eigenvalue problem

$$\Delta\phi = \lambda\phi \quad \text{on } G, \quad \phi|_{\partial G} = 0, \quad \|\phi\| \neq 0, \tag{4.15}$$

has the eigensolutions

$$\phi = \sin\frac{m\pi}{a}x \, \sin\frac{n\pi}{b}y \, \sin\frac{p\pi}{c} \tag{4.16}$$

and the eigenvalues

$$\lambda = -\pi^2 \left(\frac{m^2}{a^2} + \frac{n^2}{b^2} + \frac{p^2}{c^2} \right) \tag{4.17}$$

for $m, n, p \in \mathcal{N}$.

The smallest one is

$$\lambda_1 = -\pi^2 \left(\frac{1}{a^2} + \frac{1}{b^2} + \frac{1}{c^2} \right) \tag{4.18}$$

With (4.16) we have

$$\nabla\phi = \frac{m\pi}{a} \cos \sin \sin + \frac{n\pi}{b} \sin \cos \sin + \frac{p\pi}{c} \sin \sin \cos, \tag{4.19}$$

where the arguments of sin and cos as in the triple factors are the same as in (4.16) from left to right.

Now we define the operator $L = L(\mathbf{u})$ by

$$H(\mathbf{u}) \cdot \mathbf{w} = L\mathbf{w} + \nu \varDelta \mathbf{w}\,, \qquad (4.20)$$

i. e.

$$L(\mathbf{w}) = (\mathbf{w} \cdot \nabla)\mathbf{u} - (\mathbf{u} \cdot \nabla)\mathbf{w} = (\nabla \mathbf{u}) \cdot \mathbf{w} - (\nabla \mathbf{w}) \cdot \mathbf{u}\,. \qquad (4.21)$$

With any eigenfunction ϕ of (4.16) we see with the diagonalstructure of $\varDelta$ in (4.12) that

$$\mathbf{v} = \begin{pmatrix} \phi \\ \phi \\ \phi \end{pmatrix} \quad \text{and} \quad \varDelta \mathbf{v} \neq \lambda \mathbf{v} = 0\,, \qquad (4.22)$$

$\mathbf{v}$ is an eigenfunction of the operator matrix $\varDelta$. Since the ϕ in (4.16) form a Fourier–orthonormal basis in the Hilbertspace $\mathcal{L}^2(G)$, each $\mathbf{w} \in \mathcal{L}^2$ coming from (4.8) can be represented componentwise in the form

$$\mathbf{w}_i = \sum_n \sum_m \sum_p \sigma_{inmp}\, \phi_{nmp}$$

and the integral condition in (4.8) leads to

$$\int_G \mathbf{w}^T L\mathbf{w}\, dV + \nu \sum_{i=1}^3 \sum_n \sum_m \sum_p \sigma_{inmp}^2\, \lambda_{nmp} \leq 0\,. \qquad (4.23)$$

From

$$\|\mathbf{w}^T L(\mathbf{u})\mathbf{w}\| \leq \|\mathbf{w}^T\|\, \|L(\mathbf{u})\mathbf{w}\| = \|\nabla \mathbf{u}\|\, \|\mathbf{w}\|^2 + \|\mathbf{w}\|\, \|\mathbf{u}\|\, \|\nabla \mathbf{w}\|$$

we get the bound $\kappa := (\|\nabla \mathbf{u}\|\, \|\mathbf{w}\|^2 + \|\mathbf{w}\|\, \|\mathbf{u}\|\, \|\nabla \mathbf{w}\|)V(G)$ and

$$\left| \int_G \mathbf{w}^T L\mathbf{w}\, dV \right| = \left| \int_G \mathbf{w}^T \nabla \times (\mathbf{u} \times \mathbf{w})\, dV \right| \leq \kappa \qquad (4.24)$$

and with $\int_G \mathbf{w}^2\, dV = \sum \sum \sum \sum \sigma_{nmp}^2$ we see that (4.23) holds at least when
$(\lambda_{nmp} \leq \lambda_1 < 0 \text{ (cf. (4.18))}$

$$\kappa \leq \nu \|\mathbf{w}\|^2\, |\lambda_1| \qquad (4.25)$$

or in other form

$$\frac{\kappa}{\|\mathbf{w}\|^2\, |\lambda_1|} = \frac{\left| \int_G \mathbf{w}^T L\mathbf{w}\, dV \right| (a \cdot b \cdot c)^2}{\pi^2 \int_G \mathbf{w}^2\, dV\, (b^2c^2 + a^2c^2 + a^2b^2)} \leq \nu\,. \qquad (4.26)$$

To compare with the Reynolds number we rewrite (4.26) in the form

$$\mathrm{Re} \approx \frac{\|\mathbf{u}\| \cdot \frac{1}{\sqrt{\lambda_1}}}{\nu} \leq \frac{\|\mathbf{u}\|\, \|\mathbf{w}\|^2 \sqrt{\lambda_1}}{\kappa}\,. \qquad (4.27)$$

With (4.24) and with $V(G) = a \cdot b \cdot c$ (in our example) we get

$$\frac{\kappa}{\|\mathbf{w}\|^2} = \left(\|\nabla \mathbf{u}\| + \frac{\|\mathbf{u}\| \, \|\nabla \mathbf{w}\|}{\|\mathbf{w}\|} \right) V(G) \qquad (4.28)$$

and the condition (4.26) takes the form

$$\frac{\kappa}{\|\mathbf{w}\|^2 |\lambda_1|} = \frac{\|\mathbf{u}\| \, V(G)^3}{\pi^2 \, (a^2 b^2 + a^2 c^2 + b^2 c^2)} \left(\frac{\|\nabla \mathbf{u}\|}{\|\mathbf{u}\|} + \frac{\|\nabla \mathbf{w}\|}{\|\mathbf{w}\|} \right) < \nu . \qquad (4.29)$$

With $a = b = c$ the factor

$$\frac{V(G)^3}{(a^2 b^2 + a^2 c^2 + b^2 c^2)} = \frac{a^2}{3} = \frac{S(G)}{18}$$

is connected with the surface $S(G)$ of G. The terms $\frac{\|\nabla \mathbf{u}\|}{\|\mathbf{u}\|}$ and $\frac{\|\nabla \mathbf{w}\|}{\|\mathbf{w}\|}$ are relative condition numbers characterizing the sensibility of the flow under some small pertubations $\delta \mathbf{u}$. Since the condition (4.29) is a sufficient stability condition satisfying (4.8), we conclude that the number

$$\beta := \frac{\|\mathbf{u}\| \, S(G)}{\nu \, \pi^2 \, 18} \left(\frac{\|\nabla \mathbf{u}\|}{\|\mathbf{u}\|} + \frac{\|\nabla \mathbf{w}\|}{\|\mathbf{w}\|} \right)$$

seems to be a sharper characterisation of self-stabilisating flows than by the Reynolds number. Due to (4.29)

$$\beta < 1 \qquad (4.30)$$

is a sufficient condition for self-stabilisating laminar flows. Note that the geometry is inbuilt by the sizes of the region G.

*

The formulation (4.8) has not yet the form of a proper energy relation. In the following step we derive such a form in (4.38). From the condition

$$\int_G \mathbf{w} \frac{\partial \mathbf{w}}{\partial t} \, dV = \int_G \mathbf{u} \operatorname{rot} \frac{\partial \mathbf{w}}{\partial t} \, dV + \int_G \nabla \cdot \left(\mathbf{u} \times \frac{\partial \mathbf{w}}{\partial t} \right) dV \le 0, \qquad (4.31)$$

where we have used (2.12) and with

$$\operatorname{rot} \frac{\partial \mathbf{w}}{\partial t} = -\nu \operatorname{rot}^4 \mathbf{u} - \operatorname{rot}^2 (\mathbf{w} \times \mathbf{u}) \qquad (4.32)$$

we have

$$K(\mathbf{w}) = [(\nabla \mathbf{w}) - (\mathbf{w} \cdot \nabla)] \nabla \times -\nu \Delta^2 \qquad (4.33)$$

and thus (with Gauss theorem)

$$\int_G \mathbf{u}^T K(\mathbf{w}) \mathbf{u} \, dV + \int_{\partial G} \mathbf{n} \cdot \left(\mathbf{u} \times \frac{\partial \mathbf{w}}{\partial t} \right) dA \le 0 \qquad (4.34)$$

Under the condition that ∂G has a unique decomposition in form

$$\partial G = \partial G_1 \cup \partial G_2, \tag{4.35}$$

where with the orthonormal vector $\mathbf{n}$

$$\begin{aligned}
\mathbf{n} \times \mathbf{u} &= 0 \quad \text{on } \partial G_1 \\
\mathbf{u} &= 0 \quad \text{on } \partial G_2
\end{aligned} \tag{4.36}$$

hold (which is given in most practical problems), we have

$$\boxed{\int_{\partial G} \mathbf{n} \cdot \left(\mathbf{u} \times \frac{\partial \mathbf{w}}{\partial t} \right) dA = 0}. \tag{4.37}$$

In this case (4.34) reduces to

$$\boxed{\int_G \mathbf{u}^T K(\mathbf{w}) \mathbf{u} \, dV \leq 0} \tag{4.38}$$

which corresponds with (4.9).

To prove (4.37) we insert (2.53) and derive

$$\int_{\partial G} \mathbf{n} \cdot ([\nabla \times (\mathbf{u} \times \mathbf{w})] \times \mathbf{u} - \nu \mathbf{u} \times (\nabla \times \nabla \times \mathbf{w})) \, dA$$

$$= \int_{\partial G} \mathbf{n} \cdot [-(\mathbf{w} \cdot \nabla)\mathbf{u} + (\mathbf{u} \cdot \nabla)\mathbf{w}) - \nu \Delta \mathbf{w}] \times \mathbf{u} \, dA$$

$$= \int_{\partial G} [\mathbf{n}, \mathbf{h}, \mathbf{u}] \, dA$$

$$= \int_{\partial G_1} -[\mathbf{h}, \mathbf{n}, \mathbf{u}] \, dA + \int_{\partial G_2} [\mathbf{u}, \mathbf{n}, \mathbf{h}] \, dA = 0$$

since $\mathbf{n} \times \mathbf{u}|_{\partial G_1} = 0$ and $\mathbf{u}|_{\partial G_2} = 0$ with (4.36). ($[\mathbf{n}, \mathbf{h}, \mathbf{u}]$ is a short notation for $\mathbf{n} \cdot (\mathbf{h} \times \mathbf{u})$).

We stop here our discussion about some interesting properties concerning self-stabilisation criterions for laminar flows and proceed in the following chapter to the general instationary turbulent case.

5. The Instationary Turbulent Flow in Temporal Mean

In the stationary case the entropy minimum principle has lead to some remarkable and interesting field properties of the pressure and of the velocity. So it is obvious that the temporal mean of **instationary turbulent flows** delivers again stationarity and the entropy minimum principle delivers similar results as in the stationary case.

We denote the temporal mean (average in time) of a vector function **u** by

$$\langle \mathbf{u} \rangle = \underline{u} := \frac{1}{T} \int_0^T \mathbf{u}(x, \tau)\, d\tau, \quad \text{with } T \to \infty \tag{5.1}$$

The mean-operator $\langle \cdot \rangle$ is a linear operator with the following rules:

$$\mathbf{f} \text{ vector function}$$

$$\mathbf{f}(\mathbf{u}) = \underline{\mathbf{f}(\mathbf{u})} + (\mathbf{f}(\mathbf{u}))'$$

$$\langle \mathbf{f}(\mathbf{u}) \rangle = \underline{\mathbf{f}(\mathbf{u})} + \langle (\mathbf{f}(\mathbf{u}))' \rangle$$

$$\langle \mathbf{u}' \rangle = \underline{u}' = 0 \tag{5.2}$$

$$\langle \mathbf{u} \rangle = \langle \mathbf{u} + \mathbf{u}' \rangle = \underline{u} \tag{5.3}$$

$$\langle \delta \mathbf{u} \rangle = \delta \mathbf{u} \tag{5.4}$$

$$\langle \underline{u}'^2 \rangle = \underline{u}'^2 \tag{5.5}$$

$$\langle u^2 \rangle = \underline{u}^2 + \langle u'^2 \rangle$$

Let $\underline{a}$ be a time independent function then

$$\langle \underline{a}\, \mathbf{u} \rangle = \underline{a} \langle \mathbf{u} \rangle = \underline{a}\, \underline{u} \tag{5.6}$$

Similarly, if L is a time independent operator then

$$\langle L\mathbf{u} \rangle = L \langle \mathbf{u} \rangle = L\underline{u}. \tag{5.7}$$

A consequence of (5.2) is that for all linear functions $l(\mathbf{u}')$

$$\langle l(\mathbf{u}') \rangle = 0. \tag{5.8}$$

Let us consider

$$\langle J(\mathbf{u}) \rangle = \int_G \langle (\nabla \times \mathbf{u})^2 \rangle\, dV = \int_G (\nabla \times \underline{u})^2 + \langle (\nabla \times \mathbf{u}')^2 \rangle\, dV, \tag{5.9}$$

where $2\langle (\nabla \times \underline{u})(\nabla \times \mathbf{u}') \rangle = 0$ and the variation due to (5.4):

$$\langle J(\mathbf{u} + \delta\mathbf{u}) \rangle \quad = \quad \int_G \langle (\nabla \times \mathbf{u})^2 + 2(\nabla \times \mathbf{u}) \cdot (\nabla \times \delta\mathbf{u}) + (\nabla \times \delta\mathbf{u})^2 \rangle dV$$

$$\langle \delta J(\mathbf{u}) \rangle \quad = \quad \langle J(\mathbf{u} + \delta\mathbf{u}) - J(\mathbf{u}) \rangle = \int_G \langle (\nabla \times \underline{u})^2 + 2(\nabla \times \underline{u})(\nabla \times \mathbf{u}')$$

$$+ (\nabla \times \mathbf{u}')^2 + 2(\nabla \times \underline{u}) \cdot (\nabla \times \delta\mathbf{u}) + 2(\nabla \times \mathbf{u}') \cdot (\nabla \times \delta\mathbf{u})$$

$$+ (\nabla \times \delta\mathbf{u})^2 \rangle\, dV \tag{5.10}$$

With (5.9) we find

$$\langle \delta J(\mathbf{u}) \rangle \quad = \quad \int_G 2\langle (\nabla \times \mathbf{u}) \cdot (\nabla \times \delta\mathbf{u}) \rangle + (\nabla \times \delta\mathbf{u})^2\, dV$$

$$= \quad \int_G 2(\nabla \times \underline{u}) \cdot (\nabla \times \delta\mathbf{u}) + (\nabla \times \delta\mathbf{u})^2\, dV \tag{5.11}$$

since due to (5.6), (5.7) and (5.8) some terms in (5.10) vanish.
By the way we see that

$$\langle \delta J(\mathbf{u}) \rangle = \delta J(\langle \mathbf{u} \rangle) = \delta J(\underline{\mathbf{u}}) . \tag{5.12}$$

Now we apply the same principle as in (2.63) to the temporal mean of dissipation

$$\underline{D} = \langle D(\mathbf{u}) \rangle = \langle \nu J(\mathbf{u}) \rangle = \int_G \nu \langle (\nabla \times \mathbf{u})^2 \rangle \, dV = \min \quad \text{on } G . \tag{5.13}$$

As we see (5.13) leads in (5.12) to the same form as formulated in (2.36) of
Lemma 2.3 so that we get in the mean the same results for turbulent flows
stationary in temporal mean.

In particular Lemma 2.1, Lemma 2.2 and Lemma 2.3 hold in the temporal
mean if we consider region G independent of time, since the vectorfields
$a(\mathbf{u})$, $\mathbf{u}$ and $\mathbf{v}$ in (2.17), (2.31), (2.36) occur linearly. Thus (2.60) holds in the
temporal mean

$$\boxed{\nabla \cdot (\underline{\mathbf{u}} \cdot \nabla \underline{\mathbf{u}}) + \langle \nabla \cdot (\mathbf{u}' \cdot \nabla)\mathbf{u}' \rangle = 0} . \tag{5.14}$$

Similarly (3.4), (3.5), (3.6), (3.9), (3.11), (3.13), (3.14), (3.15), (3.16) and
(3.17) hold in the mean. A particular result therefore is

$$\boxed{\Delta(\varrho g \mathbf{z}) = 0} \tag{5.15}$$

$$\boxed{\Delta \underline{p} = 0} \tag{5.16}$$

similar to (3.21) and (3.20). From

$$\begin{aligned}
\langle \mathbf{u} \times \mathbf{w} \rangle &= \langle (\underline{\mathbf{u}} + \mathbf{u}') \times (\underline{\mathbf{w}} + \mathbf{w}') \rangle \\
&= \underline{\mathbf{u}} \times \underline{\mathbf{w}} + \langle \mathbf{u}' \times \underline{\mathbf{w}} \rangle + \langle \underline{\mathbf{u}} \times \mathbf{w}' \rangle + \langle \mathbf{u}' \times \mathbf{w}' \rangle = \nabla \underline{\gamma} + \nabla \underline{\gamma}'
\end{aligned}$$

where the 2nd and 3rd term vanish, we derive

$$\begin{aligned}
\langle \nabla \times \mathbf{q} \rangle &= \langle \nabla \times (\mathbf{u} \times \mathbf{w}) \rangle \\
&= \nabla \times (\underline{\mathbf{u}} \times \underline{\mathbf{w}}) + \nabla \times \langle \mathbf{u}' \times \mathbf{w}' \rangle = \nabla \times \nabla \underline{\gamma} + \nabla \times \nabla \underline{\gamma}' = 0
\end{aligned}$$

which proves

$$\nabla \times \underline{\mathbf{q}} = 0 \quad \text{and} \quad \underline{\mathbf{q}} = \nabla(\underline{\gamma} + \langle \gamma' \rangle) = \nabla \gamma \tag{5.17}$$

and with $\frac{\partial \underline{\mathbf{w}}}{\partial t} = 0$

$$\boxed{\Delta \underline{\mathbf{w}} = 0} \tag{5.18}$$

follows. The properties (3.31) and (3.22) with (5.18) become the form

$$\boxed{\Delta \left(\underline{\gamma} + \underline{\gamma}' \right) = 0} \quad \text{and} \quad \boxed{\Delta \left(\frac{u^2}{2} + \frac{u'^2}{2} \right) = 0} \tag{5.19}$$

6. Conclusion

In this paper we have discussed a lot of new principles and properties coming out from TDNE. In case of stationary laminar or quasistationary turbulent flows we have proved potential field properties of pressur p, kinetic energy $\frac{u^2}{2}$ and the lateral accelleration term $\mathbf{q}$. Under the assumption that our derivations hold, we have corrected Wieghardt's assumption that $\nu \operatorname{rot} \mathbf{w} = \nu \nabla \alpha$ does **not** generally exist! (cf. (3.16)).

Furthermore, in the (quasi-)stationary case we have proved that Helmholtz minimum principle of dissipation is equivalent with the minimum principle of entropy production of the TDNE. Additionally we have derived some new and mathematically approved stability conditions for flows under small perturbations. The approach in this paper seems to give new insights into the turbulent behaviour of flows and allows in future to derive new qualifications of turbulence.

References

[Gy] Gyarmati, I.: *Non-equilibrium Thermodynamics*, Springer-Verlag, Berlin 1970
[GM] de Groot, Sybren R.; Mazur, Peter: *Anwendung der Thermodynamik irreversibler Prozesse*, B.I.-Wissenschaftsverlag, 1944, Bibliographisches Institut Mannheim
[P] Prigogine, Ilya: *Vom Sein zum Werden*, Piper, München 1985
[Wi] Wieghardt, K.: *Theoretische Strömungslehre*, B. G. Teubner, Stuttgart 1969
[He] Helmholtz, H. von: *Über Integrale der hydrodynamischen Gleichungen, welche Wirbelbewegungen entsprechen*, I. reine u. angew. Math. 55 (1858), 25-55
[Mar] Martensen, E.: *Potentialtheorie*, B. G. Teubner, Stuttgart 1968
[Bff] Bischoff, H.: *Berechnungsmethoden der technischen Hydromechanik und der deterministischen Hydrologie mit Ergebnisverifikation und automatischer Sensitivitätsanalyse*, Habilitationsschrift an der Techn. Hochschule Darmstadt, 1993

Boundary Layer Turbulence and the Control by Suction

Y. Aihara

Department of Aeronautics and Astronautics, The University of Tokyo, Tokyo, Japan

Summary. Despite a large number of experiments and numerical analysis done recently, the turbulent boundary layer still remains the behaviors of the flow difficult to understand. This paper is an attempt at constituting an easy-to-comprehend story about the turbulent boundary layer on the flat plate out of the basic experimental data on the flow and the nature of the dissipative field from the thermodynamics of irreversible processes. The behavior of the turbulence after transition is interpreted to be the tendency of entropy growth but at the same time of a slow rate of entropy production. Formation of a turbulent boundary layer which changes little in the flow direction and the presence of a feed back mechanism to maintain such boundary layer are conceivable, and it is experimentally verified that the turbulence can be controlled by a relatively small amount of suction which changes the balance of the feedback mechanism.

1. Introduction

When we want to artificially control the fluid flow, typically to laminate a flow and decrease the air resistance, what is fundamentally important thing to be done is to know the natural behavior of the flow. In other words, the thing to be done must be to grasp a principle dominating the natural flow and work out a measure following the principle. The difficulty of scientifically knowing the property of the flow lies in that the behavior of the flow is essentially nonlinear and that dissipation due to viscosity, is significant. These make it difficult to discover the principle of dynamics and to establish the self-closure of the basic equation for the turbulent motion.

Nevertheless we believe that the fundamental of the natural flow must be essentially simple and rational, for instance, the natural flow would follow the less resistant and consequently less dissipative processes. If then thermodynamics of irreversible processes is linked definitely to the dissipative flow involving the turbulence, the complicated turbulence phenomena will get a fresh interest of being associated with wide areas of science[1].

It is important to see whether diverse information obtainable from the rapidly advancing technology of numerical analysis and experimentation can have the way to such an approach.

What is being tried here is to examine the meaning and effect of statistical data on the turbulent field and write a story as easy to understand as possible describing how the flow field develops or to what extent the flow can be controlled by such a conception. This work may offer one approach to grasping the flow, which is liable to be neglected in an excessively complex numerical model or experimentation.

2. Flow Change with the Development of Boundary Layer

Fig. 2.1 shows how the boundary layer develops along a flat plate. Over a certain extent from the leading edge (see Fig. 2.1 for coordinates, velocity components etc.) of $Re = Ux/\nu \lesssim 2 \times 10^5$, ν being the coefficient of dynamic viscosity, a laminar boundary layer is formed in which the momentum change from 0 on the wall surface to the main stream takes place and the molecular viscosity transmits the momentum. This is a field where presence of an object unilaterally influences a uniform flow so that a theoretical solution may be obtained with the boundary conditions on the wall and of the main stream given.

In the downstream, disturbances due to various causes are superposed on the laminar flow, and as they are amplified, the flow becomes unsteady. The area corresponding to the development of an instability of the laminar flow is the transition region.

Usually, the transition starts[2] with an amplification of a progressive small-amplitude wave, so-called Tollmien-Schlichting wave. With the amplification, the wave becomes nonlinear and three dimensional, and the transport process, except in the vicinity of wall surface, turns dominantly convective. Thereby the correlation of turbulent velocities in different directions, i.e., Reynolds stress becomes significant in the transport process; notably the correlation $\overline{u'v'}$ (—indicates time average), u' being the fluctuation in the flow direction and v' normal to the wall surface, is significant. Deformation of the mean flow and the generation of higher harmonics under the nonlinear effect due to the amplification of Tollmien-Schlichting wave in the initial stage of transition are discussed from an analytical viewpoint[3][4].

$-\overline{u'v'}\partial\bar{u}/\partial y$ indicates the rate of energy supply from the mean flow to the turbulence; in the transition area with $\overline{u'v'} < 0$, the turbulence grows.

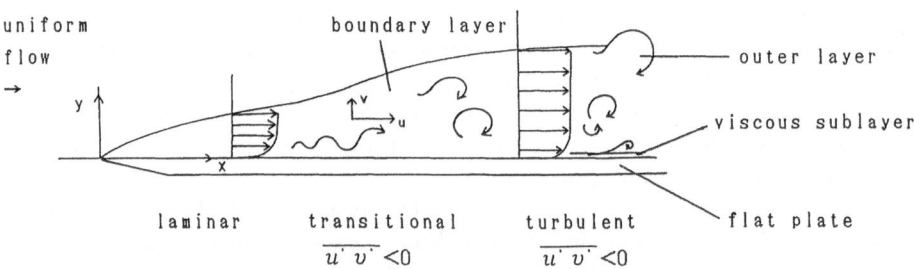

Fig. 2.1. Concept of the flow field

The condition $\overline{u'v'} < 0$ contributes to the unification of the mean velocity distribution in the boundary layer by the turbulence, because it transports a flow mass of large momentum in the main flow toward the wall surface and at the same time transports a flow mass of small momentum near the wall toward the main stream. Particularly the former effect of bringing the large momentum of the main stream close up to the wall surface and in consequence leveling the distribution of the mean velocity in the boundary layer as a whole leads to an increased wall surface friction due to a steepened gradient of the mean velocity near the wall surface, in other words, leads to the rapid increase in the frictional resistance in the transition region.

This tendency continues into the downstream turbulent boundary layer. The irregular motion in the outer layer of the turbulence boundary layer shows $\overline{u'v'} < 0$. Here a simple, fundamental doubt occurs to our mind. True, the transition to a turbulence effectively contributes to uniformity of phenomena but at the same time it causes the air resistance and the energy dissipation to increase. Thus the problem is how such changes can be interpreted as a dispensation of the nature.

Now let us remind that in a field involving dissipation the entropy invariably grows and moves in the direction of a thermodynamically uniform state. Presence of an object in the flow disturbs uniformity and it is quite natural that the flow tries to remove it or carry it away downstream on the flow. If it is an expression of the natural will for uniformity that the flow tries to blow it away by applying a large frictional force through the transition to turbulence, another doubt occurs why the flow does not enhance the turbulent motion applying a decisively large frictional force continuously.

It is known that in a fully developed turbulent boundary layer on a flat plate, while the turbulence is developing, the local frictional coefficient becomes smaller than in the transition region and it decreases in the downstream direction, though more slowly than in the laminar flow. This fact may be construed to represent the natural direction (uniformity of flow, entropy increase) and at the same time the nature's compromise for harmony with the environment (lowering of entropy production rate). We refrain from going farther to discussion about the scientific expression of the nature's dispensation at work here, but phenomenologically speaking, it can be speculated that unlike in the case of a laminar flow, in the formation of a turbulent boundary layer the turbulent motion plays an active role and our knowledge about the profundity of the natural flow is still insufficient.

It is known that the basic Reynolds averaged equation for the theoretical analysis of the turbulent boundary layer has the difficulty of closure, and as is well-known, to fill the gap various turbulence models have been worked out. Here, however, we are going to discuss the natural mechanism of suppressing the increase of resistance and entropy assumed to be taking place in the turbulent boundary layer, by the behaviors of the flow experimentally revealed so far. It is easy to imagine that this mechanism is associated with the three-dimensional structural flow which is regarded as longitudinal vortices in the viscous sublayer distinctively observed in the turbulent boundary layer after transition. It is known[5][6] that on account of the action of the longitudinal vortices, the fluid of small momentum near the wall surface lifts up as horseshoe vortices, which shed from the wall and collapse, thereby blending with the fluid of large momentum in the outer layer. Just like in the outer layer, this action together with the turbulent motion leads to $\overline{u'v'} < 0$, but it should be appreciated for its function of suppressing an increase in the mean momentum near the wall and to a certain extent stopping an increase in the frictional resistance by reducing the mean velocity gradient at the wall surface.

Thus the occurrence of the longitudinal vortex in the viscous sublayer can be considered as a sort of counterpart to the disturbance in the outer layer and we can assume its generation and development as closely linked to the intensity of the disturbance in the outer layer. The following discussion in 4 will be centered on this point.

3. Discussion on Energy

Prior to the discussion on the flow mechanism under 4, we try to consider on the flow from a viewpoint of energy balance. From the Navier-Stokes equation for an incompressible flow, we derive the following equation for the total energy $p_0 = \rho v^2/2 + p$:

$$\frac{\overline{Dp_0}}{Dt} = \nu \nabla^2 \overline{p_0} - \mu \overline{\omega^2}, \quad \omega \equiv \nabla \times v, \tag{3.1}$$

that is, the diffusion equation with energy dissipation due to the square of vorticity ω is obtained. Here v, ρ and p are the three-dimensional velocity vector, the density and the static pressure, respectively.

Meanwhile, we have an equation for entropy S as follows:

$$\rho T \frac{\overline{DS}}{Dt} = -2\nu \nabla^2 \bar{p} + \mu \omega^2, \tag{3.2}$$

where T is the absolute temperature of the fluid. Comparing (3.1) with (3.2), we see that energy dissipation of fluid and the entropy growth depend on ω^2. On the other hand, upon the integration for the volume V around the object of both equations, we see that the first terms on the right side respectively reveal the significance of the static pressure gradient on the wall surface for the energy dissipation and the entropy production, giving a clue to the flow control.

Now eliminating the term ω from the volume integration of (3.1) and (3.2) we get:

$$\rho T \int_{\Sigma} \left(\overline{vS} \right)_n d\Sigma = U\bar{D} + \nu \int_{\Sigma} \left(\nabla \bar{\mathcal{L}} \right)_n d\Sigma \tag{3.3}$$

where Σ, U, D, and $\mathcal{L}$ indicate respectively the surface of V, the relative velocity of the object and fluid, the total air resistance of the object and the Lagrangian of flow $(\rho\overline{v^2}/2 - \bar{p})$, while the index n indicates a vector normal to the surface. The left side indicates the growth rate of the total entropy, while the first term on the right side indicates the growth of the total energy, which is the work done by the air resistance to the body, and the second term is the area integration of the Lagrangian gradient on the inspected surface. $(\nabla\mathcal{L})_n$ may be equated to zero at a sufficiently remote position and accordingly (3.3) shows the relation at which the total work done by the air resistance can be equated to the total entropy growth. From (3.1) and (3.2), it is understood that in the flow field, locally the change of p_0 does not necessarily correspond to that of S. Further from (3.1) and (3.3) it follows that, p_0 and $\mathcal{L}$ are involved in the equations in the complicated forms unlike in the case of Hamiltonian or Lagrangian in the analytical dynamics. From this fact, it would be seen how difficult for us to identify the fundamental principle of the flow.

4. Self-Organization of Turbulent Boundary Layer

Common feature to the occurrence in a flow, of longitudinal vortices with their axes in the flow direction is that they are caused by a dynamically unstable balance between the distribution of the static pressure normal to the main flow and those of inertial forces. Examples are the longitudinal vortices due to thermal instability[7], Taylor vortices[8], Görtler vortices[9] and the like. It has been clarified[10] that in the turbulent boundary layer the dynamic instability between v and the distribution of static pressure normal to the wall induced by v, gives rise to longitudinal vortices in the viscous sublayer. This implies that there is a stage in which the random disturbance triggers the coherent motion. The mechanism here may be explained as follows.

The balance of momentum in the normal (y) direction to the wall in the turbulent boundary layer may be expressed as follows:

$$\frac{\partial \overline{v'^2}}{\partial y} = -\frac{1}{\rho}\frac{\partial \bar{p}}{\partial y} + \nu\left(\frac{\partial^2 \bar{v}}{\partial x^2} + \frac{\partial^2 \bar{v}}{\partial y^2} + \frac{\partial^2 \bar{v}}{\partial z^2}\right) \tag{4.1}$$

Since the viscous term on the right side may be ignored in the outer layer, the change in $\bar{p}$ is seen to be triggered by the disturbance v'. Conversely, since in the viscous sublayer the inertia term on the left side can be ignored, it turns out that the static pressure distribution formed in the outer layer penetrates the viscous sublayer and comes to balance with the viscous term on the right side. Then employing the following continuity equation

$$\frac{\partial \bar{u}}{\partial x} + \frac{\partial \bar{v}}{\partial y} + \frac{\partial \bar{w}}{\partial z} = 0 \tag{4.2}$$

and rewriting the viscous term in (4.1), we see that within the viscous sublayer the following holds;

$$0 = -\frac{1}{\rho}\frac{\partial \bar{p}}{\partial y} + \nu \left(\frac{\partial \omega_z}{\partial x} - \frac{\partial \omega_x}{\partial z} \right) \tag{4.3}$$

where

$$\omega_z = \frac{\partial \bar{v}}{\partial x} - \frac{\partial \bar{u}}{\partial y}, \quad \omega_x = \frac{\partial \bar{w}}{\partial y} - \frac{\partial \bar{v}}{\partial z}$$

(4.3) indicates how the vorticity change parallel to the wall surface takes place within the viscous sublayer under the influence of the outer layer. It suggests possibility of , say, occurrence of a periodic vorticity change within the x, z−plane, but since the change of the mean volume in the x-direction within the turbulent boundary layer is likely to be small, it is supposed that in the viscous term, $\nu \partial \omega_x / \partial z$ prevails. Accordingly, $\partial \bar{p} / \partial y$, that is, the disturbance in the outer layer generates within the viscous sublayer, ω_x, i.e., longitudinal vortices and the change in the z-direction, that is, the spacing of these longitudinal vortices is proportional to the intensity of the disturbance. When the disturbance is strong, small scale longitudinal vortices occur but the viscosity near the wall will dampen such small structures. In consequence the energy supply from the viscous sublayer to the turbulence in the outer layer will be reduced, resulting in the decaying of turbulence in the outer layer. Then $\partial \bar{p} / \partial y$ decreases, which extends the spacing of longitudinal vortices and thereby the energy supply to the outer layer is restored. Thus a feedback control between such random field disturbance and structural field, that is, a sort of self-organization supposedly performs the function of sustaining the turbulent boundary layer in an average, definite state. A more specific study revealing the relation between the disturbance in the outer layer and the formation of the longitudinal vortices in the viscous sublayer has been conducted[10].

The structure of the turbulent boundary layer revealed here may be regarded as a dynamic evidence supporting the debates[1] in the thermodynamics of irreversible processes pertaining to suppression of entropy growth rate in the nonequilibriumu dissipative system, the self-organization for formation of a structure out of irregularity, or existence of a feedback control system for co-existence of these two.

5. Experiment on Control of Turbulent Boundary Layer

What is essentially important in the interest of debates under 4 is an experimental verification to determine that in the first place the regular, three-dimensional longitudinal vortices can be generated out of the random turbulence. Reference (10) reports that when micro-particles were adhered to a flat plate near the leading edge and turbulence was generated within the laminar boundary layer, regular longitudinal vortices emerged close to the wall surface. At low velocities of the flow, both the longitudinal vortices and

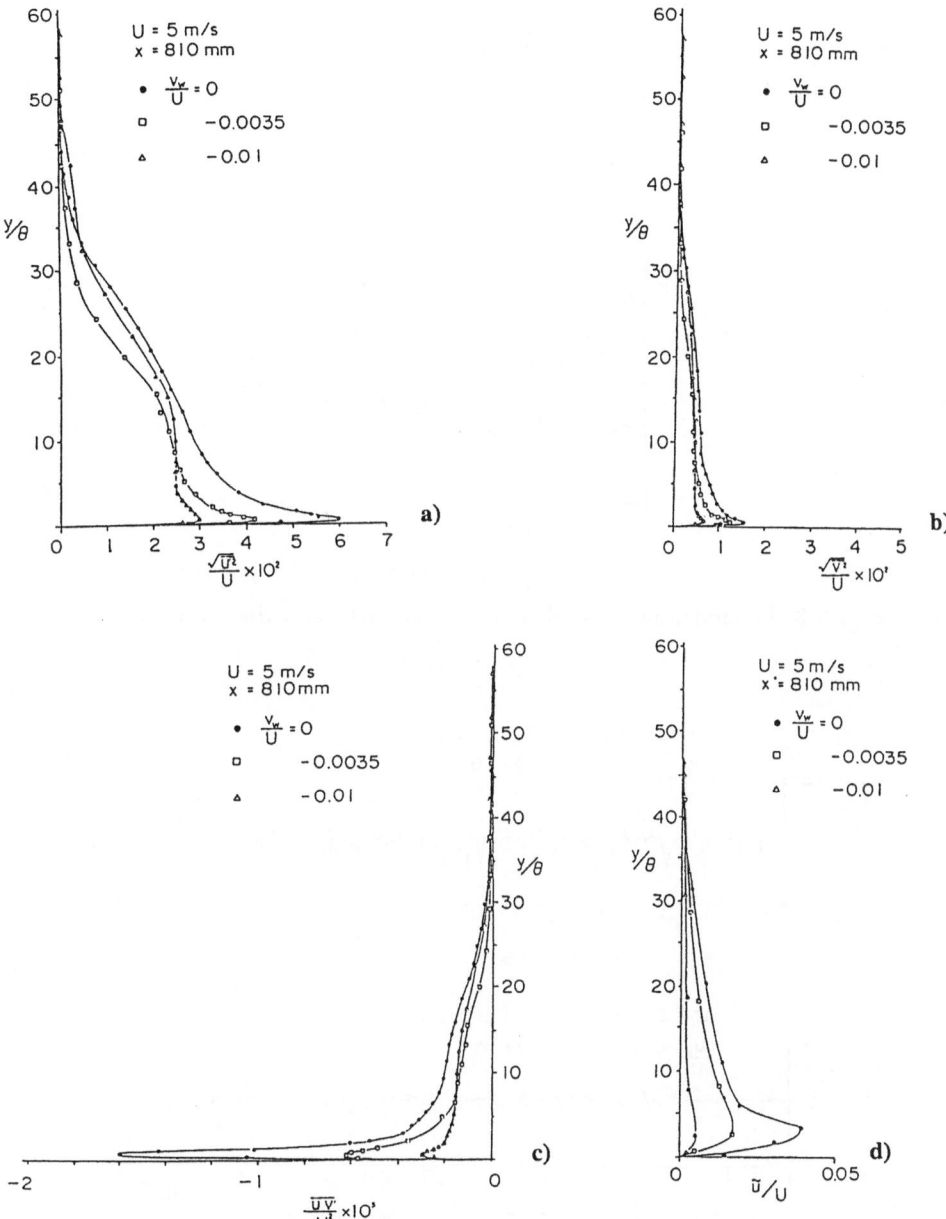

Fig. 5.1. Effects of suction on the turbulent boundary layer. θ; momentum thickness. (a) Distributions of $\sqrt{\overline{u'^2}}/U$, (b) Distributions of $\sqrt{\overline{v'^2}}/U$, (c) Distributions of $\overline{u'v'}/U^2$, (d) Distributions of the amplitude of longitudinal vortices ($\tilde{u}$ = (the peak value–the valley value in the lateral periodicity of u by longitudinal vortices)/2)

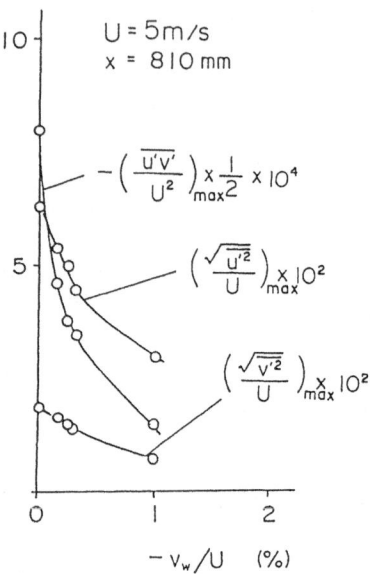

Fig. 5.2. Reduction of the peak values of the turbulence due to suction

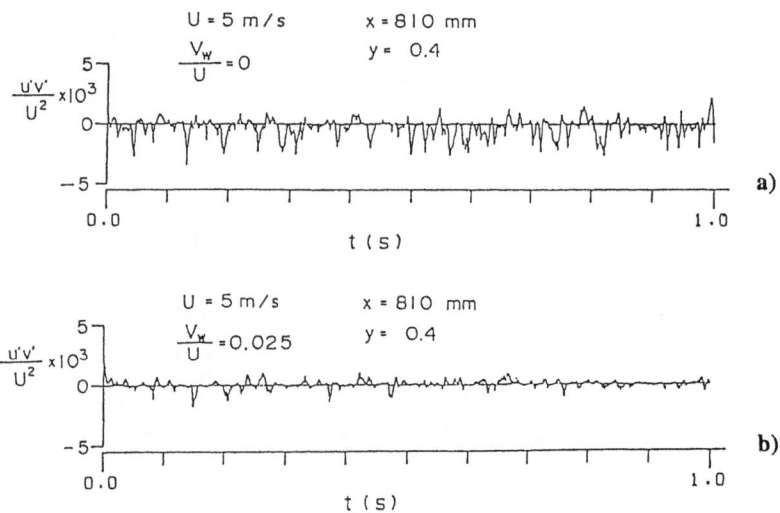

Fig. 5.3. Fluctuation of $u'v'$. (a) $v_w/U = 0$. (b) $v_w/U = 0.025$

the turbulence decayed downstream, but when the Reynolds number based on the particle diameter exceeds 100, energy was supplied from the longitudinal vortices to the disturbance, and the transition to turbulence proceeded under the linkage between the turbulence and longitudinal vortices. For the purpose of artificially controlling the above-mentioned feedback system required for generation of such a turbulent boundary layer, an experiment was conducted[11] to apply the suction to the flow from the wall surface and control the static pressure gradient $\partial \bar{p}/\partial y$ normal to the wall surface. The flat plate of 500 mm width mentioned in Ref. (10) was replaced over a length of 500 mm to 800 mm from the leading edge with a porous flat plate to suck the flow over this distance.

As $\partial \bar{p}/\partial y$ generated in the turbulent boundary layer comes from $\partial \overline{v'^2}/\partial y$ in the outer layer, the suction rate $v_w (< 0)$ required for this control is expected to be effective, even if the volume is small. Fig. 5.1 shows the variations of $\sqrt{\overline{u'^2}}/U$, $\sqrt{\overline{v'^2}}/U$, $\overline{u'v'}/U^2$ and $\tilde{u}/U$ ($\tilde{u}$: amplitude of variation in x-direction, among the three-dimensional changes in the mean velocity due to longitudinal vortices) against v_w/U. It is seen that even a small volume of suction can depress both the turbulence and the longitudinal vortices. Particularly the reduction in $|\overline{u'v'}|$ is prominent (Fig.5.2). Fig. 5.3 is a record indicating a difference in the instantaneous variation of $u'v'$ depending on presence or not of $u'v'$. The drop in the absolute value of $|\overline{u'v'}|$ due to v_w does not take place simply with the reduction of negative correlation of $u'v'$, but it is characterized by an increase in the positive correlation; namely it suggests a tendency of relaminarization of the turbulent boundary layer. This fact vividly testifies that $\partial \bar{p}/\partial y$ changes with v_w and in consequence the turbulent structure changes, which implies a loss of balance in the above-mentioned feedback loop. As for the process of the intensities of disturbance and longitudinal vortices

6. Conclusions

The following are concluded from the investigations of the turbulent boundary layer along a flat plate;

(1) Transition region is a flow field where the co-existence of the direction toward uniformity of the natural flow and its tendency toward slow rate of change can be seen.

(2) In the turbulent boundary layer, the feed back mechanism between the turbulence in the outer layer and the longitudinal vortices in the inner layer is vital to the sustenance of the flow field.

(3) The turbulence can be changed by controlling the static pressure gradient normal to the wall surface through the suction at the wall which is important for the above-mentioned feedback mechanism.

References

1. Glansdorff, P., Prigogine, I., 1951, Thermodynamic Theory of Structures, Stability and Fluctuations, Wiley-Interscience.
2. Schlichting, H., 1968, Boundary Layer Theory, McGraw-Hill.
3. Stuart, J.T., 1962, Non-linear Effects in Hydrodynamic Stability, Proc. 10th Int. Congr. Appl. Mech., Elsevier pub. Co., Amsterdam, N.Y., 63.
4. Herbert, T., 1984, Secondary Instability of Plane Shear Flows —Theory and Applicalion, Proc. IUTAM Symp. "Laminar—Turbulent Transition" Novosibirsk, USSR, 9.
5. Kline, S.J., Reynolds, W.C., Schraub, F.A. and Rundatadler, P. W., 1967, The Structure of Turbulent Boundary Layers, J. Fluid Mech., 30, 741.
6. Cantwell, B.J., 1981, Organized Motion in Turbulent Flow, Ann Rev. Fluid Mech., 13, 457.
7. Terada, T., 1928, Some Experiments on Periodic Columnar Formation of Vortices, Caused by Convection, Tokyo Imperial Univ. Aero. Res. Inst. Rep. 3(31).
8. Taylor, G.I., 1923, Stability of a Viscous Liquid Contained between Two Rotating Cylinders, Phil. Trans. R. Soc. Lond. A 223, 280.
9. Görtler, H., 1940, Über eine dreidimensionale Instabilität laminarer Grenzschichten an konkaven Wänden, Nachr. Ges. Wiss. Göttinger, New Ser. 2 no. 1.
10. Aihara, Y., 1990, Formation of Longitudinal Vortices in the Sublayer due to Boundary–Layer Turbulence, Jour. Fluid Mech. 214, 111.
11. Aihara, Y., Koyama, H., 1992, Self Organization in Turbulent Boundary Layer and the Control by Suction, Jour. Japan Soc. Aeron. Space Sci. 40, 560.

On the Quasi-Geostrophic Drag on a Rising Sphere in a Rotating Fluid

M. Ungarish

Department of Computer Science, Technion, Haifa 32000, Israel

Summary. The angular velocity profile and the drag on a spherical solid particle which rises slowly in a rapidly rotating fluid in a non-long container are estimated via the finite-difference solution of a quasi-geostrophic approximation model. In contrast with the classic geostrophic approach, the radial motion in the slightly inviscid core (including the outer Stewartson layer) is incorporated—in addition to the Ekman layers transport—and shown to affect strongly the results when $\varepsilon = \left(\frac{1}{2}HT^{-\frac{1}{2}}\right)^{\frac{1}{2}}$ is not very small, where H is the dimensionless particle-to-wall distance and T is the Taylor number. Comparisons with the disk particle and with experimental results are discussed and the limitations of the present approximation are pointed out.

1. Introduction

Consider the slow motion of a spherical particle of radius a^* in an incompressible fluid in a container of length $2H^*$ rotating with Ω^* around the axis z. Let $W^*\hat{z}$ be the velocity of the particle, and ρ^* and ν^* the density and kinematic viscosity of the fluid. Hereafter, length is scaled with a^*, velocity with W^*, pressure with $W^*\nu^*\rho^*/a^*$ and force with $W^*\nu^*\rho^*a^*$; the asterisk denotes dimensional variables. In the cylindrical r, θ, z system, rotating with Ω^* and attached to the center of the particle, the velocity components are $u, r\omega, w$. The relevant parameters are the Taylor number, $T = \Omega^*a^{*2}/\nu^*$, Rossby number, $Ro = W^*/\Omega^*a^*$, and (half) height H, assumed, respectively, large, small and moderately large (a more stringent specification follows). For example, we would like to consider $T \sim 10^4, Ro \sim 10^{-3}, H \sim 5 \div 20$, which corresponds to some accessible experiments.

The available relevant classic theory is the linear ($Ro = 0$) approach, upon which the flow field can be treated in the asymptotic $T \to \infty$ limit as a superposition of z-independent "cores", Ekman boundary layers and Stewartson shear layers on the cylinder circumscribing the particle. Recently, Ungarish & Vedensky (1994) solved this linear flow-field problem for a disk particle as a whole, for arbitrary T, by transform methods, but no extension of this improved solution to other particle shapes is available.

The drag force, D, which is of major concern in practical applications, turns out to be a puzzling, still unsolved issue, which cast doubts on the usefulness of the linear theory ($Ro = 0$) in this problem. The well-known dilemma was produced by Moore & Saffman (1968) prediction, derived under the geostrophic-flow approximation, for a rigid sphere,

$$D_0 = \frac{43}{105}\pi T^{\frac{3}{2}}; \qquad (1.1)$$

Maxworthy's (1968) experimental attempts of verification yielded considerably (typically 20%) smaller values for the attained range of large T and small Ro.

The discrepancy was attributed to the neglected non-linear terms (i.e., not sufficiently small Ro in the experiment), but Barnard & Pritchard (1975) remarked:

> A correction for the effect of the finite thickness of the free shear layers in the theory of Moore & Saffman would almost certainly reduce the discrepancy between the theory and the measurements. Whether this would then lead to satisfactory agreement must await the appropriate calculations.

Here some preliminary results of such calculations are reported.

2. Solution

The model used here is of a quasi-geostrophic core, which encompasses the "inviscid" and "outer Stewartson" shear regions. In this core u, w and p are functions of r only. The thin embedding Ekman layers (on the solid boundaries) and "inner Stewartson" $T^{-\frac{1}{3}}$ layer (on the cylinder circumscribing the particle) supply matching conditions, as given by Moore & Saffman (1969).

Moreover, we consider the simplified—yet representative—case of the particle in the mid-plane between the lids of the container, while the lateral walls are completely disregarded. We obtain the major equation

$$\varepsilon^2 \left(1 - \frac{f(r)}{H}\right)\left(\frac{d^2\omega}{dr^2} + \frac{3}{r}\frac{d\omega}{dr}\right) - \frac{1}{2}\left[1 + (1 + f'^2)^{\frac{1}{4}}\right]\omega = \frac{1}{2}T^{\frac{1}{2}}, \qquad (2.1)$$

for $0 \le r \le 1$, upstream side, subject to

$$\frac{d\omega}{dr}(r = 0) = \omega(r = 1) = 0, \qquad (2.2)$$

where

$$\varepsilon = \left(\frac{1}{2}HT^{-\frac{1}{2}}\right)^{\frac{1}{2}}, \quad f(r) = (1 - r^2)^{\frac{1}{2}}. \qquad (2.3)$$

The first term in the LHS of (2.1) represents the contribution of the co transport, and the second one the contribution of the Ekman layers. These add up to the volume displacement caused by the advance of the particle given in the RHS. In the downstream side $|\omega|$ has the same value but the o by the radial momentum equation $-2T\omega r = -dp/dr$, hence the correspondi force on the particle (the shear contribution is very small) is expressed as

198

$$D = -4\pi T \int_0^1 \omega r^3 dr. \tag{2.4}$$

Note that the geostrophic model is given by (2.1) with $\varepsilon = 0$, in which case (2.4) yields (1.1), and is expected to be relevant when $\varepsilon \ll 1$. The present model is expected to be accurate under the conditions: $(H/T)^{\frac{1}{3}} \ll 1$ and $\varepsilon/(H/T)^{\frac{1}{3}} \gg 1$.

The major task performed here was the numerical finite differences solution of (2.1), followed by the calculation of (2.4).

3. Results

Some results of angular velocity and drag calculations are displayed in Figs. 3.1 and 3.2.

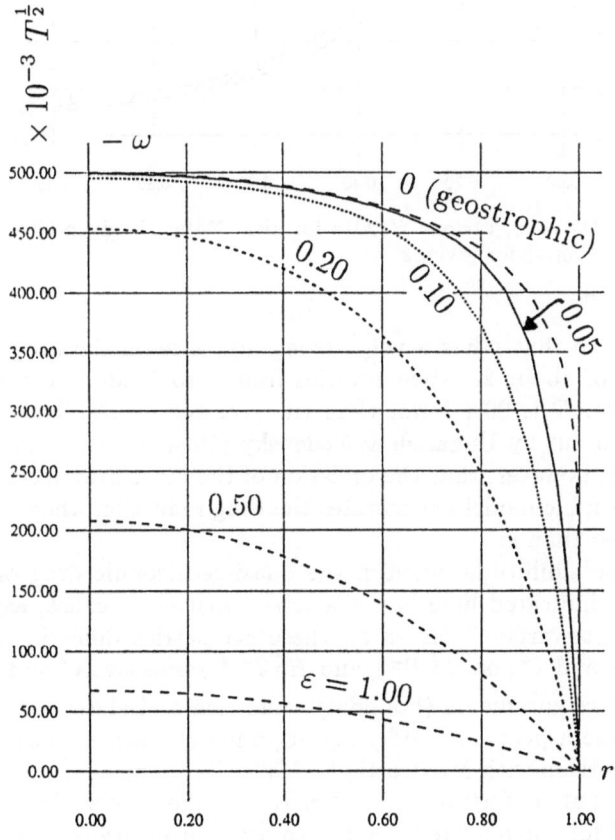

Fig. 3.1. $-\omega$ vs. r, various ε; $H = 5$

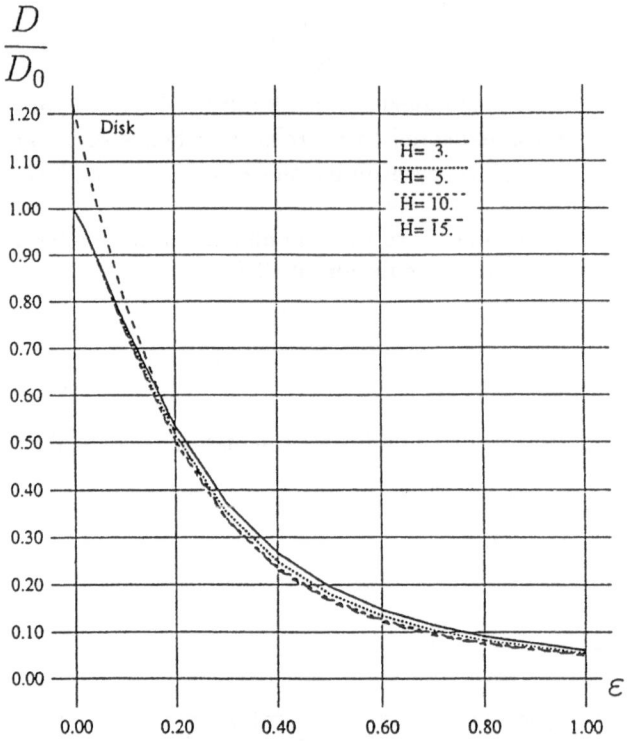

Fig. 3.2. D/D_0 vs. ε, various H; also for disk. Note: D_0 given by (1); disk results do not depend on H for given ε

We observe that ε has a very strong influence on the drag: D decreases by a factor of about 20 when ε varies from 0 to 1. Moreover, even for ε as small as 0.05, D is 20% lower than the geostrophic D_0. These trends have been pointed out by Ungarish & Vedensky (1994) for the disk particle; it is shown there, however, that the omission of the inner Stewartson layer in the quasi-geostrophic model exaggerates the drag reduction when $(H/T)^{\frac{1}{3}}$ is not much smaller than ε.

The close similitude between the quasi-geostrophic drag on a disk and on a sphere indicated here is not a trivial matter, because, asymptotically, the "outer Stewartson" layers on these geometries differ in structure and thickness (ε and $\varepsilon^{\frac{8}{7}}$, or $H^{\frac{1}{2}}T^{-\frac{1}{4}}$ and $H^{\frac{4}{7}}T^{-\frac{2}{7}}$ respectively) and are triggered by different requirements (boundary condition and shear reduction, respectively). These aspects are under investigation (Ungarish, 1994).

A comparison with Maxworthy's (1968, Fig. 3) measurements is of course interesting, but, unfortunately, restricted by the unavailability of the exact values of H for the reported points. An attempt to recover these values from the given details, produced the following relevant comparison. (The reported experimental error is ±2.5 %).

T	H	Ro	ε	D/D_0, test	D/D_0, here	$(H/T)^{\frac{1}{3}}$
$2.6 \cdot 10^4$	5.15	$\sim 5 \cdot 10^{-4}$	0.13	0.82–0.87	0.68	0.06
$2.5 \cdot 10^4$	5.15	$\sim 10^{-4}$	0.13	0.88	0.68	0.06
$2.5 \cdot 10^4$	10.50	$\sim 3 \cdot 10^{-3}$	0.18	0.73	0.54	0.07

Evidently, the present results give the proper trend of the drag, but the accuracy is low: the quasi-geostrophic drag seems to underestimate the experimental values by, roughly, the same percentage by which the geostrophic drag overestimated them. This can be attributed to two effects, still in the domain of the linear theory:

(a) The experimental value is some average for different positions of the particle along the axis, hence higher (by a few percent) than the mid-plane drag.

(b) The inner Stewartson layers, see last column in table, are not much thinner than the outer layer of width $\sim \varepsilon$. As shown in Ungarish & Vedensky (1994) for the disk geometry, the incorporati improve the agreement with the measurements. The inclusion of these layers in the sphere drag calculations, however, is a presently unsolved difficult problem.

In spite of the low predictive accuracy in the tested practical situations, the quasi-geostrophic drag calculations are valuable results, because: (a) They provide the lower limit of the linear drag, while D_0 gives the uppe (b) They rehabilitate the reliability of the linear theory. It seems that the discrepancy reported by Maxworthy is not a failure of the linear approach for $Ro \sim 10^{-3}$, rather a result of the oversimplified treatment of the Stewartson layers in the derivation of D_0. (c) They describe, at least qualitatively, the important transition region from the "short" container geostrophic case, $\varepsilon \to 0$, where D_0 applies, to the "long" container case, $\varepsilon > 1$, where $D \approx (\pi/24)T^2/H$, see Hocking, Moore & Walton (1979), Ungarish & Vedensky (1994) and Ungarish (1994).

In closing, we remark that the present discussion is also connected to the recent results of Bush, Stone & Bloxham (1992, 1994) who considered the analogous motion of drops via a geostrophic theory. The comparison with experiments indicates that the geostrophic drag in this more complex problem, again, considerably overestimates the physical results. The extension of the present investigation to non-rigid and non-spherical particles is therefore an interesting task (Ungarish, 1994).

Acknowledgments
The research was partially supported by the Fund for the Promotion of Research at the Technion.

References

1. Barnard, B.J.S. & Pritchard, W.G., 1975. The motion generated by a body moving through a stratified fluid at large Richardson numbers. J. Fluid Mech. 71, 43-64.
2. Bush, J.W.M., Stone, H.A., Bloxham, J., 1992. The motion of an inviscid drop in a bounded rotating fluid. Phys. Fluids A 4, 1142-1147.
3. Bush, J.W.M., Stone, H.A., Bloxham, J., 1994. Axial drop motion in rotating fluids. J. Fluid. Mech., submitted.
4. Hocking, L.M., Moore, D.W. & Walton, I.C., 1979. The drag on a sphere moving axially in a long rotating container. J. Fluid Mech. 90, 781-793.
5. Maxworthy, T., 1968. The observed motion of a sphere through a short, rotating cylinder of fluid. J. Fluid Mech. 31, 643-655.
6. Moore, D.W. & Saffman, P.G., 1968. The rise of a body through a rotating fluid in a container of finite length. J. Fluid Mech. 31, 635-642.
7. Moore, D.W. & Saffman, P.G., 1969. The structure of free vertical shear layers in a rotating fluid and the motion produced by a slowly rising body. Trans. Roy. Soc. Lond. A 264, 597-634.
8. Ungarish, M. & Vedensky, D., 1994. The motion of a rising disk in a rotating axially bounded fluid for large Taylor number. J. Fluid Mech., submitted.
9. Ungarish, M., 1994. On the quasi-geostrophic drag on a rising particle or drop in a rotating fluid., in preparation.

Two-Dimensional Nonlinear Saturation Behaviour of Instability Waves in a Boundary Layer at Mach 5

N.A. Adams and L. Kleiser

DLR, Institute for Theoretical Fluid Mechanics, Bunsenstraße 10, 37073 Göttingen, Germany

Summary. The two-dimensional nonlinear evolution of a second-mode instability wave in a flat plate boundary layer at a free-stream Mach number of $M_\infty = 5$ is investigated by direct numerical simulation. An explicit spectral/finite-difference scheme employing the temporal model is used. A nonlinear saturation process is found, during which initially a weak viscous shock develops near the wall. A comparison of the numerical result obtained in the shock region with a one-dimensional analytic weak-shock solution is made and a good agreement of the shock-normal velocity distribution and shock thickness is found. The nonlinear saturation is characterized by a cascade of states with alternating high and low energy levels of the higher Fourier-modes. The system evolves on a slow time scale towards a steady state which appears to be different from the undisturbed laminar state.

1. Introduction

For incompressible flows detailed experiments, theoretical efforts and numerical simulations have led to a basic understanding of the boundary-layer transition process. Due to the enormous difficulties of controlled experiments at high supersonic flow speeds hardly any experimental results revealing detailed flow phenomena at the nonlinear stages of supersonic transition have been published. Thus presently there is a strong interest in numerical simulation techniques (Kleiser and Zang [12]). Both temporal (refs. [8], [4], [1], [2], [3], [20]) and spatial (refs. [25], [7], [19]) simulation models have been successfully applied to moderate and high Mach number flows.

In a low-disturbance environment the initial stages can essentially be treated by linearized theories (Mack [17]; Herbert [11]). At higher Mach numbers the linear stability characteristics become substantially more complex (Mack [17]). A two-dimensional second-mode (Mack mode) instability moving with a low subsonic velocity relative to the freestream becomes most unstable at about $M_\infty > 3$. Much less is known about the nonlinear disturbance development. For incompressible boundary-layer flow, usually a nonlinear saturation state (cf. Koch [13]) is attained from the linearly most unstable wave for a flow which is constrained to two dimensions. The flow does not undergo transition to turbulence since the most important turbulence generation mechanism, the stretching and convection of spanwise vorticity, is absent. The mode energy of the fundamental wave and its higher harmonics increases monotonically up to a certain saturation threshold. Though the problem of the flow evolution restricted to two spatial dimensions may appear artificial, it alleviates the understanding of certain physical processes. Phenomena such as the build-up of embedded shocklets are much easier understood in two dimensions.

It is well known that in compressible mixing layer transition shock waves build up just above and beneath the spanwise vortex rollers if the flow is constrained to two dimensions. In a comparable three-dimensional flow, however, they have not been observed so far [22]. At higher Mach numbers compressibility effects may change the saturation process of a boundary layer considerably. Hence there is sufficient reason to perform a numerical simulation of the two-dimensional disturbance development in a boundary layer, which in addition is relatively cheap when compared to a three-dimensional problem as was studied in [1], [2], [3].

2. Governing Equations

We consider the problem of perturbation evolution in the boundary layer along a flat plate. The temporal simulation approach is adopted, allowing the use of periodic boundary conditions in the streamwise and spanwise directions x and y. The integration domain has the extents L_x in the streamwise and L_y in the spanwise direction and is truncated at L_z in the wall-normal direction. A sketch of the integration domain is shown in Fig. 2.1, together with the Cartesian coordinate system $\{x, y, z\}$. For ease of notation in the following

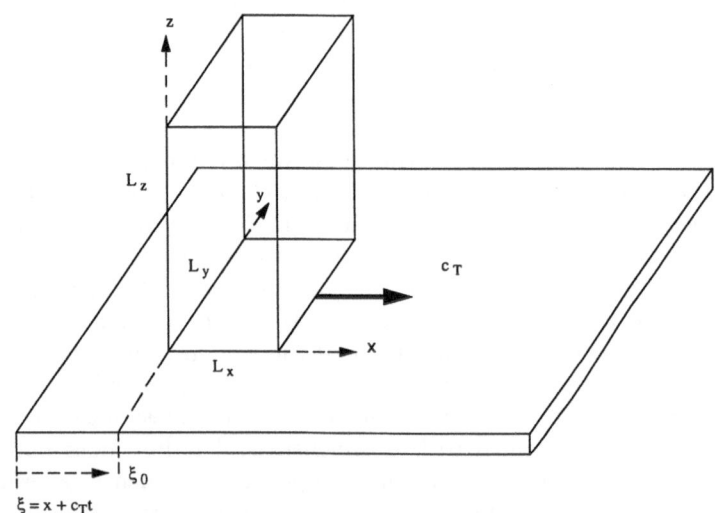

Fig. 2.1. Sketch of the computational domain

an index notation (with summation convention) is used, where the subscripts 1, 2 and 3 correspond to the coordinates x, y and z, respectively.

The basic equations to be solved are the three-dimensional unsteady compressible Navier-Stokes equations in conservative formulation. The basic flow is imposed by adding a forcing term (see e.g. ref. [4]), which in the present simulation is kept constant in time. Sutherland's law is used to give the variation of viscosity with temperature. A perfect gas with specific heat ratio $\kappa = 1.4$ is assumed. In the following dimensional quantities are denoted with an asterisk. The non-dimensionalization is done by

$$u_i = u_i^*/U_\infty^* \ , \quad \rho = \rho^*/\rho_\infty^* \ , \quad T = T^*/T_\infty^* \ ,$$

$$p = p^*/(\rho_\infty^* U_\infty^{*\,2}) \ , \quad E_T = E_T^*/(\rho_\infty^* U_\infty^{*\,2}).$$

Here u_i denotes the velocity components, ρ the density, p the pressure and E_T the total energy to be defined below. The time t is non-dimensionalized by U_∞^*/δ_1^*, where the reference length δ_1^* is taken as the displacement thickness of the undisturbed laminar basic flow. The governing equations can be cast in non-dimensional form as (e.g. ref. [5])

$$\frac{\partial \rho}{\partial t} = -\frac{\partial \rho u_i}{\partial x_i} + Z_1 \tag{2.1}$$

$$\frac{\partial (\rho u_i)}{\partial t} = -\frac{\partial (\rho u_i u_j + p\delta_{ij})}{\partial x_j} + \frac{\partial \tau_{ij}}{\partial x_j} + Z_{i+1} \tag{2.2}$$

$$\frac{\partial E_T}{\partial t} = -\frac{\partial (E_T + p)u_i}{\partial x_i} - \frac{\partial q_i}{\partial x_i} + \frac{\partial (u_j \tau_{ij})}{\partial x_i} + Z_5 \ , \tag{2.3}$$

with the total energy being $E_T = p/(\kappa - 1) + u_i u_i/2$ and $i = 1,2,3$. The shear-stress tensor obeying Newton's relation and Stokes' hypothesis is given by

$$\tau_{ij} = \frac{\mu}{Re}\left(\frac{\partial u_i}{\partial x_j} + \frac{\partial u_j}{\partial x_j} - \frac{2}{3}\frac{\partial u_k}{\partial x_k}\frac{\partial u_k}{\partial x_k}\delta_{ij}\right) \ ,$$

and the heat flux vector is

$$q_i = -\frac{\mu}{(\kappa - 1)M_\infty^2 Pr Re}\frac{\partial T}{\partial x_i} \ .$$

The Reynolds number of the flow is defined as $Re = U_\infty^* \delta_1^*/\nu_\infty^*$, and the Prandtl number $Pr = c_p^* \mu_\infty^*/\lambda_\infty^*$, where c_p is the specific enthalpy and λ is the heat conductivity. The Prandtl number is set to $Pr = 0.7$. The dynamic viscosity μ is an explicit function of the temperature only and is assumed to obey Sutherland's law

$$\mu(T) = T^{\frac{3}{2}}\frac{1+S}{T+S} \ ,$$

where $S = 110.4 \ K$. As an additional algebraic relation the perfect-gas relation holds

$$\kappa M_\infty^2 p = \rho T \ .$$

The forcing term in equations (2.1)–(2.3) is given by

$$
\begin{aligned}
Z_1 &= c_T \frac{\partial \mathrm{P}}{\partial \xi} \\
Z_2 &= c_T \frac{\partial (\mathrm{P}U)}{\partial \xi} - \frac{1}{Re} \frac{\partial}{\partial z} \left(\mathrm{M} \frac{\partial U}{\partial z} \right) \\
Z_3 = Z_4 &= 0 \\
Z_5 &= c_T \frac{1}{2} \frac{\partial (\mathrm{P}U^2)}{\partial \xi} - \frac{1}{(\kappa-1)M_\infty^2 Pr Re} \frac{\partial}{\partial z} \left(\mathrm{M} \frac{\partial \Theta}{\partial z} \right) - \\
&\quad - \frac{1}{Re} \frac{\partial}{\partial z} \left(\mathrm{M}U \frac{\partial U}{\partial z} \right)
\end{aligned}
\tag{2.4}
$$

where $\xi = x + c_T t$ is the downstream coordinate in a fixed frame of reference, while the integration domain is moving downstream with the velocity c_T. For the results presented herein we let $c_T = 0$. The basic flow profiles are denoted with $U(z)$ (streamwise velocity), $\Theta(z)$ (temperature), $\mathrm{P}(z)$ (density) and $\mathrm{M}(\Theta)$ (viscosity). It can be shown [1] that adding this kind of forcing term to the Navier-Stokes equations provides a formulation equivalent to the full nonlinear perturbation equations.

As initial condition, the laminar basic flow is superimposed with eigensolutions of the linear stability equations and additionally or alternatively random fluctuations. The basic flow is given by the compressible laminar boundary-layer similarity equations for an adiabatic wall, also employing Sutherland's viscosity law. The similarity solution is obtained numerically by a shooting method [1]. Besides the periodic boundary conditions in the streamwise and the spanwise direction, at the wall a no-slip condition is used. The wall is assumed to be isothermal with respect to the fluctuations. At the artificial boundary at the upper truncation plane time-dependent non-reflecting boundary conditions adapted from Thompson [24] are used. For details on initial and boundary conditions the reader is referred to Adams [1].

3. Numerical Method

The Navier-Stokes equations (2.1)–(2.3) are discretized using the method of lines. The periodic streamwise and spanwise directions x and y are treated by using Fourier expansions. In these directions the pseudo-spectral Fourier-collocation approach is employed (see Canuto et al. [6]). In the wall-normal direction z a compact finite-difference scheme (Lele [15]) is used, which is of 6th order at inner points and has 3rd to 4th order boundary closures for the first and second derivative operators. The (Lax- and asymptotic) stability of the scheme was proved for the linear convective and the linear diffusive limits using a normal mode and an eigenvalue analysis [1]. Time advancement is made with an explicit third-order compact-storage Runge-Kutta scheme (Wray [27]). An analytic mapping function between the evenly spaced computational interval $[0, 1]$ and the physical wall-normal interval $[0, z_{max}]$ is used [1]. This allows for a condensation of grid points near the wall and some plane $z_c \in [0, z_{max}]$ which is chosen to be the critical layer in our simulation. The mapping parameters are optimized for resolving the linear instability wave and are kept constant throughout the simulation.

The code has been carefully validated by computing the growth rates and shapes of linear eigenfunctions. Two classes of tests were made: (a) tests where eigenfunctions from linear stability analysis were used as the initial disturbances, and (b) tests where linear instabilities were allowed to grow from random background disturbances. In all cases very good agreement with linear theory was obtained. A more detailed description of the mathematical model, the numerical method, validation procedure and test results can be found in refs. [4] and [1].

4. Simulation Results

The nonlinear development of a Mack mode in a Chapman-Rubesin boundary-layer [23] at $M_\infty = 5$, a free-stream temperature of $T_\infty = 290K$ and a Reynolds number of $Re = 10000$ is investigated, extending the nonlinear investigation presented in ref. [4]. (Throughout this paper variables are non-dimensionalized with the laminar displacement thickness δ_1 and the respective freestream values.) The most unstable second mode with wavenumber $\alpha = 2.217$, frequency $\omega_r = 2.0233$ and growth rate $\omega_i = 0.0430$ is excited initially with a maximum disturbance amplitude $u'_{max} = 0.001$ by superimposing the laminar base flow with the eigensolutions from linear stability theory. The discretization throughout the simulation was $N_x = 64$, ($N_y = 4$, since the 3D-code was used,) $N_z = 129$. By increasing the resolution up to $N_x = 96$ and $N_z = 161$ the former discretization was shown to be sufficient.

5. Perturbation Evolution

D. Henningson [10] proposed a formulation of a compressible perturbation energy, which takes into account the fluctuation of the internal energy as well as the fluctuation of the kinetic energy and is positive definite. From this a *total mode energy* can be defined as

$$E(k_x; t) \quad = \quad \underbrace{\int_0^{z_{max}} \bar{\rho}(z) \hat{u}_j(k_x; t) \hat{u}_j^\dagger(k_x; t) +}_{E_{kin}}$$

$$\underbrace{\int_0^{z_{max}} \frac{1}{\kappa M_\infty^2} \frac{1}{\bar\rho^2(z)} \hat\rho(k_x;t)\hat\rho^\dagger(k_x;t)dz}_{E_{int_1}} +$$

$$\underbrace{\int_0^{z_{max}} \frac{1}{\kappa(\kappa-1)M_\infty^2} \bar\rho^2(z)\hat T(k_x;t)\hat T^\dagger(k_x;t)dz}_{E_{int_2}} . \qquad (5.1)$$

where k_x denotes the Fourier-wavenumber, $\hat u_j$ are the Fourier-transformed velocity components and $\dagger$ indicates a complex conjugate. The first term of the right-hand side is essentially equivalent to the content of kinetic energy in each Fourier mode. The sum of the second and the third term is the content of internal energy of each Fourier mode.

The evolution of the total mode energy is given in Fig. 5.1 for a span of more than 380 periods $T_\omega = 3.1056$ of the primary wave. After about $52 T_\omega$ one can observe the dominant modes reaching a first maximum of the fluctuation energy at the expense of the mean-flow energy. More than 80% of the initial total energy is contributed by the internal energy $E_{int_1} + E_{int_2}$. Fig. 5.2 shows the respective fractions of energy separated for terms E_{kin}, E_{int_1} and E_{int_2} from equation (5.1) for mode 1. The mode-energy peaks in Fig.

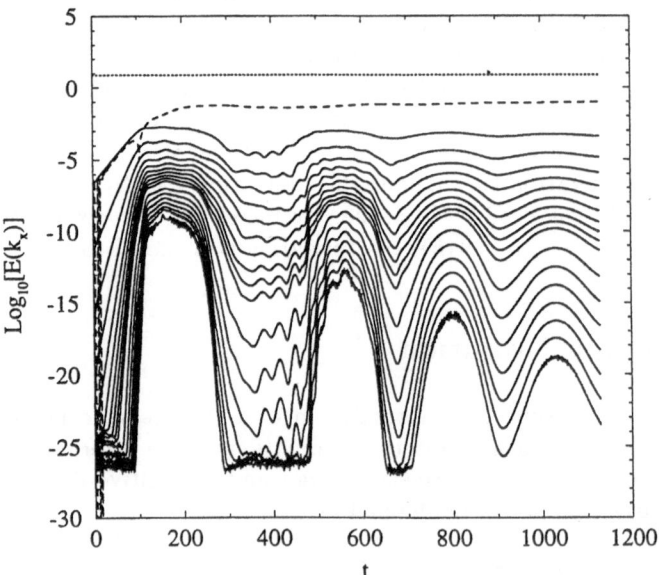

Fig. 5.1. Evolution of mode energy $E(t)$ for the 0-mode (dotted line), for the 0-mode perturbation (dashed line) and, with decreasing energy, Fourier modes $k_x = 1, 2, 3, 4, 5, 6, 7, 8, 9, 10, 13, 16, 19, 22, 25, 28, 31$ (solid lines)

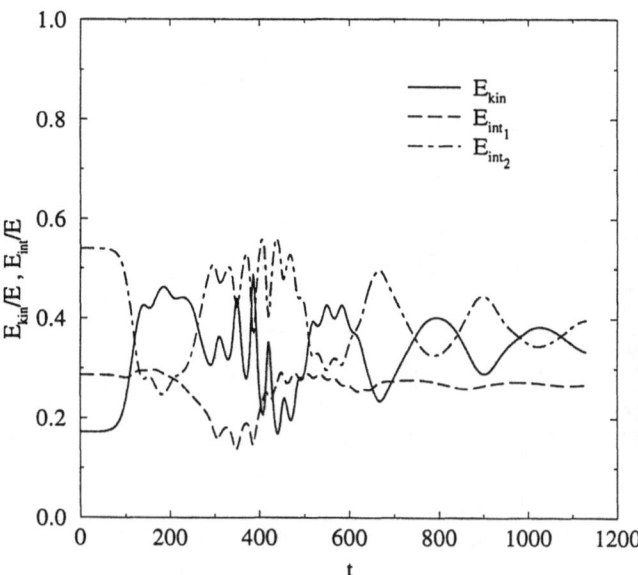

Fig. 5.2. Evolution of the different energy fractions of the total mode energy for mode $k_x = 1$

5.1 are accompanied by a maximum of the kinetic energy fraction of mode 1. For comparison the fractions of kinetic energy and internal energy for the higher mode 9 are shown in Fig. 5.3. After the initial transient during which the energy content of the higher harmonics is very low, they show a similar behaviour as mode 1. Thus in general two processes of energy redistribution take place :

(a) drain of total energy from the mean flow to the perturbations ,
(b) redistribution of the internal and kinetic energy content of each Fourier mode.

With increasing time the oscillating redistribution process ceases and is expected finally to reach a quasi-steady state with constant total mode energy in each Fourier mode and constant fractions of kinetic and inner energy. Since the 0-mode perturbation does not vanish at the final stage the new mean flow is expected to be different from the laminar basic flow. This can be inferred from Fig. 5.4 which shows the evolution of the mean flow profiles and the corresponding root-mean-square fluctuations (the tilde denotes a mass-weighted or Favre average).

It should be noted that the above behaviour is:

(a) Independent of the initial disturbance amplitude, as has been verified by two additional runs, one with an amplitude of $A = 10^{-4}$ and one with and amplitude of $A = 10^{-2}$. Once the second mode has reached a certain threshold amplitude the observed damped limit-cycle behaviour sets in.

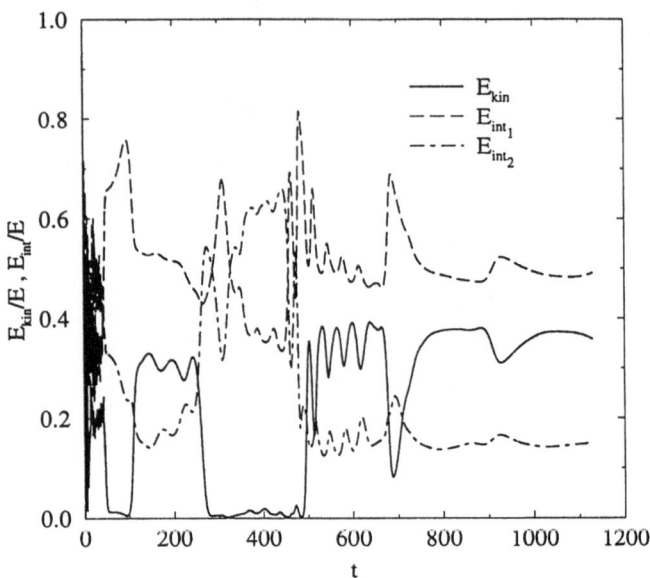

Fig. 5.3. Evolution of the different energy fractions of the total mode energy for mode $k_x = 9$

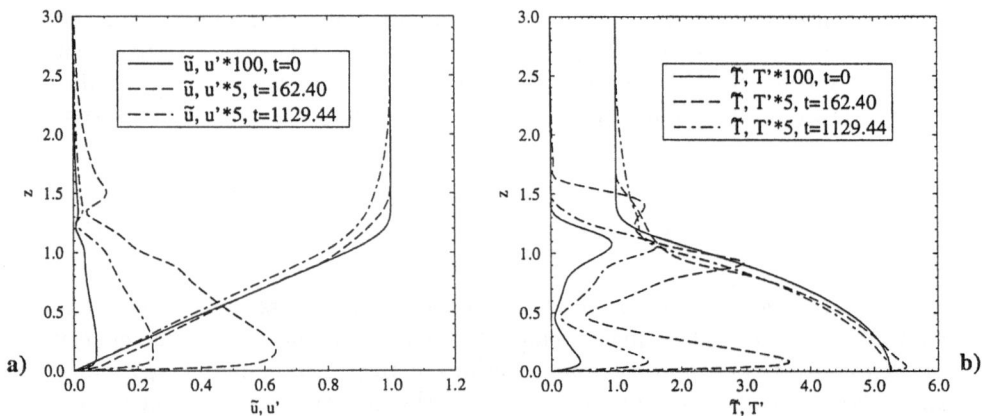

Fig. 5.4. Evolution of the mean velocity profiles (a) and the mean temperature profiles (b) and the corresponding RMS-fluctuations

(b) Independent of the grid resolution. A simulation with a resolution in x increased to $N_x = 96$ and another one with a resolution in z increased to $N_z = 141$ have been performed covering the first cycle of the cascade and essentially no differences in the flow behaviour have been observed.

6. Evolution of the Flow Field

A strong local pressure gradient develops near the wall after about $52T_w$, Fig. 6.1(a). It is located between a large-scale spanwise vortex, Fig. 6.1(b), at $z = 1.3$, near the critical layer, and the near-wall vorticity layer. It will be shown in the following subsection that the flow in this strong pressure gradient region corresponds to a viscous weak-shock solution. In the shock region high momentum fluid is convected towards the wall. Due to the strong vortex near the critical layer the flow in a frame of reference moving with the phase speed of the primary wave is deformed in a similar way as in a Laval nozzle. To visualize this in Fig. 6.2 a set of streamlines of the flow in a frame of reference moving with the phase velocity of the primary wave are shown together with the pressure field. Depending on the state upstream of the most narrow cross section a shock may develop. In a comparable three-dimensional flow situation a build-up of a similar embedded shocklet has not been observed so far. Before the spanwise vortices reach a sufficient strength they have already undergone a secondary instability and are three-dimensionally deformed [1] so that the situation becomes completely different.

Fig. 6.3 shows the evolution of the shape factor

$$H = \frac{\delta_1}{\delta_2}$$

and the skin-friction coefficient

$$C_f = \frac{2}{Re} \overline{\mu \frac{\partial u}{\partial z}}\bigg|_{z=0} \,,$$

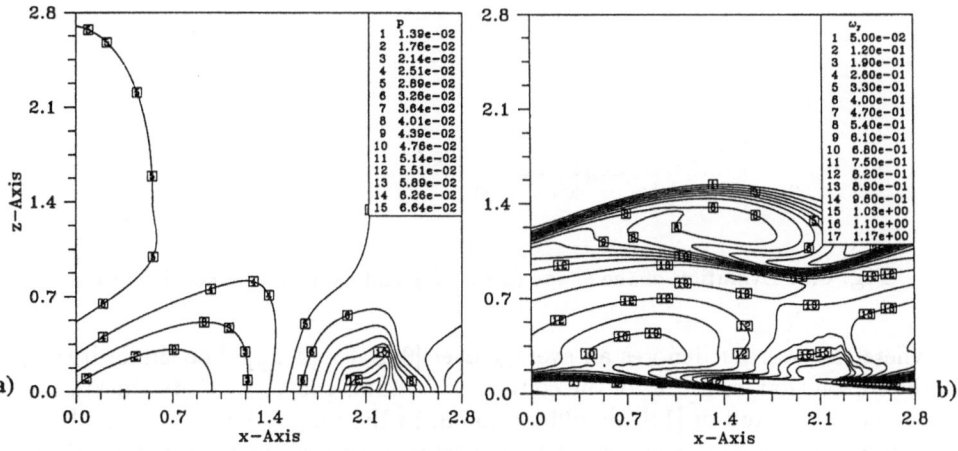

Fig. 6.1. Distribution of (a) pressure and (b) spanwise vorticity (only part of the total ω_y interval shown: $max = 12.73$, $min = -1.67$) at $t = 162.40$

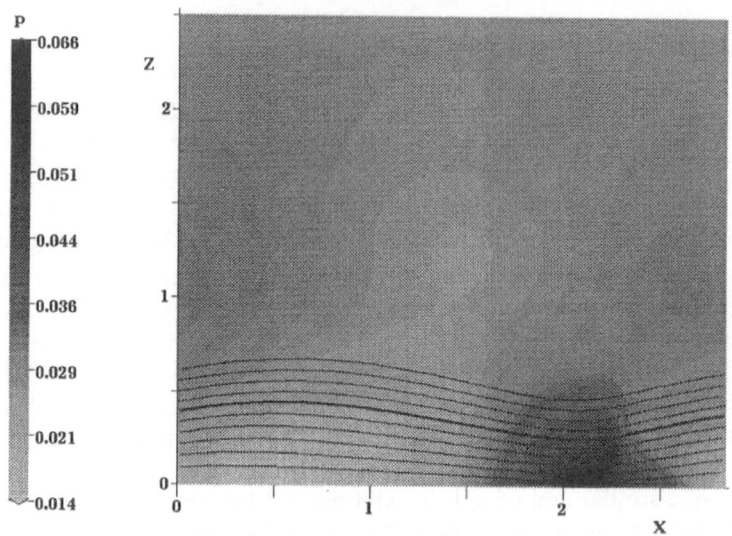

Fig. 6.2. Pressure distribution at $t = 162.40$ and instantaneous streamlines of the flow in the moving reference frame. The streamline which passes the maximum local pressure gradient is highlighted

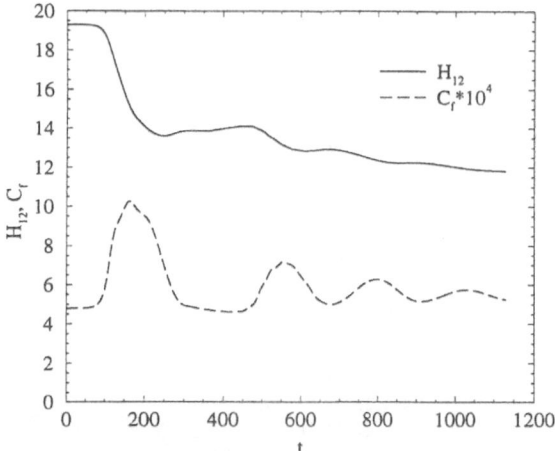

Fig. 6.3. Evolution of the shape factor H_{12} and skin friction coefficient C_f

where the overbar denotes an average over $[0, L_x] \times [0, L_y]$. The similarity of their evolution with that of the compressible mixing layer vorticity thickness, which is reported in [14], should be noted. In the mixing layer flow Lele [14] observed an energy exchange between spanwise vortices and mean flow. This oscillating energy exchange process causes the vortices to change their shape periodically. During this process a cascade of compression and decompression waves are generated which diminish the potential energy contained in the mi-

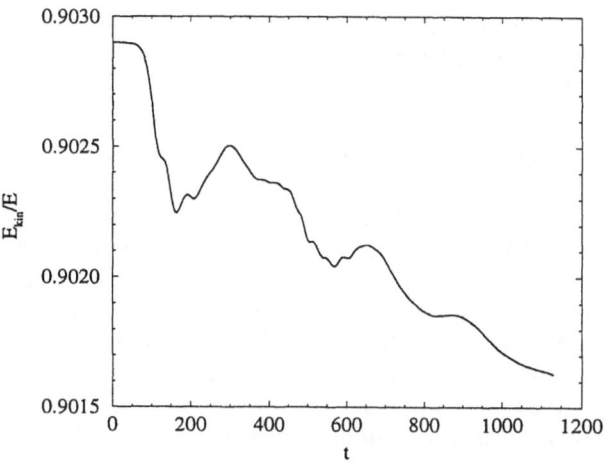

Fig. 6.4. Evolution of fraction of kinetic energy of the mean flow

xing layer. Though we did not investigate the drain of potential (acoustic) energy, we observe that the mean flow, besides its loss of total energy, periodically feeds kinetic energy into the higher Fourier modes. Also the fraction of kinetic energy contained in the mean flow decays while evolving towards an equilibration level like a damped oscillator, Fig. 6.4. Correspondingly the fraction of internal energy of the mean flow increases. A peak of this energy redistribution coincides with the appearance of the near-wall shocklet which will be identified in the following subsection. Since the spanwise vortices essentially fix the amount of kinetic energy in the Fourier mode 1 the energy cascade corresponds to a pulsation of the strength or shape of the spanwise vortex. This behaviour appears to be similar to the observations for the compressible mixing layer, where a connection between the oscillating changes in the vorticity thickness and the change of vortex shape has been reported [14].

7. Shock Identification

One possible criterion for identifying shocklets in homogeneous turbulence was used for example in [16]. The pressure ratio close before and behind the shock was compared with the pressure ratio from the Rankine-Hugoniot equations. These give the conservation of mass, momentum and total energy across a shock for a flow which is isentropic except for the shock discontinuity and thus are strictly valid only for inviscid flow (the shock solution is a weak solution of the Euler equations). For a viscous flow they are also correct (in a one-dimensional environment) in the limit far upstream and far downstream of the shock. That means that in contrast to the inviscid flow the domain to be considered not just surrounds the discontinuity but extends infinitely upstream and downstream. Away from the shock, however, the approximate one-dimensionality is strongly violated and makes the jump relations difficult to apply for shock verification. In our case the mean flow field changes on a scale which is comparable to the extent of the domain which is necessary to apply the Rankine-Hugoniot equations within the boundary layer. Hence it is difficult to estimate the correct pre- and post-shock states which give the according jumps of the variables of state. Another possibility is to compare the change of the flow variables along a streamline which passes the shock with a one-dimensional viscous shock solution (a genuine solution of the Navier-Stokes equations). Taking into account the diffusive terms the Rankine-Hugoniot equations can be extended as is shown e.g. in [26] to a system of ordinary differential equations along a streamline crossing the shock. Indeed these equations can be derived directly from the Navier-Stokes equations by introducing certain simplifications. In case of a low pre-shock Mach number M_1 an approximate analytic solution to these equations can be found. Although strictly valid only for $M_1^2 - 1 \ll 1$, this solution turns out to be applicable also at larger M_1 as remarked in [26]. The normalized distribution of the velocity across the shock is defined as

$$\Phi = \frac{q - q_2}{q_1 - q_2} ,$$

where q denotes the shock-normal velocity and the indices 1 and 2 indicate the pre-shock and post-shock conditions, respectively. For the idealized 1D weak-shock solution Φ can be written as

$$\Phi_{1D}(\xi) = \left[1 + \mathrm{Exp}\left(\frac{G}{\bar{C}}(q_1^* - q_2^*)\xi \right) \right]^{-1}$$

[26]. In this equation the following definitions are used :

$$\xi = \frac{x^*}{\lambda^*} , \quad \bar{C} = \sqrt{\frac{8kT_1^*}{\pi m_L^*}} , \quad \lambda^* = 2\nu_1^* \frac{1}{\bar{C}} , \quad G = \frac{\kappa + 1}{\frac{4}{3} + \frac{\kappa+1}{Pr}} , \quad q = \sqrt{u^2 + w^2} .$$

The asterisk denotes dimensional quantities, λ^* is the mean free path, ν^* the kinematic viscosity and $\bar{C}$ is the average molecular speed, which is calculated from the temperature T_1^*, the Boltzmann constant k and the molecular mass of air m_L^*. Thus the shock solution can be obtained from the known pre-shock quantities using the above equations.

As is evident from Fig. 6.2 the shock is oblique. Thus the change of the flow state has to be investigated along a streamline which crosses the shock. By interpolating all variables along the highlighted streamline of Fig. 6.2, which passes the shock where it is most intense, all variables of state along the streamline can be obtained. Table 7.1 lists their values in front of and behind the shock. From iso-contour plots of the pressure at two successive time levels it is found that the shock moves downstream almost with the phase speed of the primary wave, $c_{Ph} = 0.9126$. Thus the velocities which

Table 7.1. Pre-shock and post-shock flow variables from DNS on a streamline crossing the shock.

Variable	pre-shock	post-shock
x	2.3608	2.2433
z	0.2885	0.2650
p	0.0324	0.0500
T	5.1574	5.8409
u	-0.5672	-0.4448
w	-0.1254	-0.0693
ρ	0.2199	0.2998
θ^*	12.47°	8.86°
$\xi = \tau^*/\lambda^*$	-15.809	16.110

are relevant for the fluid state are to be taken in a reference frame moving downstream with c_{Ph}, which explains the negative entries for u in table 7.1. The arc-length along the streamline is denoted by τ. The deflection across the shock is small, $\Delta\theta = 3.61°$. The pre-shock Mach number is $M_1 = 1.28$ and the mean free path is calculated as $\lambda_1^* = 3.8031 \cdot 10^{-7}[m]$. The shock thickness from the DNS is $\delta_{Sh}^* = 31.82\lambda_1^*$. The displacement thickness of the unperturbed laminar flow is $\delta_1^* = 270.85\lambda_1^*$. A so-called velocity-gradient thickness is defined by [26]

$$\delta_{vg} = \frac{q_1 - q_2}{\frac{dq}{d\xi}\big|_{max}} \ .$$

For the DNS solution we obtain approximately $\delta_{vg}^*|_{DNS} = 22.78\lambda_1^*$ and for the ideal solution $\delta_{vg}^*|_{1D} = 14.78\lambda_1^*$.

A comparison of Φ_{1D} and Φ_{DNS} along the highlighted streamline of Fig. 6.2 is made in Fig. 7.1. Additionally the normalized pressure $\Pi(\xi) = (p(\xi) - $

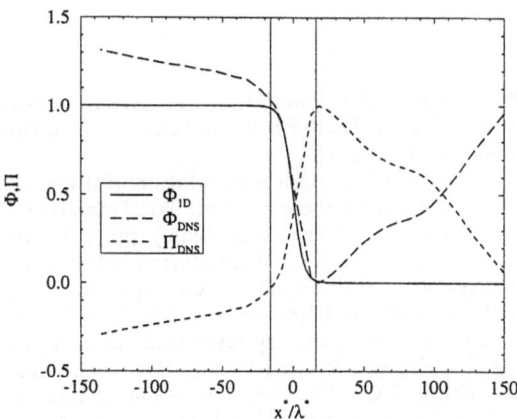

Fig. 7.1. Comparison of idealized 1D analytic weak shock solution Φ_{1D} and DNS solution Φ_{DNS} at $t = 162.40$

$p_2)/(p_1 - p_2)$ is shown. The vertical lines approximately mark the shock thickness as it is found from the DNS. Since the pre-shock and post-shock states are known from the DNS solution, the condition tested is the agreement of the slopes of Φ_{DNS} and Φ_{1D} across the shock. From Fig. 7.1 it is evident that the slopes are very close. Thus it is confirmed that the flow field can locally be described by a one-dimensional flow, which for the given data develops a weak viscous shock.

8. Conclusions

The two-dimensional nonlinear evolution of a second-mode instability wave in a laminar boundary layer at Mach 5 has been investigated. A characteristic phenomenon observed is a cascade of quasi-saturated states in which the kinetic energy of higher Fourier modes acquires alternating minima and maxima. During the first of these saturated states a weak shock near the wall has been observed. The shock was identified by a comparison with an analytic viscous weak shock solution. The energy cascade is interpreted as the relaxation of an energy imbalance of the total energy between the mean flow and the perturbation and between the fractions of kinetic and internal energy of the mean flow and the respective perturbation modes. Starting from an unstable situation the flow evolves towards a new equilibrium steady state in a similar way as a damped oscillator.

Acknowledgements
The authors wish to acknowledge the use of the post-processing package COMADI by H. Vollmers (DLR) and the linear stability solver COSMET by Dr. M. Simen (DLR).

References

1. Adams, N.A., 1993 Numerische Simulation von Transitionsmechanismen in kompressiblen Grenzschichten, DLR-FB 93-29 (also doctoral thesis, Technische Universität München) (in German).
2. Adams, N.A., Kleiser, L., 1993 Numerical simulation of transition in a compressible flat plate boundary layer, ASME FED-Vol. 151, Transitional and Turbulent Compressible Flows, Kral, L.D., Zang, T.A. (eds.), pp. 101–110.
3. Adams, N.A., Kleiser, L., 1993 Numerical simulation of fundamental breakdown of a laminar boundary-layer at Mach 4.5, AIAA-paper 93-5027.
4. Adams, N.A., Sandham, N.D., and Kleiser, L., 1992 A method for direct numerical simulation of compressible boundary-layer transition, NNFM 35, Vos, J.B., Rizzi, A., Ryhming, I. (editors), Vieweg-Verlag, Braunschweig, pp. 523–532.
5. Anderson, D.A., Tannehill, J.C., Pletcher, R.H., 1984 Computational Fluid Mechanics and Heat Transfer, Hemisphere Publ. Corp., New York.
6. Canuto, C., Hussaini, M.Y., Quarteroni, A., Zang, T.A., 1988 Spectral Methods in Fluid Dynamics, Springer-Verlag, Berlin.

7. Eißler, W., Bestek, H., 1993 Spatial numerical simulations of nonlinear transition phenomena in supersonic boundary layers, ASME FED-Vol. 151, Transitional and Turbulent Compressible Flows, Kral, L.D., Zang, T.A. (eds.), pp. 69–76.

8. Erlebacher,G., and Hussaini, M.Y., 1990 Numerical experiments in supersonic boundary-layer stability, Phys. Fluids A 2, pp. 94–104.

9. Fasel, H., Thumm, A., Bestek, H., 1993 Direct numerical simulation of transition in supersonic boundary layers: oblique breakdown, ASME FED-Vol. 151, Transitional and Turbulent Compressible Flows, Kral, L.D., Zang, T.A. (eds.), pp. 77–92.

10. Henningson, D., 1992 Private communication.

11. Herbert, T., 1988 Secondary instability of boundary layers, Ann. Rev. Fluid Mech. 20, pp. 487–526.

12. Kleiser, L., and Zang, T.A., 1991 Numerical simulation of transition in wall-bounded shear flows, Ann. Rev. Fluid Mech. 23, pp. 495–537.

13. Koch, W., 1992 On a degeneracy of temporal secondary instability modes in Blasius boundary-layer flow, J. Fluid Mech. 243, pp. 319–351.

14. Lele, S.K., 1989 Direct numerical simulation of compressible free shear flows, AIAA-paper 89-0374.

15. Lele, S.K., 1992 Compact finite difference schemes with spectral-like resolution, J. Comp. Phys. 103, pp. 16–42.

16. Lee, S., Lele, S.K., Moin, P., 1991 Eddy shocklets in decaying compressible turbulence, Phys. Fluids A 3(4), pp. 657–664.

17. Mack, L.M., 1984 Boundary-layer linear stability theory, in Special Course on Stability and Transition of Laminar Flow, AGARD Report No. 709, pp. 3-1-3-81.

18. Ng, L., and Erlebacher, G., 1992 Secondary instabilities in compressible boundary layers, Phys. Fluids A 4, pp. 710–726.

19. Pruett, C.D., Chang, C.L, 1993 A comparison of PSE and DNS for high-speed boundary-layer flows, ASME FED-Vol. 151, Transitional and Turbulent Compressible Flows, Kral, L.D., Zang, T.A. (eds.), pp. 57–67.

20. Pruett, C.D., and Zang, T.A., 1992 Direct numerical simulation of laminar breakdown in high-speed, axisymmetric boundary layers, Theor. Comp. Fluid Dyn. 3, pp. 345–367.

21. Sandham, N.D., and Adams, N.A., 1993 Numerical simulation of boundary-layer transition at Mach two, Applied Scientific Research 51, pp. 371–375.

22. Sandham, N.D., Reynolds, W.C., 1991 Three-dimensional simulations of large eddies in the compressible mixing layer, J. Fluid Mech. 224, pp. 133–158.

23. Stewartson, K., 1964 The Theory of Laminar Boundary Layers in Compressible Fluids, Oxford University Press, Amen House, London, pp. 33–60.

24. Thompson, K.W., 1987 Time dependent boundary conditions for hyperbolic systems, J. Comp. Phys. 68, pp. 1–24.

25. Thumm, A., Wolz, W., and Fasel, H., 1990 Numerical simulation of spatially growing three-dimensional disturbance waves in compressible boundary layers, in Laminar-Turbulent Transition, D. Arnal and R. Michel (eds.), Springer, pp. 303-308.

26. Vincenti, W.G., Kruger C.H., 1965 Introduction to Physical Gas Dynamics, John Wiley & Sons Inc., pp. 412–424.

27. Wray, A.A., 1986 Very low storage time-advancement schemes, Internal Report, NASA Ames Research Center, Moffet Field, CA.

Inertial Convection in Turbulent Rayleigh-Bénard Convection at Small Prandtl Numbers

G. Grötzbach and M. Wörner

KfK, Institut für Reaktorsicherheit, Postfach 3640, D-76021 Karlsruhe, Germany

Summary. Inertial convection is a two-dimensional flow mechanism effectively transporting heat. It was predicted theoretically to exist in Rayleigh-Bénard convection in liquid metals at Rayleigh numbers above 10^4. In numerical simulations it was found in this range and at smaller Rayleigh numbers. Here, the method of direct numerical simulation is used to study the details of inertial convection in the fully turbulent regime in liquid sodium, $Pr = 0.006$, at $Ra = 24,000$. Application of a semi-implicit time integration scheme and of a fast elliptic solver make such simulations possible. The results show the inertial convection still exists in this range of Ra, but it occurs only locally and over certain time intervals in that areas in which the flow is roughly two-dimensional at large scales. In an aperiodic oscillation this flow mechanism is in competition to a more irregular, three-dimensional flow state. The Nusselt numbers at both walls oscillate with the changing flow structure. They show larger values during the occurrence of the inertial convection and smaller ones with the irregular flow.

1. Introduction

In new designs of liquid metal cooled nuclear reactors the removal of decay heat in accident situations is achieved by pure natural convection. Experiments are performed in scaled reactor models with water to analyse the corresponding flow phenomena and temperature transients [1]. The interpretation of the experiments and the transfer of the results to real reactor conditions is mainly done by computer codes [1,2]. The turbulence models used in such codes have to be adapted to be applicable to purely buoyant flows and to liquid metals. One effort in this context is to determine turbulence data in simple flows to calibrate existing models, e.g. from experiments with jets [3] and heated spheres [4], or from direct numerical simulations for Rayleigh-Bénard convection [5,6]. From those simulation results it is deduced that common transport equation models are incomplete in a sense that terms are neglected which turn out to be dominant in pure natural convection. For some of the missing terms no models are known. Therefore, a complementary effort is to gain a better understanding of the physical mechanisms in natural convection of liquid metals to form a basis for the development of improved turbulence models.

Rayleigh-Bénard convection is the upward directed convective heat transfer through a large horizontal fluid layer enclosed between two plane walls. For common fluids with Prandtl numbers $Pr = \nu/\kappa$ (ν = viscous diffusivity,

κ = thermal diffusivity) around one the flow development and the mechanisms are well known [7]. The flow develops from no motion at small heating rates, that is at small Rayleigh numbers ($Ra < Ra_{cr} = 1,708$, with Ra = g $\beta \Delta T_w D^3 / (\nu/\kappa)$, g = gravity, β = volume expansion coefficient, ΔT_w = temperature difference between both walls, D = distance between walls), through steady two-dimensional and three-dimensional flows, through time-dependent three-dimensional flows, to a fully turbulent flow at large heating rates. Some uncertainty exists regarding the turbulent regime. This was subdivided in a regime with soft and hard turbulence, each one showing its own statistical features [8], but recent simulation results showed that the differences between hard and soft turbulence occur only as a consequence of different aspect ratios of the channels used [9]. Rayleigh-Bénard convection is per definition the heat transfer through an infinite horizontal fluid layer, and therefore a kind of hard turbulence found at small aspect ratios is not a separate flow status of this flow.

Liquid metals have very small Prandtl numbers; their thermal diffusivity is much larger than their viscous diffusivity. Therefore, the heat transfer by molecular conduction is dominant up to much larger heating rates than in fluids with small thermal diffusivity. The transition from no motion to irregular flow extends in experiments only over a range of Rayleigh numbers being apart less than a factor of two [7]. This very different behaviour was also predicted theoretically [10]. Thus one finds in liquid metals already at small Ra very irregular or turbulent convection, but the heat transfer rate only shows considerable convective contributions at much larger Ra ($Ra > 10^4$), see e.g. the experiments of [11,12]. In this range a fly-wheel type of convection, called inertial convection, is predicted by two-dimensional methods to occur [13], but experimentally indications for the inertial convection are found even at smaller Rayleigh numbers [14]. With three-dimensional simulations of Rayleigh-Bénard convection in liquid sodium, which gave the first field data on the velocity field in this type of flow, we could show that inertial convection definitely exists at $Ra = 3,000$ and $6,000$ and that the flow becomes more irregular with increasing Rayleigh number [6,15]. Indicators for the inertial convection seem to be small secondary vortices near the walls between the larger rolls and a more regular, roughly two-dimensional arrangement of the rolls.

In this paper we use the method of direct numerical simulation to analyse Rayleigh-Bénard convection in liquid sodium, $Pr = 0.006$, at a Rayleigh number of 24,000. The simulation model TURBIT [16] had to be extended by a semi-implicit time-integration scheme for the energy equation to allow for these simulations [6,17]. The simulation results are analysed regarding the large scale features of the flow, especially regarding the existence of the inertial convection when three-dimensional methods are used in this range of Rayleigh numbers, and regarding the dynamics of the flow. The mechanisms of the special flow features found are explained and their influences on the heat transfer capabilities are investigated.

2. Simulation Method

The simulation model used is the TURBIT code [16]. It is based on the complete three-dimensional time-dependent conservation equations for mass, momentum, and energy for a Newtonian fluid. The fluid is considered to be incompressible; for the buoyancy term the validity of the Boussinesq approximation is assumed. The equations are made dimensionless by the length scale D, by the velocity scale $(g\beta\Delta TD)^{1/2}$, and by the temperature scale $\Delta T = T_{w1} - T_{w2}$. The code uses a finite volume scheme on a staggered grid. The pressure is calculated by Chorin's projection method. The resulting Poisson equation can be solved efficiently in simple geometries like in infinite plane channels by direct methods [18].

The time integration scheme used originally in the code is the second order explicit Euler-leapfrog scheme. Application of this scheme to natural convection in liquid metals leads to enormous CPU-time requirements. The stability criterion of this explicit scheme can be written by using Einstein's summation convention [19]:

$$\Delta t \leq \left(\frac{|u_i|_{max}}{\Delta x_i} + 4 \frac{Max(\nu,\kappa)}{\Delta x_i^2} \right)^{-1} \tag{2.1}$$

where u_i , $i = 1,2,3$, denotes the components of the velocity vector and Δx_i denotes the grid widths. The temperature field in natural convection of liquid metals is governed by the large thermal conductivity allowing only for large scale structures in the temperature field and for thick thermal boundary layers. In contrast, the velocity field has not only large scales but also very small spatial structures and very thin viscous boundary layers near walls. For direct numerical simulations of turbulence it is essential to choose grids which resolve all relevant length scales of the flow. Therefore, the velocity field requires very fine grid widths. Thus, the diffusion terms become dominant in the stability criterion 2.1 and liquid metals require very small time steps for numerical stability of the thermal diffusion. These time steps are much smaller than those required to ensure stability of even the highest frequencies in the convective time scales. Substantially larger time steps can be achieved by time integration schemes which do not have the thermal diffusivity in the stability criterion. Thus, the diffusive terms of the energy equation have to be treated implicitly, whereas all other terms may still be treated explicitly.

Recently the code was extended by two semi-implicit time integration schemes [6,17] which can be used alternatively. They were selected to be suitable for diffusion dominated problems and to be consistent with the explicit scheme used for the velocity field. Both semi-implicit schemes treat the diffusive terms $L = \kappa\nabla^2 T$ by the implicit Crank-Nicolson method, CN, whereas for the non-linear convective terms $N = \underline{u}\nabla T$ the explicit Adams-Bashforth, AB, Eq. 2.2, or the Leapfrog scheme, LF, Eq. 2.3, is used, respectively:

$$\frac{T^{n+1} - T^n}{\Delta t} = \frac{1}{2}(3N^n - N^{n-1}) + \frac{1}{2}(3L^{n+1} + L^n) \tag{2.2}$$

$$\frac{T^{n+1} - T^{n-1}}{2\Delta t} = -N^n + \frac{1}{2}(L^{n+1} + L^{n-1}) \tag{2.3}$$

Here n denotes the time level. In the code both schemes are realized in a way that, like the fully explicit scheme, both are started by an Euler step, and after about 40 to 60 time steps an averaging step is used to damp spurious oscillations which might develop in the solution.

A von Neumann stability analysis for the linearized problem indicates the LFCN scheme should be stable for Courant numbers $C = u_{max}\Delta t/\Delta x_{min} < 1$ and for any value of the diffusion number $D = \kappa\Delta t/(\Delta x_{min}^2)$, whereas the stability criterion of the ABCN scheme depends on the diffusion number and always needs some diffusion for stability. Practical tests showed that both schemes become unstable in some applications for diffusion numbers of about six and greater. To avoid these stability problems a maximum value for the diffusion number was chosen, $D_{max} = 4$. This gives an upper limit for the time step width calculated from Eq. 2.1 with $\kappa = 0$.

The sets of linear equations resulting from the implicitly treated diffusion terms are those from a Helmholtz equation. Thus, in principle these sets can be solved with general Helmholtz (or most Poisson) solvers. With direct numerical simulations, the coefficients of the Helmholtz equation are space dependent, but their space dependence can be separated; therefore very efficient direct Poisson solvers, like from [18], can be used. More serious difficulties are due to the manifold of combinations of wall conditions used in the thermal diffusion term (von Neumann and / or Dirichlet conditions), which are usually not all available with common fast solvers. Correspondingly, the boundary conditions of the package from [18] were extended. As a result of using modified direct Poisson solvers to solve the set of equations resulting from the semi-implicit time integration scheme, the additional CPU-time is only 10 - 20% of that of the fully explicit scheme. The storage requirement is about the same as that of the fully explicit scheme. As the time step can be increased, e.g. for sodium at least by one order of magnitude, this method provides the efficiency required to make these simulations possible.

3. Case Specifications and Initial Data

For simulation of Rayleigh-Bénard convection a plane channel should be considered which is infinite in both horizontal directions. This is achieved in the simulation by using periodic boundary conditions in both horizontal directions. The periodicity lengths are X_i, with $i = 1, 2$, Fig. 3.1. They have to be chosen large enough to resolve all large scale phenomena. In earlier simulations for common fluids with Prandtl numbers around $Pr = 1$ these lengths turned out to have a strong influence on the heat transfer through the fluid layer [9,20]. Even our recent simulations with $X_i/D = 7.92$ [15,21] seem not to fulfil all requirements of a complete statistical decoupling in the horizontal directions. In liquid metals the large thermal diffusivity is responsible for an

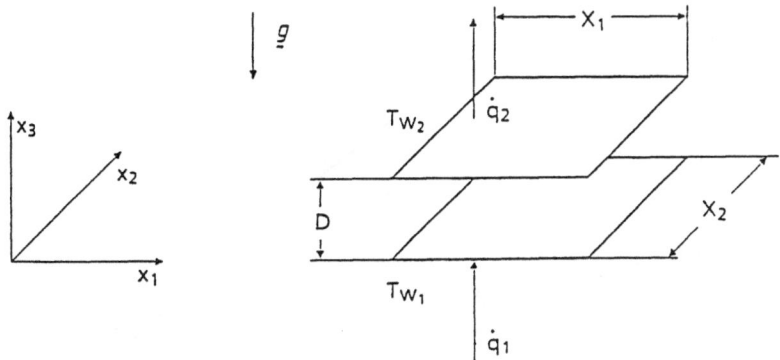

Fig. 3.1. Geometry and definitions for the Rayleigh-Bénard problem

even farer horizontal coupling and therefore for larger scales, but there are no data available from experiments to select adequate periodicity lengths. From simulations on a coarse grid for $Ra = 6,000$ using horizontal extensions X_i/D from 4 to 16 we found, that the large scale features of the flow simulated for aspect ratios of 8 and greater are almost independent on the aspect ratio [6]. Therefore we used $X_i = 8$ for the convection in sodium also at the Rayleigh number considered here, $Ra = 24,000$, which is in the fully turbulent regime. In direct numerical simulations not only the large scales have to be resolved adequately, but also the small scales. The smallest scales in liquid metals occur in the velocity field. The size of these can be calculated on several ways [20], e.g. by means of the Kolmogorov length scale and by the Grashof analogy [22]. The latter means the statistical features of the small scales in the velocity field were found to be similar in different fluids for comparable Grashof numbers, $Gr = Ra/Pr$, and therefore the mean dissipation profile needed to calculate the Kolmogorov length is similar. The profile can be taken from simulations or experiments for other fluids. The finest grids using $N_1 = N_2 = 250$ mesh cells in each horizontal direction, Table 3.1., are finer than the Kolmogorov scale at any position in the channel.

Table 3.1. Case specifications and simulation times on a SIEMENS/FUJITSU VP400-EX. Each case is started from that one on the line above

Ra	Gr	$N_1 = N_2$	N_3	Δx_{3w}	T_{max}	N_t	$CPU/$ $VP400$
12,000	$2 * 10^6$	128	19	0.03	358.8	16,000	
12,000	$2 * 10^6$	160	25	0.02	400.1	19,040	
12,000	$2 * 10^6$	160	31	0.01	410.1	22,000	
12,000	$2 * 10^6$	200	35	0.008	411.5	22,640	
12,000	$2 * 10^6$	250	39	0.005	444.4	61,440	55h
24,000	$4 * 10^6$	250	39	0.005	471.5	84,000	60h

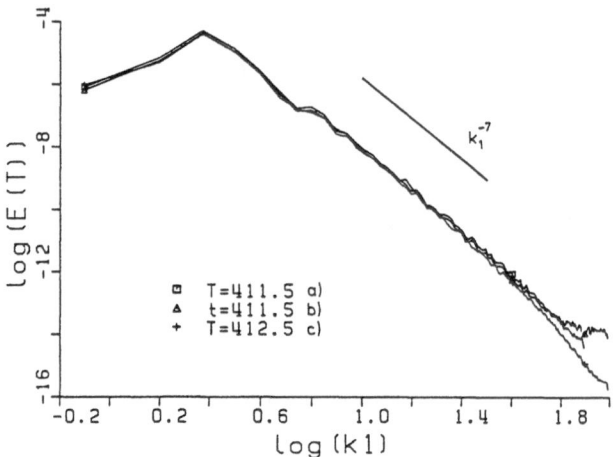

Fig. 3.2. One-dimensional energy spectra of temperature fluctuations over wave number k_1 at channel mid plane, $x_3 = 0.5$, for $Ra = 12,000$, a) coarse grid, $200^2 * 35$, $t = 411.5$, b) same result interpolated to fine grid, $200^2 * 39$, c) fine grid, $250^2 * 39$, after integration to $t = 412.5$

Linear wall approximations are applied for the diffusive terms in the mesh cells next to the walls. Therefore, the viscous and thermal boundary layers have to be resolved sufficiently by the grid. In liquid metals the viscous layer is the finer one. The Grashof analogy is used to specify the vertical resolution by the vertical grid width Δx_{3w} near the wall. The criterion of [20] to use at least 2 to 3 cells inside the boundary layer is sufficient from an engineering point of view, e.g. to get sufficient accuracy in the global energy balance or in the Nusselt number, but in studying terms from turbulence models based on second order transport equations, a much finer resolution is required near the walls for analysing purposes. In using a non-equidistant vertical distribution the number of vertical nodes can be limited to $N_3 = 39$, Table 3.1.

Initial conditions are gained from the last run of a series of simulations for $Ra = 12,000$, Table 3.1. The first case used a coarse grid, zero velocities, and random temperature fluctuations superimposed to a schematic piecewise linear vertical profile of the mean temperature which roughly represents a realistic initial vertical mean energy distribution. The other simulations on the finer grids for $Ra = 12,000$ are started from the simulation results for the same Rayleigh number gained from the next coarser grid using parabolic spatial interpolation.The interpolated data on the fine grid show increased fluctuation amplitudes at small scales, Fig. 3.2, but after integration over a time period between 0.1 and 1 dimensionless time units the spectrum is as steep again as the theoretically expected k_1^{-7} slope. For $Ra = 24,000$ the simulation results for the smaller Rayleigh number on the finest grid are used as initial data; this is possible without any transformation of the data because with the scaling, especially of the velocities, chosen here the dimensionless velocity values do not change with Ra at large Rayleigh numbers. Both latter methods of creating initial conditions turned out to allow for reductions of CPU-times by one order of magnitude or more compared to runs starting from more or less random data on the required fine grid. The CPU-times given in Table 3.1. are those on the finest grids only. The number of time steps N_t is consecutively counted through the complete sequence of runs.

4. Results

4.1 Verification

Problems in verifying these simulation results occur due to the complete lack of detailed turbulence data on the velocity field in Rayleigh-Bénard convection of liquid metals and due to the existence of only very few data on the temperature fields. In [6,23] all available data are used to perform a verification. As the uncertainties of the experimental results are rather large, additional recalculations of experiments for air [22] and the recalculation of GAMM benchmarks [17] were used to verify the implementation of the semi-implicit time integration scheme.

A very crude verification, especially of the periodicity lengths chosen, can be deduced from horizontal cuts through the instantaneous temperature fields, Fig. 4.1. Despite a fully turbulent state these figures show band-like structures. According to [24] this indicates the existence of remainders of regular vortex systems. The wavelength of this structure is about 2.7. This is,

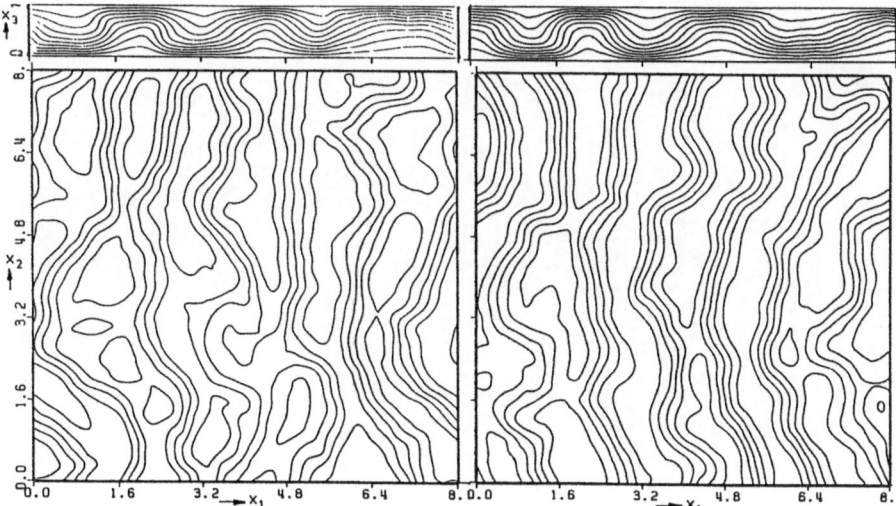

Fig. 4.1. Contourlines of instantaneous temperature fields at $Ra = 24,000$; isoline increment 0.1, top vertical cuts at $x_2 = 6.816$, bottom horizontal cuts at $x_3 = 0.5$, left $t = 478.7$, right $t = 488.3$

as to be expected, above the value at the critical Rayleigh number [10] and it shows that the periodicity lengths chosen can represent about 3 vortex pairs. This resolution for large scale structures is better than in other published simulations for other fluids, e.g. in [9,25]. From the vertical cuts it gets obvious that convection only weakly contributes to the heat transfer. Accordingly, the Nusselt number analysed is with 1.36 near one. This is in reasonable agreement with experiments [11,12].

4.2 Flow Mechanisms and Dynamics

According to current knowledge the Rayleigh number $Ra = 24,000$ of this simulation is in the turbulent regime. Nevertheless the instantaneous temperature fields at both times give large band like structures which indicate the existence of irregular roll systems, Fig. 4.1. Vortex bands are indeed found in the velocity field, for example in u_3, Fig. 4.2. There exist three large vortex pairs which are locally disturbed in a three-dimensional manner. In the simulations for smaller Ra it was found that the inertial convection coexists with small secondary currents near the walls [6,24]; here those parts of the

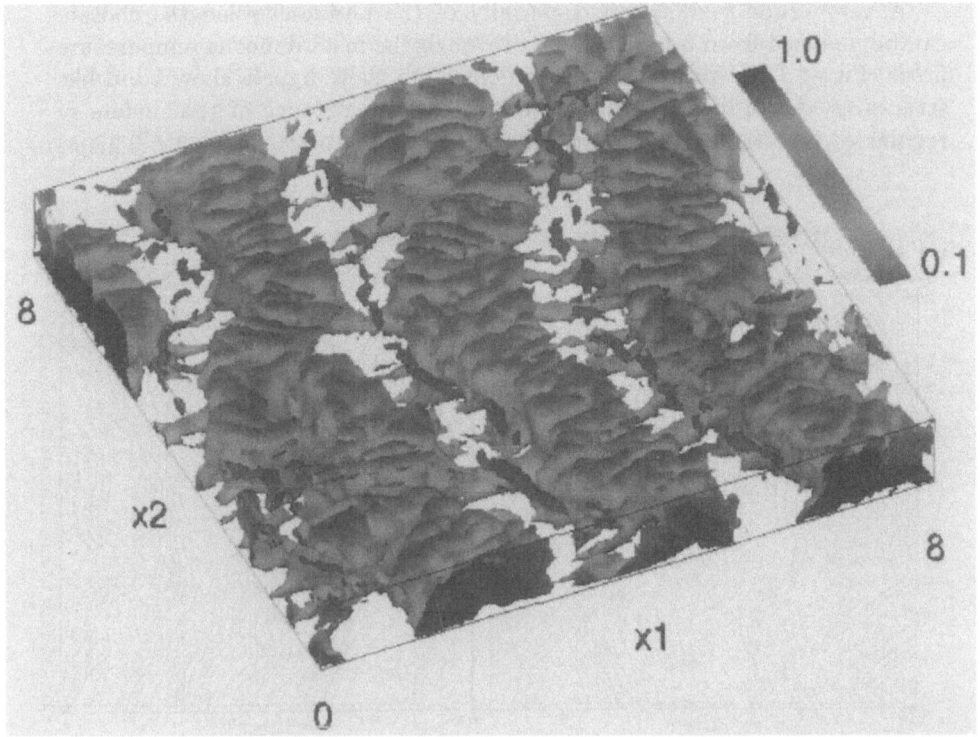

Fig. 4.2. Isosurface for a small positive value of the vertical velocity, $u_3 = 0.05$, colour code for the temperature. $Ra = 24,000$, $t = 488.3$. Inside the isosurface the flow is upward, outside it is mainly downward

secondary currents which have positive vertical velocities are found as small clouds in the downdraft area. Thus, there are indicators for the existence of remainders of the inertial convection. Indeed, the analysis of a time sequence of such figures in the form of a computer generated movie shows there is always a competition between the formation of two-dimensional structures, which are necessary to form inertial convection [13], and their destruction by meandering into more irregular, three-dimensional ones.

A more detailed search for the inertial convection is performed in that areas and at that times in which a pronounced two-dimensional character of the flow field is found. Scanning through the velocity data for $t = 488.3$ shows nearly everywhere a highly irregular, turbulent flow field, Fig. 4.3. However, only near $\{x_1 = 2, x_2 = 6.8\}$ one pair of vortices is found that shows an arrangement in the channel which is symmetrical to the mid plane and that has a regular radial velocity distribution. It extends from about $x_1 = 1.2$ to about $x_1 = 3$, and in the axial direction it has a length between one and two D. The life time is only a few time units. The corresponding profile for the vertical velocity component in that plane confirms, this vortex pair has a velocity profile which is linear over a larger area. That means, each of these vortices rotates at least in the inner part like a solid body. Outside the range

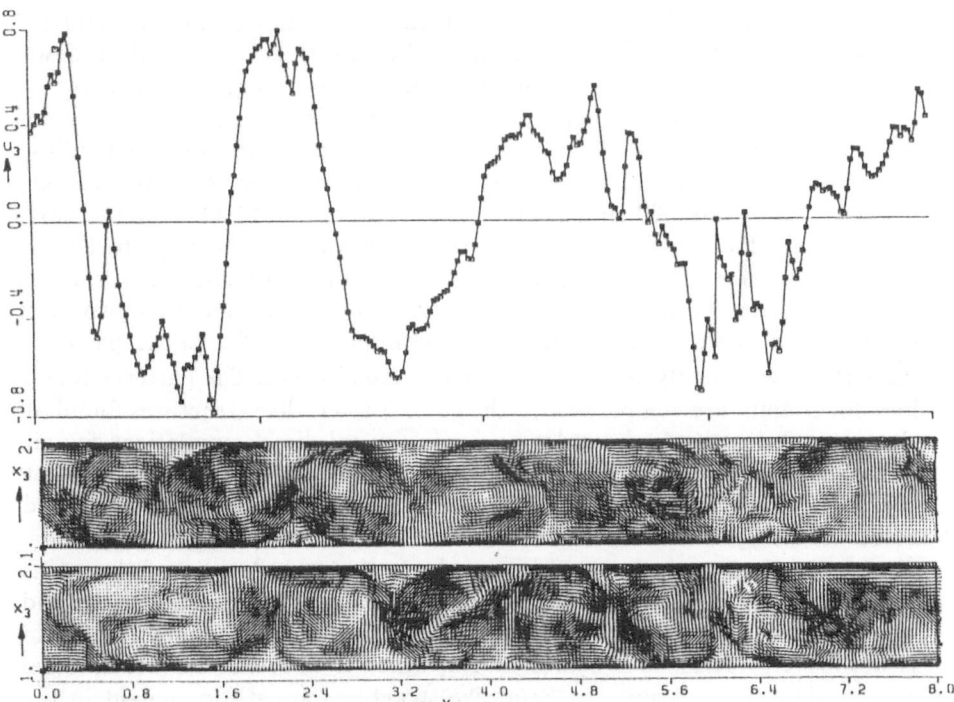

Fig. 4.3. Instantaneous profile of vertical velocity u_3 at $\{x_2 = 6.81, x_3 = 0.5, t = 488.3\}$, top, and vector plots for (u_1, u_3) at $\{x_2 = 6.81, t = 488.3\}$, middle, and at $\{x_2 = 2.81, t = 488.3\}$, bottom

of these two vortices, that is below $x_1 = 1.5$ and above $x_1 = 2.5$, the velocity profile is highly irregular or turbulent, as it is to be found in all other parts of the channel.

The corresponding temperature field shows in the horizontal sections wide areas with roughly two- dimensional character, that is not only in the area around $\{x_1 = 2, x_2 = 6.8\}$, but also in a much larger surrounding, Fig. 4.1. In the vertical sections one finds according to the small value of the Nusselt number only slight distortions by convection, except for $t = 488.3$ in that area in which the inertial convection was found in the velocity field. There we find stronger distortions of the isolines toward the upper wall in the centre of the vortex pair and towards the lower wall in both outer areas of the vortex pair. This means the inertial convection is locally and for short times responsible for an intensive vertical heat transfer by convection. With these results the aperiodic production and destruction of the inertial convection can be explained like follows: Vortices rotating like solid bodies obey no internal dissipation. Therefore, the rotation speed grows to large values, and thus also the convective heat transfer is augmented. Increasing rotation speeds lead to shear instabilities near the walls and therefore to irregular, three-dimensional structures. These are more dissipative and do not allow for such large velocities. The rotation speed is decreasing and the convective heat transfer is reduced. The molecular heat transfer becomes dominant. It filters off the small scale fluctuations produced by the irregular, turbulent flow. Thus, the flow relaminarizes locally and starts to form regular rolls again.

Considering the small scales of the velocity field, an other phenomenon is found in liquid metals. Thin spoke pattern like structures exist extending across the rolls, Fig. 4.2. These spoke patterns were not observed at the Rayleigh numbers $3,000$ and $6,000$ analysed earlier [6, 24]. They do not only exist near the lower wall, but also near the upper wall as is indicated by a very rugged isosurface. These structures are formed by thin bands with upward moving fluid in the downdraft area near the lower wall and by downward moving fluid in the updraft area near the upper wall. The spoke patterns show the flow is highly three-dimensional at small scales. The patterns found here for sodium are comparable to the spoke-pattern like structures found at the edge of the viscous boundary layer in several direct simulations for air at comparable Grashof numbers [21,24,25]. In general, the velocity field at this Rayleigh number obeys much smaller scales as the temperature field and correspondingly it has also smaller time scales.

A statistical analysis at midplane shows that the flatness of the vertical velocity component is near 3, which indicates Gaussian distributions and therefore turbulent features, whereas the flatness for the temperature field is around 1.7. Thus, it is nearer to 1.5, which is the value for a sinusoidal distribution. This means, the band-like structures are still dominant in the temperature field, despite they can be identified in the contourline presentations of horizontal cuts through the temperature field, Fig. 4.1, only over short times and only locally.

4.3 Heat Transfer Statistics

The integral heat transfer capabilities are analysed by means of a time history for plane averaged Nusselt numbers at both walls, Fig. 4.4. As these data are not time-averaged, the consequences of the irregular oscillations found in the flow become obvious. The Nusselt numbers change within a 10%-band. The corresponding oscillations are very slow. Therefore, immense averaging times are required in experiments to determine accurate results for the Nusselt numbers.

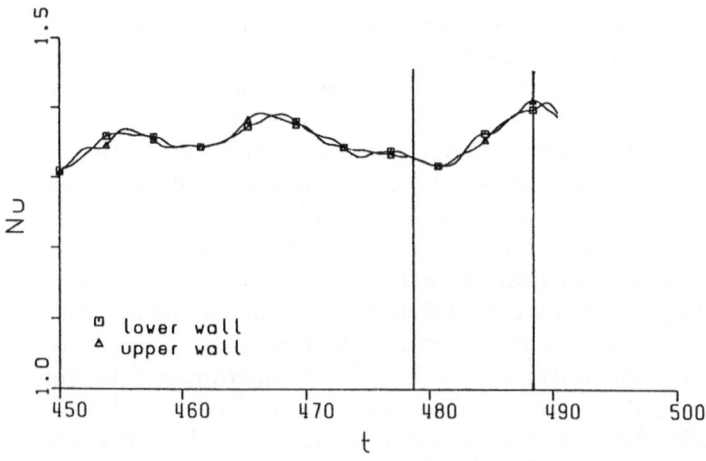

Fig. 4.4. Time dependent Nusselt numbers at lower and upper walls averaged horizontally. $Ra = 24,000$. The vertical lines mark the times for which the temperature fields are given in Fig. 4.1

Correlating the time-dependent Nusselt numbers to the instantaneous temperature fields, Fig. 4.1, shows the wall heat fluxes always reach a larger value at those times at which the flow field is more regular or two-dimensional. They are large especially in those time intervals in which the inertial convection occurs; compare the times used in Figs. 4.1 and 4.3. The Nusselt numbers exhibit smaller values at times with more irregular or three-dimensional flow fields. From Fig. 4.3 it can be deduced that the inertial convection is a flow with rather large rotation velocities and that it occurs only over short times extending over small parts of the channel. This means the inertial convection is a type of flow which has a strong influence on the overall heat transfer in the channel. It maximises the convective heat transfer.

5. Conclusions

The method of direct numerical simulation is used to investigate Rayleigh-Bénard convection in liquid metals. The large thermal conductivity of these fluids requires an implicit treatment of the thermal diffusion term. This, and using extended direct Helmholtz solvers, ensures efficient simulations and make such simulations possible at all. The long time periods to be simulated need additional measures like using not too fine grids for the development phase of the flow and interpolation of results from coarser grids to finer grids, on which the final time interval is simulated which is finally used for analysis.

The inertial convection was predicted by theoretical two-dimensional methods to occur in Rayleigh-Bénard convection of liquid metals at Rayleigh numbers above 10^4. With former direct simulation results it was shown that in the expected range of Ra indications for this flow mechanism, like a locally two-dimensional flow field and small secondary currents, are found, but that pronounced forms of inertial convection exist only at smaller Rayleigh numbers. This is in agreement with indications from experiments. Simulation results for $Ra = 24,000$ are analysed here in more detail regarding the instantaneous local velocity distributions. Small areas are found in the channel only for short times in which large scale vortex pairs exist which rotate like a solid body. These vortices are found only in small parts of those regions in which the flow is roughly two-dimensional and in which small counter rotating secondary vortices occur near the walls. This means, local two-dimensionality and secondary vortices are no sufficient indicators for the inertial convection because these indicators occur also outside those areas. The velocity field also shows the inertial convection at these Ra is superimposed by highly three-dimensional spoke patterns. These patterns consist of small and fast scales which clearly show the highly turbulent nature of these flows. Thus, with increasing Rayleigh number the inertial convection does not vanish abruptly, but it is still occurring locally and for short time periods. It is in competition to three-dimensional flow structures and forms an aperiodic oscillation. Large computational domains are required to find such changes in flow structures. In analysing the integral heat transfer through the fluid layer it is found, the predominantly two-dimensional vortices, that is the inertial convection, transport heat much more efficient than those flow structures that are at large scales more three-dimensional and irregular.

References

1. H. Hoffmann, D. Weinberg, W. Baumann, K. Hain, W. Leiling, K. Marten, H. Ohira, G. Schnetgöke, K. Thomauske, Scaled model studies of decay heat removal by natural convection for the European fast reactor, Proc. Sixth Int. Topical Meeting on Nuclear Reactor Thermal Hydraulics, Oct. 5-8, 1993, Grenoble, France, Vol. 1, pp. 54-62.
2. H. Ninokata, Advances in computer simulation of fast breeder reactor thermalhydraulics, Proc. Conf. on Supercomputing in Nuclear Applications, Mito, Japan, March 12-16, 1990, pp. 80-85.
3. J.U. Knebel, L. Krebs, U. Müller, Experimental investigations on the velocity and temperature field in axisymmetric jets of sodium, Ninth symposium on turbulent shear flows, Kyoto, Japan, August 16-18, 1993.

4. D. Suckow, Experimentelle Untersuchung turbulenter Mischkonvektion im Nachlauf einer beheizten Kugel, Dr. thesis, University Karlsruhe, KfK 5174, 1993.
5. M. Wörner, G. Grötzbach, Turbulent heat flux balance for natural convection in air and sodium analysed by direct numerical simulations. Fifth Int. Symposium on Refined Flow Modelling and Turbulence Measurements, Sept. 7-10, 1993, Paris, IAHR-Proc. pp. 335 - 342.
6. M. Wörner, Direkte Simulation turbulenter Rayleigh-Bénard-Konvektion in flüssigem Natrium, Dr. thesis, University Karlsruhe, KfK 5228, 1994.
7. R. Krishnamurti, Some further studies on the transition to turbulent convection, JFM 60, 1973, pp. 285-303.
8. F. Heslot, B. Castaing, A. Libchaber, Transition to turbulence in helium gas, Phys. Rev. A36, 1987, pp. 5870-5873.
9. S. L. Christie, J. A. Domaradzki, Numerical evidence for nonuniversality of the soft/hard turbulence classification for thermal convection, Phys. Fluids A 5, 1993, pp. 412-421.
10. R.M. Clever, F.H. Busse, Convection at very low Prandtl numbers, Physics of Fluids, A2, 1990, pp. 334-339.
11. V. Kek, Bénard Konvektion in flüssigen Natriumschichten, Dr. thesis, University Karlsruhe, KfK 4611, 1989.
12. V. Kek, U. Müller, Low Prandtl number convection in layers heated from below, Int. J. Heat Mass Transfer 36, 1993, pp. 2795-2804.
13. R.M. Clever, F.H. Busse, Low Prandtl number convection in a layer heated from below, JFM 102, 1981, pp. 61-74.
14. A. Chiffaudel, S. Fauve, B. Perrin, Viscous and inertial convection at low Prandtl number: Experimental study, Europhysics Letters 4, 1987, pp. 555-560.
15. G. Grötzbach, M. Wörner, Flow mechanisms and heat transfer in Rayleigh-Bénard convection at small Prandtl numbers, ERCOFTAC-Workshop on DNS and LES, Guildford, UK., March 27-30, 1994, Proc. by Kluwer, Dordrecht, Ed. P. Voke, L. Kleiser, J.P. Chollet (1994).
16. G. Grötzbach, Direct numerical and large eddy simulation of turbulent channel flows, Encyclopaedia of Fluid Mechanics, Gulf Publ. Houston, Vol. 6, 1987, pp. 1337-1391.
17. M. Wörner, G. Grötzbach, Analysis of semi-implicit time integration schemes for direct numerical simulation of turbulent convection in liquid metals, Notes on Numerical Fluid Mechanics, Ed. J.B. Vos, A. Rizzi, I.L. Ryhming, Vol. 35, pp. 542-551, Verlag Vieweg, Braunschweig, 1992.
18. H. Schmidt, U. Schumann, H. Volkert, Three-dimensional, direct and vectorized elliptic solvers for various boundary conditions, DFVLR-Mitt. 84, 15, August 1984.
19. U. Schumann, Linear stability of finite difference equations for three-dimensional flow problems, J. Comp. Phys. 18, 1975, pp. 465-470.
20. G. Grötzbach, Spatial resolution requirements for direct numerical simulation of the Rayleigh-Bénard convection, J. Comp. Phys. 49, 1983, pp. 241-264.
21. G. Grötzbach, Simulation of turbulent flow and heat transfer for selected problems of nuclear thermal hydraulics, Proc. Conf. on Supercomputing in Nuclear Applications, Mito, Japan, March 12-16, 1990, pp. 29-35.
22. M. Wörner, G. Grötzbach, Contributions to turbulence modelling of natural convection in liquid metals by direct numerical simulation, Proc. Mathematical Methods and Supercomputing in Nuclear Applications, Karlsruhe, Germany, April 19-23, 1993, Vol. 1, pp. 224-235.
23. M. Wörner, G. Grötzbach, Analysis of diffusion of turbulent kinetic energy by numerical simulations of natural convection in liquid metals, Proc. Sixth Int. Topical Meeting on Nuclear Reactor Thermal Hydraulics, Oct. 5-8, 1993, Grenoble, France, Vol. 1, pp. 186-193.

24. G. Grötzbach, M. Wörner, Analysis of flow mechanisms in Rayleigh-Bénard convection at small Prandtl numbers, Proc. Mathematical Methods and Supercomputing in Nuclear Applications, Karlsruhe, Germany, April 19-23, 1993, Vol. 1, pp. 236-247.
25. C.H. Moeng, R. Rotuno, Vertical velocity skewness in the buoyancy driven boundary layer, J. Atmos. Sc. 47, 1990, pp. 1149-1162.

The GRP Treatment of Flow Singularities

M. Ben-Artzi [1, 2], A. Birman [3], and J. Falcovitz [4]

[1] Institute of Mathematics, Hebrew University, Jerusalem 91904, Israel.
[2] Partially supported by the Robert Szold Institute for Applied Science of the P.E.F. Israel Endowment Funds.
[3] Dept. of Physics, Technion – Israel Institute of Technology, Haifa 32000, Israel.
[4] Dept. of Aerospace Engineering, Technion – Israel Institute of Technology, Haifa 32000, Israel.

The GRP (Generalized Riemann Problem) method was introduced and developed in [1–7, 9–10], following the pioneering work of van Leer [13].

Basically, the GRP is an "analytic high-resolution" (second-order) extension of the classical Godunov [11] scheme, designed to solve numerically systems of conservation or "quasi-conservation" laws. One can also describe the method as a sort of "hybrid" scheme, incorporating the detailed analysis of the characteristic structure of singularities (jumps, edges of rarefaction, imposed boundaries, etc.) into a robust "shock capturing" method, based on conservative differencing.

The fundamental GRP strategy can be summarized as follows (for quasi 1-D problems).

(a) Use piecewise linear distribution of flow variables at a given time-level.
(b) Apply the GRP-analysis to evaluate fluxes at "singularities". These include, naturally, all jumps at "cell boundaries", but, optionally, also at selected strong shocks, material interfaces and so on.
(c) Use the fluxes in a straightforward time-marching of flow variables, via conservative differencing.
(d) Apply once again the GRP-analysis to determine the piecewise linear (i.e., constant slopes in cells) distribution of flow variables at the new time level.

The above four steps have been used as the *underlying basis* of a *unified approach* towards the study of a wide range of fluid dynamical phenomena. It is always expected that the (necessary) additional technicalities (such as monotonicity algorithms) are kept at the simplest possible level.

The purpose of this paper is to illustrate the application of the GRP-method to treatment of flow singularities. We shall deal here with the following three cases.

(a) Tracking singularities by adjustable meshes.
(b) Dealing with geometrical singularities (e.g., origin in spherical coordinates).
(c) Treatment of very narrow reaction zones in chemical flows.

233

Before proceeding to a discussion of these three cases, we outline briefly the basic ideas of the GRP-method. For this purpose, we shall use the system of Euler equations that model the time dependent flow of an inviscid, compressible fluid through a duct of smoothly varying cross-section. We denote by r the spatial coordinate and by $A(r)$ the area of the cross-section at r. The equations can be written in the following form.

$$A\frac{\partial}{\partial t}U + \frac{\partial}{\partial r}[AF(U)] + A\frac{\partial}{\partial r}G(U) = 0$$

(1)

$$U = \begin{pmatrix} \rho \\ \rho u \\ \rho E \end{pmatrix}, \quad F(U) = \begin{pmatrix} \rho u \\ \rho u^2 \\ (\rho E + p)u \end{pmatrix}, \quad G(U) = \begin{pmatrix} 0 \\ p \\ 0 \end{pmatrix}.$$

Here ρ, p, u, E are, respectively, density, pressure, velocity and total specific energy, where $E = e + \frac{1}{2}u^2$, e being the internal specific energy. In general, the thermodynamic variables p, ρ, e are related by an "equation of state". We shall refer to the most common case, that of an ideal "γ-law" gas, where,

$$p = (\gamma - 1)\rho e, \qquad \gamma > 1.$$

(2)

We recall the approach, originated by Godunov [11], for a conservative ("upwind") difference scheme for (1). Thus, suppose that we use equally spaced grid-points $r_i = i\Delta r$ along the r-axis and the (numerical) solution is sought at equally spaced time-levels $t_n = n\Delta t$. By "cell i" we shall refer to the interval extending between the "cell-boundaries" $r_{i\pm1/2} = (i\pm\frac{1}{2})\Delta r$. We label by Q_i^n the average value of a quantity (flow variable)Q over cell i at time-level t_n.

Similarly, $Q_{i+1/2}^{n+1/2}$ is the value of Q at the cell boundary $r_{i+1/2}$, averaged over the time interval (t_n, t_{n+1}). Generally speaking, a "quasi-conservative" difference scheme for (1) is given by,

$$U_i^{n+1} - U_i^n = -\frac{\Delta t}{\Delta V_i}\left\{\left[A(r_{i+1/2})F(U)_{i+1/2}^{n+1/2} - A(r_{i-1/2})F(U)_{i-1/2}^{n+1/2}\right]\right.$$
$$\left. + A(r_i)\cdot\left[G(U)_{i+1/2}^{n+1/2} - G(U)_{i-1/2}^{n+1/2}\right]\right\},$$

(3)

where $\Delta V_i = \int_{r_{i-1/2}}^{r_{i+1/2}} A(r)dr$ is the volume of cell i.

In (3), we must still give an appropriate interpretation to the "flux" values $F(U)_{i+1/2}^{n+1/2}$, $G(U)_{i+1/2}^{n+1/2}$. To do this, Godunov proposed to solve (at every cell-boundary $r_{i+1/2}$) the Riemann Problem (RP),

$$\frac{\partial}{\partial t}U + \frac{\partial}{\partial r}[AF(U)] + A\frac{\partial}{\partial r}G(U) = 0,$$

(4)

$$U(r,0) = \begin{cases} U_{i+1}^n, & r > r_{i+1/2}, \\ U_i^n, & r < r_{i+1/2} \end{cases}.$$

If the solution thus obtained (for r in the vicinity of $r_{i+1/2}$ and $t \geq 0$) is denoted by $U(r,t)$, then the desired flux values are obtained by,

$$U_{i+1/2}^{n+1/2} = \lim_{t \to 0+} U(r_{i+1/2}, t), \ F(U)_{i+1/2}^{n+1/2}$$

$$= F(U_{i+1/2}^{n+1/2}), G(U)_{i+1/2}^{n+1/2} = G\left(U_{i+1/2}^{n+1/2}\right). \tag{5}$$

Note that if $A(r) \equiv 1$ (planar symmetry), the solution to the RP(4) is "self-similar" (i.e., depending only on the similarity coordinate $r - r_{i+1/2}/t$) and the classical Godunov scheme is obtained. It is well-known that the resulting (first order) scheme is stable and robust, but also that jump discontinuities are poorly resolved by it.

In order to obtain a second-order variant of the Godunov scheme, van Leer [13] proposed the following idea. Assume that all flow variables are *linearly distributed* in cells (with jumps at cell boundaries). Let $U_i^n(r) = U_i^n + \frac{(\Delta U)_i^n}{\Delta r}(r - r_i)$ be the linear distribution of U in cell i, at time t_n. instead of (4) let us now solve the GRP, which is the initial value problem given by,

$$A\frac{\partial}{\partial t}U + \frac{\partial}{\partial r}[AF(U)] + A\frac{\partial}{\partial r}G(U) = 0,$$

$$\tag{6}$$

$$U(r,0) = \begin{cases} U_i^n(r), & r < r_{i+1/2}, \\ U_{i+1}^n(r), & r > r_{i+1/2}. \end{cases}$$

Let $U(r,t)$ be the solution to (6) near $(r,t) = (r_{i+1/2}, 0)$.

The key idea of the GRP method is to solve the system (6) analytically and then replace (5) by,

$$U_{i+1/2}^{n+1/2} = U_{i+1/2}^n + \frac{\Delta t}{2}\left[\frac{\partial}{\partial t}U\right]_{i+1/2}^n, \tag{7}$$

where,

$$U_{i+1/2}^n = \lim_{t \to 0+} U(r_{i+1/2}, t),$$

$$\left[\frac{\partial}{\partial t}U\right]_{i+1/2}^n = \lim_{t \to 0+} \frac{\partial}{\partial t}U(r_{i+1/2}, t). \tag{8}$$

Once $U_{i+1/2}^{n+1/2}$ is evaluated, one proceeds to compute the fluxes as in (5).

The solution to the GRP is given in [1–2, 5, 7], where the numerical implementation is also presented. Indeed, once (7) is evaluated, the algorithm (3) is applied without any modifications, except for a very simple monotonicity algorithm.

We conclude this outline of the method with the following two remarks.

(1) The evaluation of $U_{i+1/2}^n$ by (8) requires the solution of a RP of the type used in the classical Godunov scheme, namely, (4) with $A = 1$. However,

U_i^n, U_{i+1}^n are replaced by the interface values $U_i^n(r_{i+1/2})$, $U_{i+1}^n(r_{i+1/2})$, respectively.

(2) In order to get a second-order upgrading of Godunov's scheme, it suffices to determine $\left[\frac{\partial}{\partial t}U\right]_{i+1/2}^n$ with an $0(\Delta t)$ error, since then, in view of (7), the fluxes are computed with an $0(\Delta t^2)$ error. It turns out that such an approximation is extremely easy to get. Indeed, the solution to the GRP is then reduced to a modification of the solution to the corresponding RP (i.e., $U_{i+1/2}^n$ in (8)), obtained by solving a couple of (algebraic) linear equations. This constitutes what was labeled as the "E_1-scheme" and is, by definition, the simplest upgraded (to second-order) Godunov scheme. This modification can be easily installed in any existing code based on the Godunov scheme. We refer the reader to [6] for full details.

We now turn to a discussion of the three cases mentioned above.

1. Tracking of Singularities

In the difference scheme (3), assume that the cells are allowed to vary with time, so that the cell boundaries are now functions of time, $r_{i\pm1/2}(t)$. We set $r_{i+1/2}^n = r_{i+1/2}(t_n)$, etc. Instead of (3) we obtain now,

$$U_i^{n+1} = \frac{(\Delta V)_i^n}{(\Delta V)_i^{n+1}}U_i^n + \frac{\Delta t}{(\Delta V)_i^{n+1}}\left\{[(\Lambda U - F(U))\,A]_{i+1/2}^{n+1/2}\right.$$
$$\left. - [(\Lambda U - F(U))A]_{i-1/2}^{n+1/2} \right.$$
$$\left. - 1/2\left[G(U)_{i+1/2}^{n+1/2} - G(U)_{i-1/2}^{n+1/2}\right]\cdot\left(A_{i+1/2}^{n+1/2} + A_{i-1/2}^{n+1/2}\right)\right\}, \tag{9}$$

where $A_{i+1/2}^{n+1/2} = A(r_{i+1/2}^{n+1/2})$, $(\Delta V)_i^n = \int_{r_{i-1/2}^n}^{r_{i+1/2}^n} A(r)dr$, and Λ is the speed of the moving boundary relative to the Eulerian mesh (e.g., $\Lambda_{i+1/2}^{n+1/2} = r'_{i+1/2}\left(\frac{t_n+t_{n+1}}{2}\right)$).

In order to implement (9), one takes full advantage of the GRP solution, which yields complete information about the wave structure of the flow near $r_{i+1/2}$, including directional derivatives along any trajectory emanating from the singularity. The trajectories $r_{i+1/2}(t)$ can be selected to follow shock or contact discontinuities, or even characteristic curves, e.g. edges of rarefaction. Consequently, a "singularities tracking" scheme, of second-order accuracy, is naturally obtained.

As a numerical example, we consider the well-known shock tube problem proposed by Sod [12]. The tube extends from $r = 0$ to $r = 100$ (with planar symmetry, $A(r) \equiv 1$) and is divided into 100 equal cells. The fluid is a perfect gas satisfying (2) with $\gamma = 1.4$. The initial conditions are $u = 0$, $p = \rho = 1$ for $0 < r < 50$; $u = 0$, $p = 0.1$, $\rho = 0.125$ for $50 < r < 100$.

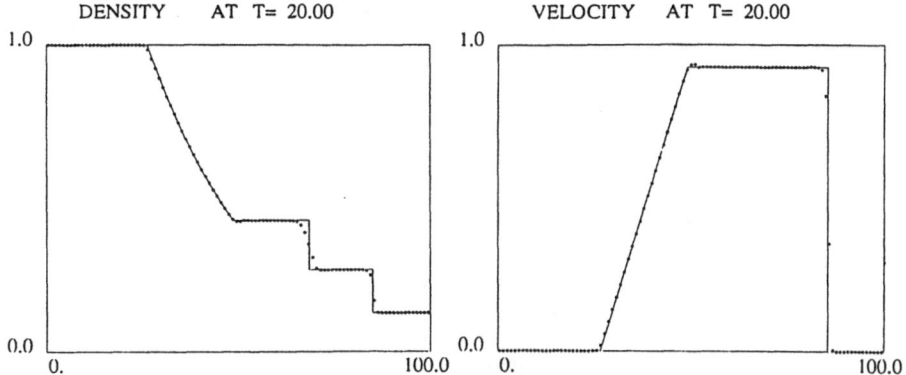

DENSITY   AT T= 20.00   VELOCITY   AT T= 20.00

Fig. 1a. Sod's problem. GRP without tracking

Figure 1a shows the velocity and density profiles at $t = 20$ (the exact self-similar solution is given by the solid curve), where the basic GRP scheme (3), (8), is used without any tracking. Observe that the contact discontinuity and the tail of the rarefaction wave are not as well resolved as the shock and the head of the rarefaction.

In Fig. 1b we show the velocity and density profiles for the same problem at the same time. However, the contact discontinuity is now "tracked" as described above. Clearly, the contact discontinuity is now sharply resolved.

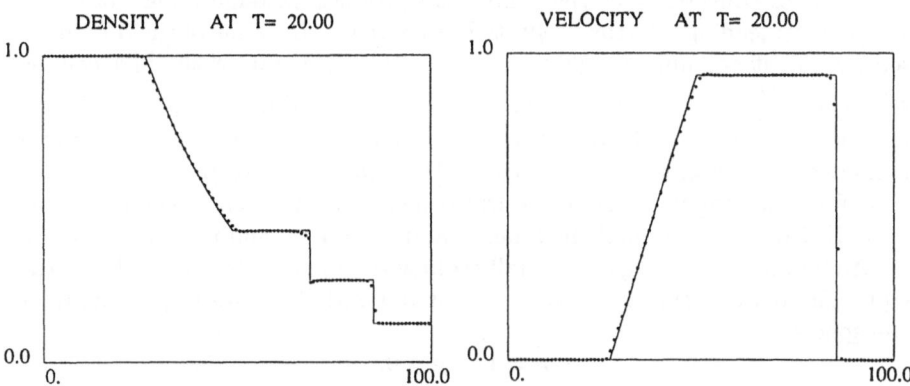

DENSITY   AT T= 20.00   VELOCITY   AT T= 20.00

Fig. 1b. Sod's problem. GRP with tracking of contact

Finally, in Fig. 1c we show the same profiles, tracking the following three singularities: Shock wave, contact discontinuity and tail of the rarefaction wave. The calculation now is in excellent agreement with the exact solution. In fact, observe that by adding the shock and the tail of rarefaction as tracked

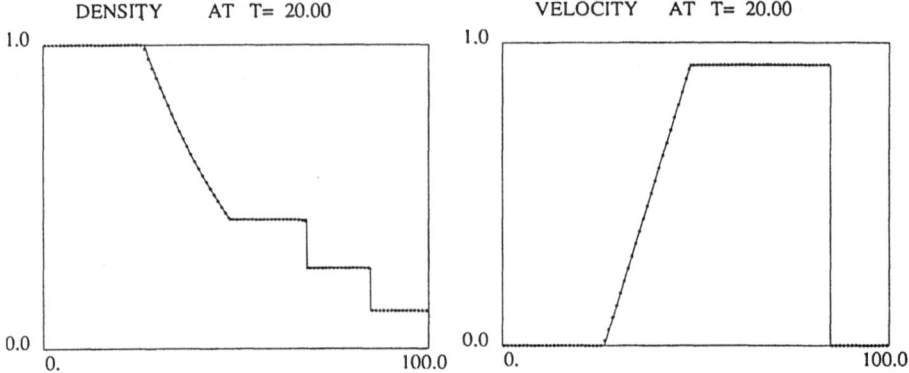

Fig. 1c. Sod's problem. GRP with tracking of shock, contact and tail of rarefaction

points, we have managed to improve the quality of the calculated profile near the contact discontinuity.

2. Geometrical Singularities

Consider the case of compressible flow with spherical symmetry. The governing equations are given by (1), with $A(r) = r^2$. The point $r = 0$, where $A(r) = 0$, is a singular point for the flow, with $u = 0$. As is well-known, shock waves converging to the center can produce extremely high pressure and density, leading to strong reflected shock waves. In terms of the difference scheme (3), it amounts to taking $r_{i-1/2} = A(r_{i-1/2}) = 0$ for the first cell, as well as $u_{i-1/2}^{n+1/2} = 0$ for all n. To provide boundary values (at $r_{i-1/2} = 0$) for p, ρ one uses appropriate reflections. We refer to [6–7] for more details and numerical results of various test cases. Here we choose to present one such case, for which we have also an exact (explicit) solution. This example is due to W.R. Noh, in an unpublished memorandum (see [4] and references there).

An infinite sphere of gas is initially cold with pure kinetic energy. Thus, the equation-of-state (2) is used with $\gamma = 5/3$ and initially, uniformly throughout the sphere,

$$\rho = 1, \quad p = 0.$$

At the initial time $t = 0$ the gas is uniformly imploding with velocity $u = -1$. The initial conditions imply that at $t = 0$ an "infinite shock" (i.e., a shock of maximal compression rate) is reflected from the origin and brings the incoming gas to rest. The compression ratio here is $\frac{\gamma+1}{\gamma-1} = 4$, and it follows from the Rankine-Hugoniot jump conditions that the (uniform) speed of the reflected shock is $W = \frac{1}{3}$. At time $t = 3$ it reaches the point $r = 1$ where it encounters fluid particles that originated at $r = 4$. Hence, the uniform density behind the shock is $\bar{\rho} = 64$. It follows again from the jump conditions that the uniform pressure behind the shock is $\bar{p} = \frac{64}{3}$. The velocity and pressure profiles ahead of the shock are equal to their initial values, while the density varies due to geometrical compression. It is easy to see that the density profile, for $r \geq 4t/3$, is given by,

$$\rho(r - t), t) = \frac{r^2}{(r - t)^2}.$$

Figure 2 shows the results obtained by using the GRP scheme with 100 points, equally spaced with $\Delta r = 1$. A uniform time step $\Delta t = 0.25$ was taken, and the profiles are shown after 900 time steps. The solid line gives the profile for the exact solution.

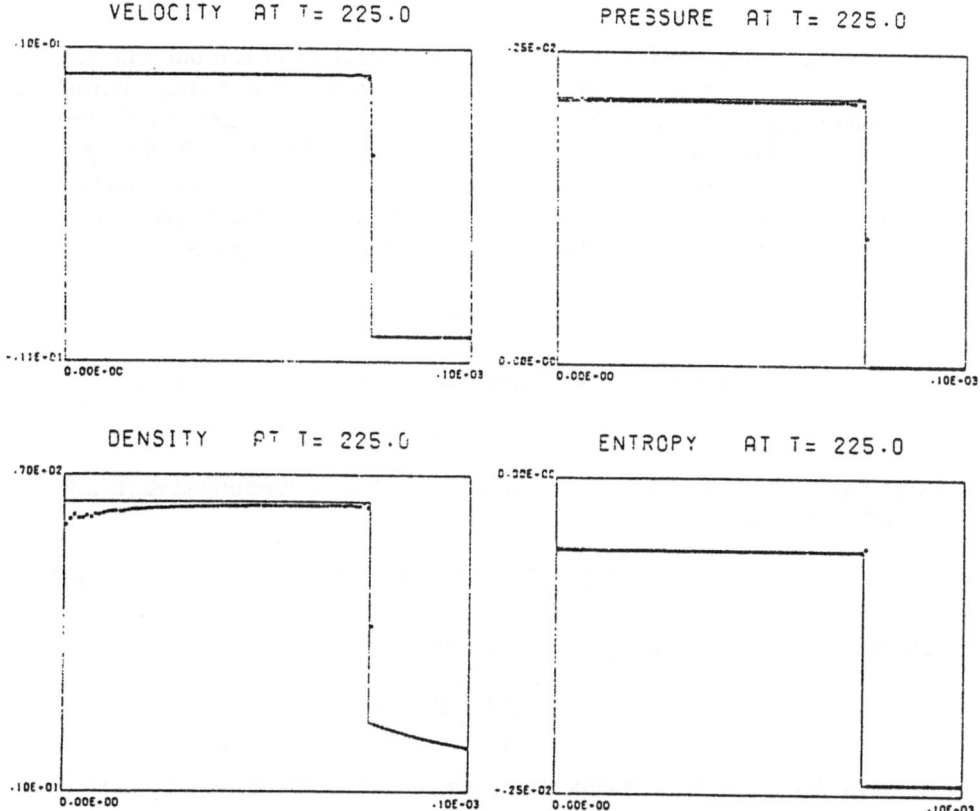

Fig. 2. Reflection of infinite shock from spherical singularity

239

3. Narrow Reaction Zones in Combustion Calculations

In this case we consider the Euler equations that model the time-dependent flow of an inviscid, compressible, reacting gas. Assuming planar symmetry, and denoting by x the spatial coordinate, the equations can be written in "quasi-conservative" form as,

$$\frac{\partial}{\partial t}U + \frac{\partial}{\partial x}F(U) = G(U),$$

(10)

$$U = \begin{pmatrix} \rho \\ \rho u \\ \rho E \\ \rho z \end{pmatrix}, \quad F(U) = \begin{pmatrix} \rho u \\ \rho u^2 + p \\ \rho u E + pu \\ \rho z u \end{pmatrix}, \quad G(U) = \begin{pmatrix} 0 \\ 0 \\ 0 \\ -k(\rho, p, z)\rho \end{pmatrix}.$$

Here ρ, p, E, u are as in (1), and z is the mass fraction of the unburnt gas. Thus, $z = 1$ (resp. $z = 0$) represents the completely unburnt (resp. burnt) gas. The total specific energy $E = e + \frac{1}{2}u^2$ contains also the "internal chemical energy", included in e. The assumed equation-of-state here is of the form $p = p(e, \rho, z)$. In addition, the system (10) involves the "reaction rate" function $k(\rho, p, z)$. Note that the fourth function ("chemical") equation of (10) can be written using total derivates ("Lagrangian") along particle-paths, as,

$$\frac{dz}{dt} = -k.$$

(11)

We specialize here to a "γ-law" gas, where, instead of (2), one has,

$$p = (\gamma - 1)\rho(e - q_0 z), \quad \gamma > 1, \quad q_0 > 0,$$

(12)

expressing an "exothermic" process. Also, we shall use a "simplified Arrhenius model", given by,

$$k = KzH(T - T_c), \quad K > 0, \quad T = \frac{p}{\rho},$$

(13)

where the Heaviside function $H(y)$ is given by,

$$H(y) = \begin{cases} 1, & y > 0, \\ 0 & y \leq 0. \end{cases}$$

In particular, there is no chemical process below the "ignition temperature" T_c.

We refer the reader to [2], where the coupling between the "fluid dynamical" and the "chemical" phases of (10) is discussed. Recall the structure of a C-J wave in the Z-N-D model for solutions of (10) (see [8] for details). The wave consists of a reaction zone of finite width moving at the speed determined by the jump conditions. Across this zone the mass fraction decreases from $z = 1$ to $z = 0$. The edge of the reaction zone facing the unburnt gas is a fluid-dynamical shock wave which raises the pressure and density to values significantly higher than those attained in the fully burnt gas. The profile is therefore referred to as the "Z-N-D spike". As the coefficient K in (13) is increased, the whole process is faster and the reaction zone becomes narrower. The same effect is achieved by increasing the size of the computatioal cell, since then the Z-N-D spike (whose width is independent of the mesh size) is "absorbed" in a small portion of the cell and the averaging process clips it almots entirely. Thus, for a study of "fast reactions" and their coupling with the underlying fluid dynamical structure it suffices to increase the mesh size. However, such an increase has a very interesting impact on the solution, which can be best described in terms of the simple stiff equation $\frac{dy}{dt} = -Ky$, $K \gg 1$. Consider the explicit scheme

$$\frac{y_{(e)}^{n+1} - y_{(e)}^n}{\Delta t} = -Ky_{(e)}^n,$$

and the implicit scheme

$$\frac{y_{(i)}^{n+1} - y_{(i)}^n}{\Delta t} = -Ky_{(i)}^{n+1}.$$

Clearly, the explicit scheme forces Δt to be of the order of magnitude of K^{-1}, while the implicit scheme does not restrict Δt. However, if $y_{(i)}^n = 1$ and if Δt is not small, then $y_{(i)}^{n+1} \approx 0$. In terms of our combustion calculations, this amounts to the burning of "one cell per one time step". As Δt is decreased, the burning process races faster! Such a combustion wave is indeed possible in the context of the system (10) – it is called "weak detonation" [2, 8]. Such a weak detonation is supersonic with respect to the flow behind it, so that it can be followed by a slower shock wave which raises further the pressure and density. This gives rise to a one-parameter family of solutions (non-physical) to (10) – and the computational scheme picks one of them by adjusting the speed of the detonation wave to $\frac{\Delta x}{\Delta t}$. As Δt is decreased, the numerical solution moves away from the correct solution!

Figure 3a shows the results of the GRP calculation for a detonation wave using the analogue of the "explicit" scheme (thus restricting Δt substantially). Observe that the Z-N-D spike exists, even though it is considerably truncated (the site of the reaction zone is roughly equal to one computational cell). Figure 3b shows profiles for the same problem, using an analogue of the "implicit" scheme. Note the smoothness of the results – despite the fact that they yield the "wrong" solution. We refer the reader to [2] for full details of this example.

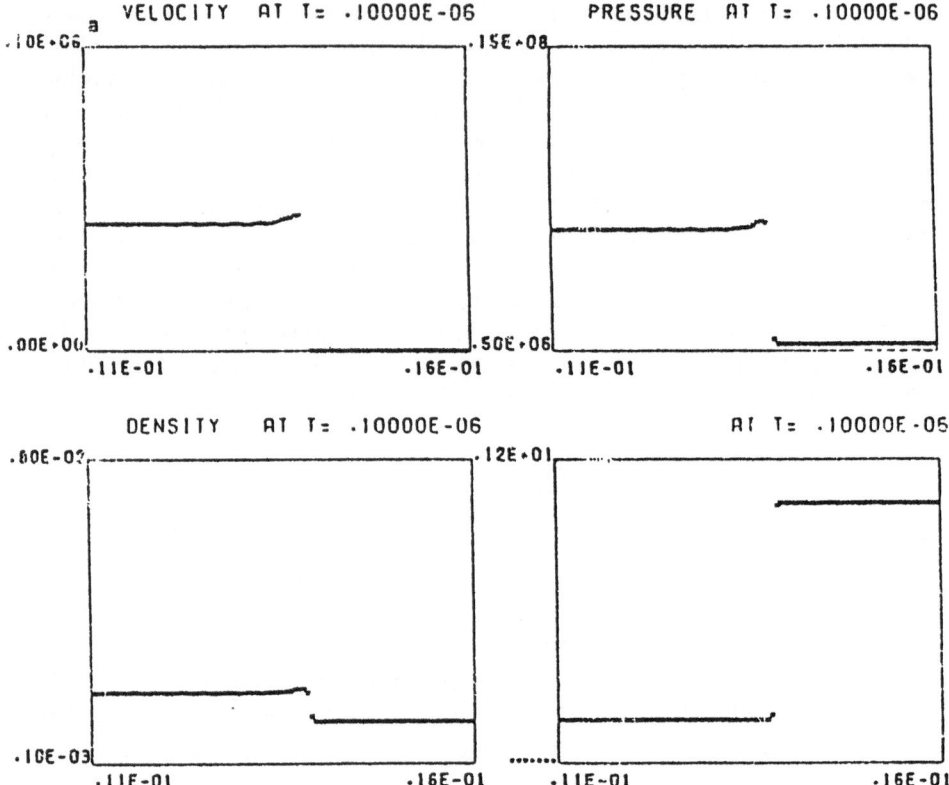

Fig. 3a. Detonation wave, explicit GRP scheme

References

1. M. Ben-Artzi: The Generalized Riemann Problem in compressible duct flow, Contemporary Mathematics **60** (1987), 11–18 (Ed. B. Keyfitz)
2. M. Ben-Artzi: The Generalized Riemann Problem for reactive flows, J. Comp. Phys. **81** (1989), 70–101
3. M. Ben-Artzi, A. Birman: Application of the "Generalized Riemann Problem" method to 1-D compressible flows with material interfaces, J. Comp. Phys. **65** (1986), 170–178
4. M. Ben-Artzi, A. Birman: Computation of reactive duct flows in external fields, J. Comp. Phys. **86** (1990), 225–255
5. M. Ben-Artzi, J. Falcovitz: A second-order Godunov-type scheme for compressible fluid dynamics, J. Comp. Phys. **55** (1984), 1–32
6. M. Ben-Artzi, J. Falcovitz: A high-resolution upwind scheme for quasi 1-D flows, in "Numerical Methods for the Euler Equations of Fluid Dynamics", Eds. F. Angrand, A. Dervieux, J.A. Desideri, R. Glowinski, SIAM Publ. 1985, pp. 66–83
7. M. Ben-Artzi, J. Falcovitz: An upwind second-order scheme for compressible duct flows, SIAM J. Sci. Stat. Comp. **7** (1986), 744–768

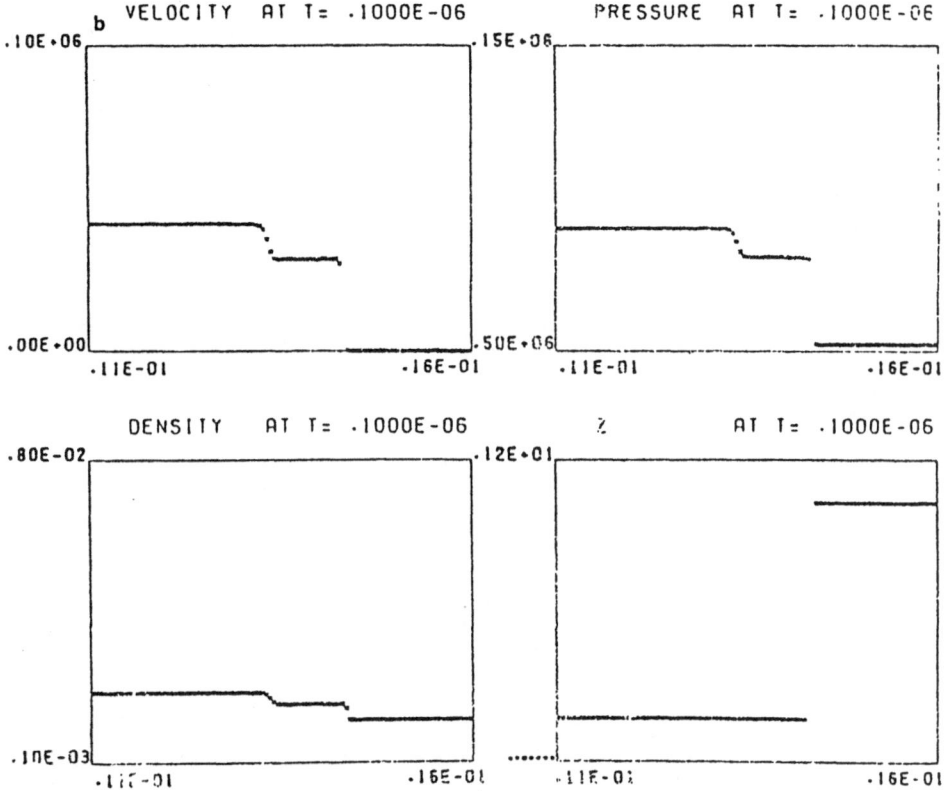

Fig. 3b. Detonation wave, implicit GRP scheme

8. R. Courant, K.O. Friedrichs: Supersonic flow and shock waves, Springer-Verlag, New York 1976
9. J. Falcovitz, A. Birman: A singularities tracking conservation laws scheme for compressible duct flows, in "Shock Waves: Proc. of the 18th Int. Symp. on Shock Waves" (Ed. K. Takayama), Springer-Verlag 1991, pp. 1107–1112
10. J. Falcovitz, A. Birman: A singularities tracking conservation laws scheme for compressible duct flows, J. Comp. Phys. (in press)
11. S.K. Godunov: A finite difference method for the numerical computation of discontinuous solutions of the equations of fluid dynamics, Mat. Sbornik **47** (1959), 271–295
12. G.A. Sod: A survey of several finite difference methods for systems of nonlinear hyperbolic conservation laws, J. Comp. Phys. **27** (1978), 1–31
13. B. van Leer: Towards the ultimate conservative difference scheme, V, J. Comp. Phys. **32** (1979), 101–136

Application of the Multidomain Local Fourier Method for CFD in Complex Geometries

L. Vozovoi[1], M. Israeli[1], and A. Averbuch[2]

[1] Faculty of Computer Science, Technion, Haifa 32000, Israel
[2] School of Mathematical Sciences, Tel Aviv University, Tel Aviv 69978, Israel

Summary. A low communication parallel algorithm is developed for the solution of time-dependent non-linear PDE's. Our particular interest is in the application of this algorithm to fluid dynamics problems. The parallelization is achieved by domain decomposition. The discretization in time is performed via a third order semi-implicit stiffly stable scheme. The elemental solutions in subdomains are constructed by using a high-order method with the Local Fourier Basis (LFB). It results in elliptic equations of Helmholtz and Poisson types, which have to be solved repeatedly at each time step. The continuity of the global solution is accomplished using a point-wise matching of the local subsolutions on the interfaces. The matching relations are derived in terms of the jumps on the interfaces. The LFB transformation enables the splitting a 2-D problem with global coupling of the interface unknowns into a set of uncoupled 1-D differential equations. Localization properties of an elliptic operator, resulting from the discretization in time of a time-dependent problem, are utilized in order to simplify the matching relations. In effect, only local (neighbor-to-neighbor) communication between the processors becomes necessary. The present method allows the treatment of problems in various complex geometries by the mapping of curvilinear domains into simpler (rectangular or circular) regions with subsequent matching of local solutions. The operator with non constant coefficients, obtained in the transformed domain, is preconditioned by an appropriate constant coefficient operator, easily inverted by the LFB. The problem is then solved with spectral accuracy by (a rapidly convergent) conjugate gradient iteration. [1]

1. Introduction

Within the last two decades, domain decomposition technique becomes an efficient tool for parallelizing algorithms, particularly, due to its ability to provide an equal loading of the computational work to different processors. The last feature is critical to avoid synchronization bottlenecks and thus, to achieve high parallel efficiency of a parallel algorithm.

Domain decomposition can be successfully combined with spectral methods, using series expansions into orthogonal bases. These methods are particularly suitable for computation of incompressible the Navier-Stokes equations at high Reynolds numbers, since they converge exponentially fast as the number of degrees of freedom N increases. For problems in multidomains polynomial (Chebyshev or Legendre) bases are mostly used. Spectral methods

[1] This research is supported partly by a grant from the French-Israeli Binational Foundation for 1993-1994 and by grant of Israel Ministry of Science for 1992-1993.

with the Fourier basis seem to be of doubtful utility for the construction of *non-periodic* local solutions in subdomains. The reason is that the truncated Fourier series of a non-periodic function (having discontinuous periodic extension at the boundaries) converges very slowly, like $1/N$, inside the region, and gives rise to $O(1)$ spurious oscillations near the boundaries, known as Gibbs phenomenon.

However, the Gibbs phenomenon can be avoided if a smooth decomposition of functions is performed. In [1]a spectrally accurate multidomain Fourier method was developed. This method employs a particular version of the projection technique, introduced by R. Coifman and Y. Meyer [4] for the construction of the Local Fourier Basis (LFB). The multidomain local Fourier method (MDLF) of [1] is especially efficient for large problems, using high resolution in space. In this case the relative amount of operations, required to perform the projection, is insignificant in comparison with that needed to execute the Fast Fourier Transform (FFT).

An important advantage of this method, when compared to other multidomain spectral techniques, is that it enables a great simplification of the matching relations for the interface unknowns. The matching of local solutions, constructed independently in each subdomain, is necessary to obtain the continuous global solution. For problems with constant coefficients and in simple domains, the use of the Fourier basis enables us to fulfill the matching of each harmonic separately, and thus, to eliminate the global coupling of the interface unknowns.

The matching relations can be simplified furthermore by utilizing localization properties of an elliptic operator, resulting from the discretization in time of a time-dependent problem [2]. Eventually, only *local communication* between the neighboring subdomains turns out to be necessary. Thus, the MDLF algorithm is completely scalable and with high parallel efficiency [3].

In this paper we generalize our previous multidomain Fourier approach in order to treat complicated geometries. The new algorithm incorporates the MDLF method along with several additional techniques, such as mapping and conjugate gradient iterations. A rapidly convergent (by $O(1)$ iterations) algorithm, using a spectral preconditioner, is constructed for the solution of full-matrix problem in a mapped domain. In effect, the spectrally accurate solution in complex geometry can be accomplished by a little more work than required to solve a constant coefficient equation in the simplest rectangular domain, using the FFT.

The structure of this paper is the following. In section 2 we describe the MDLF approach in 1-D and in 2-D for the simplest rectangular geometry. In section 3 we set up the problem in complex geometry. The numerical algorithm for solving this problem is introduced in section 4. Finally, we apply this method to a fluid dynamics problem in the channel with wavy boundaries.

2. Multidomain Local Fourier Method (MDLF)

In this section we discuss briefly the MDLF method in 1-D and its extension to 2-D for the simplest rectangular domains, decomposed into parallel strips. The detailed analysis of this method can be found in [1], [2], [3].

We describe the numerical algorithm as applied to the boundary value problem of diffusion type

$$u'' - \lambda^2 u = f(x), \qquad x \in [0, L], \tag{2.1}$$

with periodic boundary conditions

$$u(0) = u(L), \qquad u'(0) = u'(L) \tag{2.2}$$

We divide the computational interval $[0, L]$ into P pieces (subdomains) of an arbitrary size and discretize the local subproblems on the uniform grid of collocation points.

Following [4], we cover the interval with the collection of smooth overlapping bell functions $b_n(x)$, which satisfy certain properties to allow the projection in L_2. In our particular version of the LFB method the bell functions $b_n = 1$ inside the subdomains $l_n = [x_{n-1}, x_n]$ and smoothly decay outwards over the intervals 2ϵ. This choice of $b_n(x)$ is particularly convenient for the solution of differential equations using the LFB technique. The algorithm consists of two steps.

1.**Construction of the local solutions**. For each subdomain we perform:

• Extension of a local source function $f^{(n)}(x)$ beyond the interfaces over the distance 2ϵ (from each side). It can be done by exchanging data with two neighboring subdomains. For periodic problems Eq.(2.2) the continuous extension beyond the boundaries, $x < 0$ and $x > L$, can be obtained by overlapping the first and the last subdomains.

• Projection of $f^{(n)}(x)$ with the help of the bell functions b_n. The projection procedure results in a smooth function $\tilde{f}^{(n)}(x)$, localized on the extended interval $\bar{l}_n = [\bar{x}_{n-1} - \epsilon, \bar{x}_n + \epsilon]$. which can be represented by rapidly convergent Fourier series. Due to the choice of b_n, this function coincides with the original one, $f^{(n)}(x)$, on the range l_n of a subdomain.

• Integration of Eq.(2.1) with the smooth source function, using the Discrete Fast Fourier Transform (DFFT) on the extended interval $\bar{l}_n$.

The resulting solutions $\tilde{u}_p^{(n)}(x)$ will be the true solutions of Eq.(2.1) on the intervals of the subdomains l_n, where the projection procedure does not change the source function $f_n(x)$. As for the "extra" parts (on the extended intervals ϵ outside the subdomains, where the source function was distorted by projection), the most economic way is not to reconstruct the solutions there, but rather not to utilize them at all. The computational expenses for these external intervals will be the extra work to do in order to avoid the Gibbs phenomenon when the Fourier method is applied to non-periodic problems.

It can be shown that for large problems, using high resolution in space, this extra work is a small percentage of the whole computational work.

The global solution $u_p = \bigcup_{n=1}^{P} u_p^{(n)}$ will be piecewise continuous, as the local solutions have, in general, jumps at the interfaces.

2. **Matching step.** In order to obtain the continuous global solution, we correct the particular solutions, constructed at the previous step, by properly weighted local homogeneous solutions of Eq.(2.1):

$$u = \bigcup_{n=1}^{P} u^{(n)}, \quad u^{(n)} = u_p^{(n)} + A_n h_+^{(n)} + B_n h_-^{(n)} \qquad (2.3)$$

These solutions are two exponential functions $h_-^{(n)} = e^{-\lambda x}$, $h_+^{(n)} = e^{-\lambda(l_n - x)}$, $0 < x < l_n$, which decay inward the subdomain. The coefficients A_n, B_n have to be found in order to enforce the continuity of $u(x)$ and $u'(x)$ at the interfaces. They can be expressed, in closed form, in terms of jumps of particular solutions u_p, u_p' at the interfaces.

It seems that computation of matching coefficients requires the global communication between the processors since all the interfaces contribute to the matching relations. We observe, however, that if the parameter λ in Eq.(2.1) is much greater than 1 (see section 5), then the functions $h_\pm(x)$ decay rapidly away from the interfaces. Then, for each particular location x, the influence of the remote interfaces becomes negligible. As a result, the matching relations decouple in such a way, that the computation of each pair of matching parameters A_n, B_n requires only the knowledge of jumps at two nearest interfaces.

To conclude, the computation of particular local solutions and the matching procedure use only local, neighbor-to-neighbor, inter-processor communication, that makes the whole algorithm completely scalable.

The present algorithm can be extended to solve non-periodic problems. For such problems the continuous extension of functions beyond the boundaries is not available. Therefore the boundary subdomains should be treated differently. An alternative approach consists in reexpanding the Fourier partial sums, burden by the Gibbs phenomenon, into the rapidly converging Gegenbauer series [7].

The extension of this method to rectangular domains, decomposed into parallel strips, is straightforward. After applying the DFFT along the strips, we arrive at a collection of *uncoupled* 1-D ODEs for the Fourier coefficients. These equations are of the same type as Eq.(2.1). Therefore, they can be solved by using the 1-D MDLF routine.

3. Problems in Complex Geometries

Consider a second-order elliptic equation

$$\nabla^2 u - \lambda^2 u = F(x, y) \qquad (3.1)$$

in a region $\Omega = \{0 \le x \le L, \; y_1(x) \le y \le y_2(x)\}$. Here $y_1(x)$, $y_2(x)$ are smooth L- periodic functions of x, having the mean values $\bar{y}_1 = 0$, $\bar{y}_2 = 1$ on the period L. The boundary conditions are:

$$y = y_{1,2}, \qquad\qquad u = \phi_{1,2}(x),$$

$$u(0, y) = u(L, y). \qquad (3.2)$$

The curvilinear region Ω can be mapped onto the rectangle $\omega = \{0 \le \xi \le L, \; 0 \le \eta \le 1\}$ by means of a simple stretching transformation:

$$\xi = x, \qquad\qquad \eta = \frac{y - y_1(x)}{y_2(x) - y_1(x)} \qquad (3.3)$$

The complication of using a coordinate transformation (3.3) appears in the coefficients of the differential operator in the transformed domain ω. After defining

$$\alpha = y_1, \qquad \beta = y_2 - y_1, \qquad\qquad \eta_y = 1/\beta, \qquad \eta_x = -(\alpha' + \eta\beta')/\beta,$$

$$\eta_{xx} = -(\alpha'' + \beta'' + 2\eta_x\,\beta')/\beta, \qquad\qquad J(\xi, \eta) = \eta_x^2 + \eta_y^2$$

$$a = 2\eta_x/J, \quad b = 1/J, \quad c = \eta_{xx}/J, \quad d = -\lambda^2/J, \quad f = F/J \qquad (3.4)$$

(here subscripts "x" and "y" denote the derivatives with respect to corresponding variables, whereas $'$ is the derivative of a function of one variable) we can rewrite Eq.(3.1) as follows:

$$\mathcal{L}u = f \qquad\qquad \text{in } \omega \qquad (3.5)$$

where

$$\mathcal{L}u = \frac{\partial^2 u}{\partial \eta^2} + a(\xi, \eta)\frac{\partial^2 u}{\partial \xi \partial \eta} + b(\xi, \eta)\frac{\partial^2 u}{\partial \xi^2} + c(\xi, \eta)\frac{\partial u}{\partial \eta} + d(\xi, \eta)u \qquad (3.5.1)$$

The boundary conditions (3.2) take now the form:

$$\eta = 0, 1 \qquad\qquad u = \phi_{1,2}(\xi),$$

$$u(0, \eta) = u(L, \eta). \qquad (3.6)$$

Spectral approximations to the boundary value problem Eqs. (3.5), (3.6) lead to full $N \times N$ matrix equations for the N expansion coefficients. The solution of this algebraic system requires $O(N^2)$ storage locations and $O(N^3)$ operations to invert the matrix (using Gauss elimination or other direct methods). Obviously, for multi-dimensional problems of realistic size (typically, with $N \sim 10^6$) the direct solution of such a system is prohibitively expensive.

The straightforward implementation of iteration methods to the solution of large linear algebraic systems is also inefficient. The reason is that the typical number of iterations, required to reduce the error by an order of magnitude (it can be characterized by the *condition number* κ) grows with the size N of a system. For the second order operator (3.5), $\kappa \propto N^2$ whatever discretization method is used.

The convergence of iterations can be improved substantially, if an appropriate *preconditioner* to the original differential operator is used [6]. In the next section we construct an efficient *spectral* preconditioner to $\mathcal{L}$ to obtain a rapidly convergent iteration algorithm.

4. Preconditioned Iteration Method with Spectral Preconditioner

The idea of preconditioning is to solve the equation

$$H^{-1}\mathcal{L}u = H^{-1}f \qquad (4.1)$$

rather than Eq. (3.5). The preconditioned iterations (for the particularly simple Richardson scheme) look as follows:

$$Hu^{n+1} = f - (\mathcal{L} - H)u^n. \qquad (4.2)$$

Both Eqs (4.1) and (3.5) are equivalent in view of solution, but the iteration method may converge much faster for Eq.(4.1) if the preconditioner H approximates $\mathcal{L}$ in the sense that the condition number of the "compound" operator $H^{-1}\mathcal{L}$ be bounded (independent of N).

Starting with the work by S. Orszag [5], a low order (finite-difference or finite-element) approximation to $\mathcal{L}$ is commonly used in the capacity of H. The use of a higher order finite-difference preconditioner improves the rate of convergence, but it is partially off set by the additional cost of inverting a denser matrix (representing an operator, discretized on a larger stencil). Therefore, high-order finite-difference preconditioners are not useful.

We propose to construct H in the same form as $\mathcal{L}$, but with *constant* coefficients in the place of variable coefficient in $\mathcal{L}$. For instance, the preconditioner to the operator Eq.(3.5) will be

$$Hu = \frac{\partial^2 u}{\partial \eta^2} + \bar{a}\frac{\partial^2 u}{\partial \xi \partial \eta} + \bar{b}\frac{\partial^2 u}{\partial \xi^2} + \bar{c}\frac{\partial u}{\partial \eta} + \bar{d}u \qquad (4.3)$$

The coefficients $\bar{a}, \bar{b}, \bar{c}$ and $\bar{d}$ can be obtained, for example, by averaging of the corresponding non-constant coefficients in Eq.(5.5) over the region ω. The operator H, defined in a rectangular domain ω, is trivially inverted by the Fourier method.

If the coefficients in $\mathcal{L}$ vary in small ranges, then the *spectral* preconditioner Eq.(4.3), obviously, provides a better approximation to $\mathcal{L}$ than any low-order preconditioner. However, if the coefficients vary substantially on the computational region, then Eq.(4.3) is going to be a poor approximation to $\mathcal{L}$.

The situation will change if the preconditioned iteration method is combined with the domain decomposition technique. Domain decomposition provides us with the new possibility to use *different* preconditoners in subdomains, each one being a good constant-coefficient approximation to $\mathcal{L}$ on a small enough patch of the region. As a result, the convergence rate of iterations can be improved considerably.

We illustrate this conclusion by the following 1-D example

$$u'' + \frac{1}{r}u' - \frac{m^2}{r^2}u - \lambda^2 u = f(r), \qquad r \in [r_0, 1]$$

$$u'(r_0) = u(1) = 0, \qquad (4.4)$$

which has obvious applications to the modified Helmholtz equation on the unit circle. The exact solution is chosen to be $u_{ex}(r) = (r - r_0)^2 (r - 1)$ (the forcing function $f(r)$ is computed accordingly), the minimal radius $r_0 = 0.1$. On the interval $0.1 \le r \le 1$ the coefficients $1/r$ and $1/r^2$ change by one and two orders of magnitude correspondingly.

Following our approach, we devide the computational interval $r \in [r_0, 1]$ into a number of subintervals $[r_{n-1}, r_n]$, $n = 1, ..., P$ and define a set of constant coefficient operators H_n as follows:

$$H_n = u'' + \bar{a}_n u' + \bar{b}_n u - \lambda^2 u, \qquad r \in [r_{n-1}, r_n], \qquad (4.5)$$

where

$$\bar{a}_n = \frac{1}{l_n} \int_{r_{n-1}}^{r_n} \frac{dr}{r}, \qquad \bar{b}_n = \frac{1}{l_n} \int_{r_{n-1}}^{r_n} \frac{dr}{r^2}, \qquad l_n = r_n - r_{n-1}.$$

Fig. 4.1 shows the variance of the first coefficient $a(r) = 1/r$ on the interval $r \in [r_0, 1]$, and its step-wise approximation $\bar{a}_n$ for $P = 8$.

The convergence of the preconditioned conjugate gradient (PCG) iterations with the spectral preconditioners Eq.(4.4) is shown on Fig. 4.2 for different numbers of subdomains P (parameters $m = 1, \lambda = 0$). We can see that even in this "tough" case, when coefficients change considerably (by $1 - 2$ orders of magnitude) on the computational interval, a better domain decomposition strategy allows us to obtain solution with the maximum accuracy after a few CG iterations.

5. Demonstration for the Navier-Stokes System

In this section we apply the generalized MDLF method to the solution of the complete Navier-Stokes system in a region of a complicated shape. As an

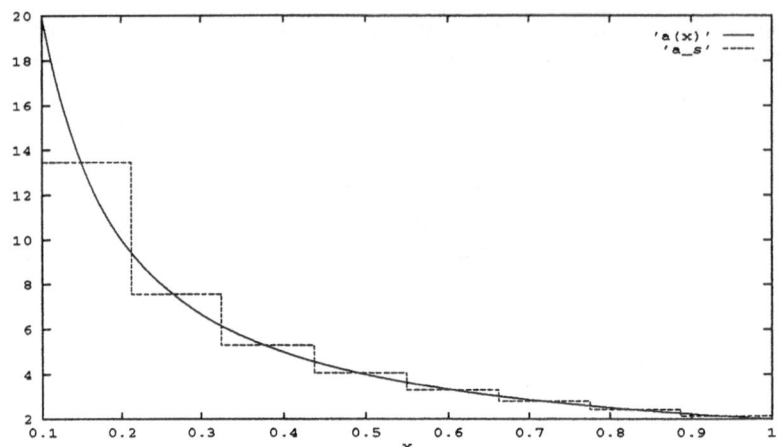

Fig. 4.1. Step-wise approximation of the variable coefficient $\frac{1}{r}$ in Eq. (4.5)

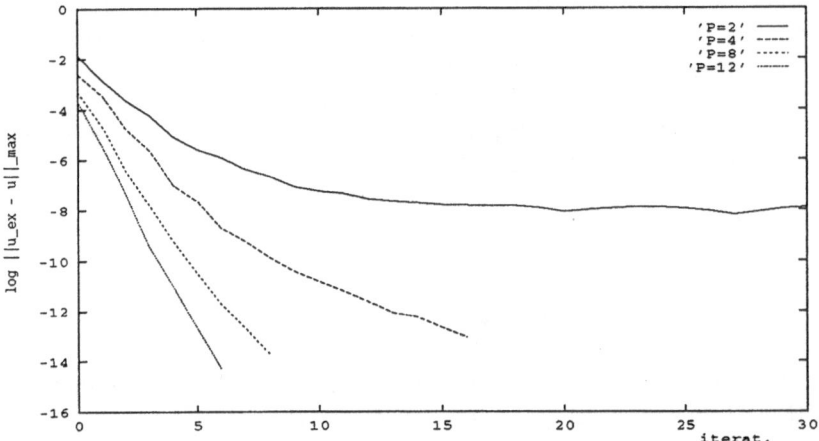

Fig. 4.2. Convergence history of the PCG iterations, using local spectral preconditioners Eq. (4.5), with $P = 2,4,8$ and 12 sub-divisions

example, we consider the problem of natural convection in a vertical channel
with sinusoidal boundaries:

$$y_1(y) = -(1 + \delta cos\ k_w x), \qquad\qquad y_2(y) = 1 + \delta cos\ k_w x. \qquad (5.1)$$

(this problem was investigated earlier by the first author [9] using a finite-
difference method). The computational domain comprises one undulation
period $L = 2\pi/k_w$ in the x direction: $\Omega = \{0 < x < L, y_1 < y < y_2\}$.

The governing equations are written in the $\psi - \phi$ formulation:

$$\frac{\partial \phi}{\partial t} + \frac{\partial \psi}{\partial x}\frac{\partial \phi}{\partial y} - \frac{\partial \phi}{\partial x}\frac{\partial \psi}{\partial y} = \Delta\phi + Gr\frac{\partial T}{\partial y}, \qquad (5.2)$$

$$\Delta\psi = -\phi \qquad (5.3)$$

$$\frac{\partial T}{\partial t} + \frac{\partial \psi}{\partial x}\frac{\partial T}{\partial y} - \frac{\partial T}{\partial x}\frac{\partial \psi}{\partial y} = Pr^{-1}\Delta T. \qquad (5.4)$$

Here ψ, ϕ, T are correspondingly stream function, vorticity and temperature; Gr, Pr are the Grashof and Prandtl numbers.

Functions ψ and T are subject to the following boundary conditions:

$$y = y_{1,2}, \qquad\qquad \psi = \psi_\nu = 0, \qquad T = \mp1, \qquad (5.5)$$

$$\psi(0,y) = \psi(L,y), \qquad\qquad T(0,y) = T(L,y). \qquad (5.6)$$

where ψ_ν denotes the normal derivative of ψ on the boundaries.

The discretization in time of Eqs.(5.2),(5,4) is performed by using the implicit, 2-d order stiffly stable scheme of [8]. It results in two elliptic equations of the type Eq.(3.1), with the parameter $\lambda \propto 1/\sqrt{\Delta t}$, where Δt is the time stepping increment. These elliptic equations, along with Eq.(5.3), are solved repeatedly at each time step by the generalized MDLF algoritm, as described in previous sections.

For typical Δt, using in computations, $\lambda \gg 1$ that justifies the implementation of the local matching procedure for the Helmholtz type equations (3.1). As for the Poisson equation (5.3), it implies the use of the global matching procedure. However, it is shown in [2], that even in this case the global matching is needed only for a few (low-frequency) harmonics, while the most (high-frequency) harmonics can be treated locally.

A particular issue, raised by the $\psi - \phi$ formulation, is the computation of ϕ on the rigid boundaries. Low order approximations, like Tom or Pearson formulas, are not appropriate for a high order method. A spectrally accurate approximation to ϕ on the boundaries can be obtained by using the algorithm by [10]. It consists in iterating the system Eqs.(3.1), (5.3) along with the following (penalty-like) procedure on the boundaries:

$$y = y_{1,2} \qquad\qquad \phi^{s+1} = \phi^s - K\psi_\nu^s \qquad (5.7)$$

The iteration scheme Eq.(5.7) compels the normal derivative ψ_ν to vanish on the boundaries, in agreement with the second b.c. of Eq.(5.5). In the simplest case of the straight boundaries, the optimal value of the iteration parameter is $K = \lambda$. The computation of $K(x)$ for general geometries is described in [10]. Since K is large, the iterations Eq.(5.7) converge very fast. The typical dependence of ψ_ν^s on s is shown on Fig. 5.1.

The algorithm has been tested at first for the plane channel ($\delta = 0$); the parallel computation was simulated on a serial machine. The well known critical Grashof number $Gr_* = 497$, which is the stability threshold for the main flow with the cubic velocity profile, has been found with the spectral accuracy, see Fig. 5.2, 5.3.

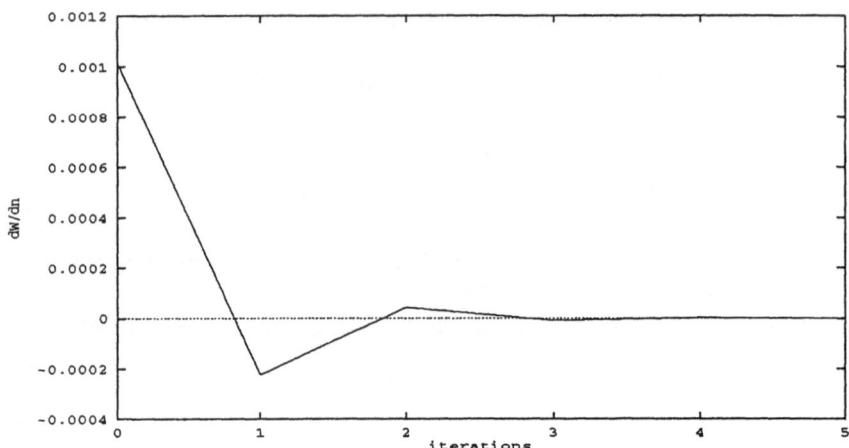

Fig. 5.1. Convergence of the iteration procedure Eq. (5.7)

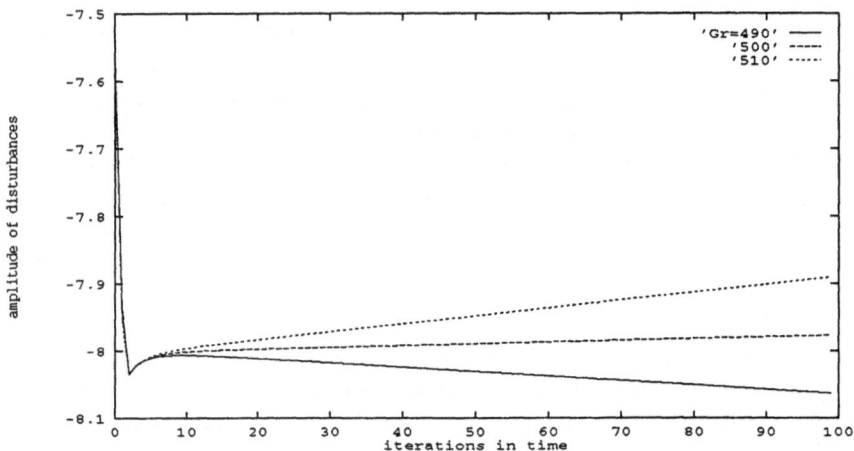

Fig. 5.2. Exponential grow (*decay*) of the disturbances at $Gr > G_*$ ($Gr < G_*$)

The steady-state flow pattern in a wavy channel, computed for the parameters $\delta = 0.1, k_w = 1.4, Gr = 400, Pr = 0$, is shown on Fig. 5.4, 5.5.

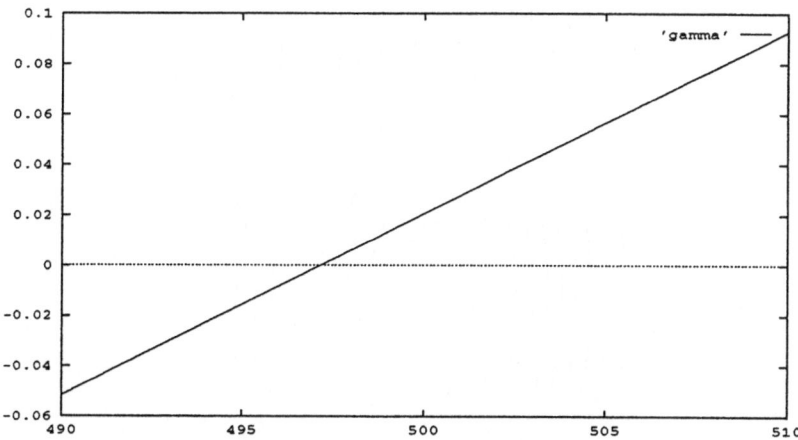

Fig. 5.3. The rate of decay γ as a function of Grashof number ($\gamma = 0$ corresponds to G_*)

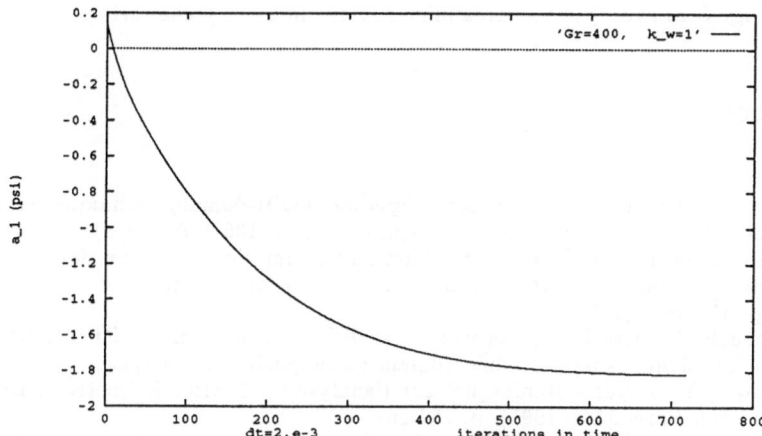

Fig. 5.4. Evolution in time of the stream function (the amplitude of the main harmonic $k = k_w$)

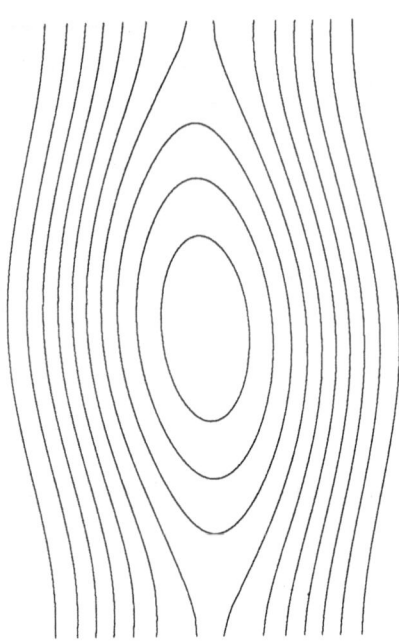

Fig. 5.5. Steady-state streamline pattern in a wavy channel

References

1. M. Israeli, L. Vozovoi, A. Averbuch - Spectral multi-domain technique with Local Fourier Basis, *Journ. Scient. Comp.*, **8**, N 2, p. 135-149, (1993).
2. M. Israeli, L. Vozovoi, A. Averbuch - Parallelizing implicit algorithm for time-dependent problems by parabolic domain decomposition, *Journ. Scient. Comp.*, **8**, N 2, p. 151-165, (1993).
3. A. Averbuch, M. Israeli, L. Vozovoi - Parallel implementation of non-linear evolution problems using parabolic domain decomposition, to appear.
4. R. Coifman, Y. Meyer - Remarques sur l'analyse de Fourier à fenêtre, série, *C.R.Acad.Sci.* Paris **312** (1991), p. 259-261.
5. S. A. Orszag - Spectral methods for problems in complex geometries. *Numerical methods for PDE's*, 1979.
6. C. Canuto, M.Y. Hussaini, A. Quarteroni, T.A. Zang - Spectral methods in fluid dynamics, *Series in Computational Physics*, Springer-Verlag, (1989)
7. L. Vozovoi, M. Israeli, A. Averbuch - Analysis and application of Fourier-Gegenbauer method to stiff differential equations, to appear in *SIAM, J.Numer.Anal.*
8. G.E. Karniadakis, M. Israeli, S. Orszag - High-order splitting methods for the incompressible Navier-Stokes equations, *J.Comp.Phys.*, **97**, No. 2, (1991), p. 414-443
9. L. Vozovoi - Convection in a vertical layer with undulating boundaries, *Fluid Dynamics*, **11**, N 2, (1976), p. 202-206
10. M. Israeli - A Fast implicit numerical method for time dependent viscous flows, *Studies in Applied Mathematics*, **49**, N 4, p. 327-349, (1970)

Confined Swirling Flows–A Continuing Challenge

Pinhas Bar-Yoseph

Computational Mechanics Laboratory, Faculty of Mechanical Engineering, Technion, Haifa 32000, Israel

1. Introduction

Secondary flows occurring in rotating machinery demonstrate potential instability and multiple flow modes regimes. A better understanding of the secondary flow structure can lead to improvement in the performance, reliability and cost of many existing rotating machines, as well as may serve to stimulate innovations.

Taylor vortex flow between eccentric coaxial rotating spherical annuli (Fig. 1.1b) with medium gap [6], exhibits a transition with hysteresis into a flow with one pair of Taylor vortices (denoted as $0 \to 1$). As the eccentricity increases, the magnitude of the hysteresis decreases until it disappears. A new nonsymmetric flow mode with *three* pairs of vortices was explored in the case of concentric spheres with thermal effects [5]. Recently, a new $0 \to 1 \to$ "0" transition caused by thermo-mechanical effects was found [8]. For a wide gap, *five* different flow modes exist for the same set of physical parameters [5]. The possibility of a vortex breakdown in a spherical gap was numerically and experimentally observed for the first time by the author and his colleagues [1, 2, 3, 4]. Recently, *steady state* transitions from vortex breakdown to multiple flow regimes were explored [8]. A numerical investigation of the stability, onset of oscillatory instability, and slightly supercritical unsteady regimes of the secondary flow confined in a disk-cylinder system (Fig. 1.1a) is presented in [12, 13]. It is shown that the appearance of vortex breakdown is not caused by instability. More recently, the influence of shear -thinning and -thickening of fluid properties on the vortex breakdown phenomenon was investigated [7].

In this chapter we describe the numerical study of Newtonian and non-Newtonian flow and heat transfer in several axisymmetric enclosures (Fig. 1.1) which show interesting features of secondary flows.

2. Statement of the Problem

The laminar, axisymmetric flow patterns for a rotating fluid contained in an enclosure formed between the rotor and the inner surface of the container have

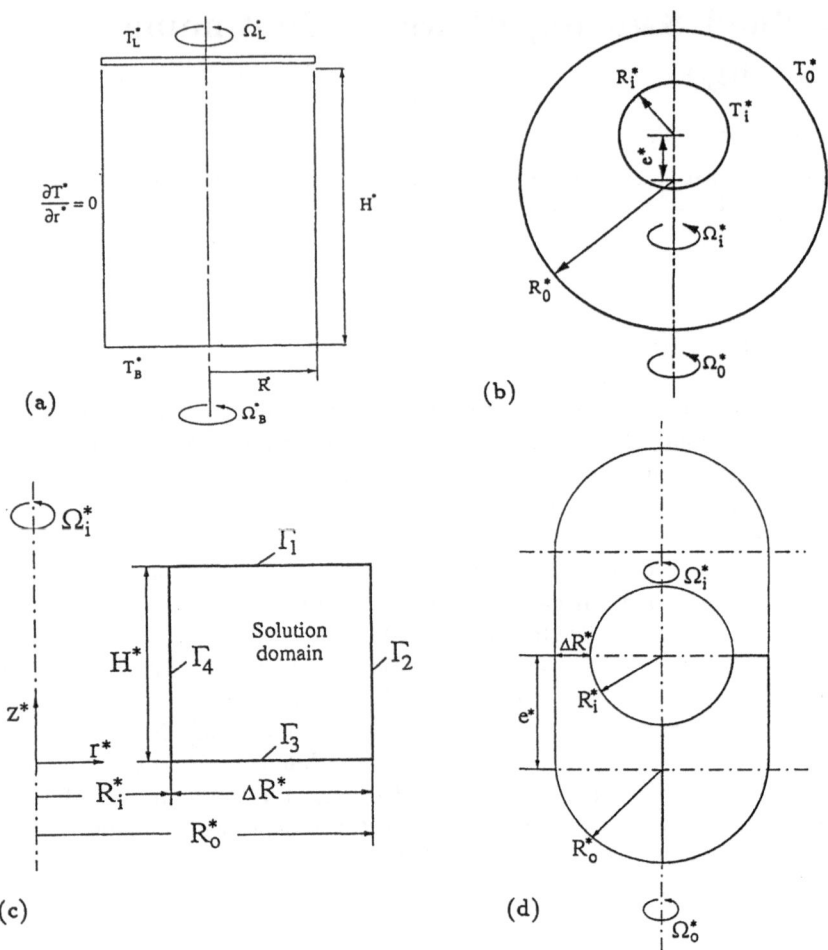

Fig. 1.1. Rotor-container systems: Geometry and definitions

been studied numerically. The flow field is mathematically described by the solution of the continuity, momentum and energy equations. The body forces considered here are those due to gravitational buoyancy, centrifugal, and Coriolis accelerations [1, 10, 14]. It is assumed that in the energy equation, the dissipation of mechanical energy and work of compression are negligible. Each rigid boundary is assumed to be impermeable and isothermal (different boundaries have different temperatures), either rotating at a constant angular velocity or is stationary. There are different ways to nondimensionalize the variables and therefore, in each case, the characteristic parameters are differently defined. Use of the laminar flow equations and Boussinesq's equation of state limits the Reynolds number to a value of the order of 10^3 and the Rayleigh number to a value of the order of 10^6. All the calculations were performed for Prandtl number $Pr = 0.7$ ($Pr := \frac{\nu^*}{\alpha^*}$; an asterisk denotes a dimensional quantity), corresponding to air.

2.1 Disk-Cylinder System

Consider a circular cylinder, of radius R^* and height H^* completely filled with a Newtonian incompressible fluid (Fig. 1.1a). The lid rotates at a constant angular speed Ω_L^* and the cylindrical container is stationary. The scale factors are: R^* for length, $R^* \Omega_L^*$ for velocity. The physical parameters are: $Re := \frac{\Omega_L^* R^{*2}}{\nu^*}$ (Reynolds number); $\kappa := \frac{H^*}{R^*}$ (aspect ratio).

2.2 Spherical Annulus

The laminar, axisymmetric flow of a Boussinesq fluid contained within *concentric* spherical annuli is investigated for the case where the outer spherical shell rotates with a constant angular velocity Ω_o^*, about a vertical axis in an axial gravitational field, and is kept at a given temperature T_o^*. The inner sphere rotates with a constant angular velocity Ω_i^* and is kept at a lower temperature T_i^* (Fig. 1.1b). The scale factors are: R_o^* for length, $\Omega_i^* R_o^*$ for velocity , and $\Delta T^* = T_o^* - T_i^*$ for the temperature difference. The nondimensional temperature is defined as $\Theta := (T^* - T_i^*)/(T_o^* - T_i^*)$. The physical parameters are: $Ra := \frac{g^* \beta^* \Delta T^* R_o^{*3}}{\nu^* \alpha^*}$ (Rayleigh number); $Re := \frac{\Omega_i^* R_o^{*2}}{\nu^*}$ (Reynolds number); $\xi := \frac{\Omega_o^*}{\Omega_i^*}$ (rotation ratio); $\delta := \frac{R_i^*}{R_o^*}$ (radii ratio) or, alternatively, the dimensionless gap $s = 1 - \delta$ (difference of radii divided by the outer radius) .. α^* is the thermal diffusivity ; β^* is the thermal expansion coefficient; ρ_0^* is a reference density; ν^* is the kinematic viscosity. Non-Newtonian effects are investigated for fluids obeying the Carreau-Yasuda constitutive relationship [9]:

$$\frac{\eta^* - \eta_\infty^*}{\eta_0^* - \eta_\infty^*} = [1 + (\lambda^* \, \dot{\gamma}^*)^a]^{(n-1)/a} \tag{2.1}$$

where η_0^* is the zero-shear-rate viscosity, η_∞^* is the infinite-shear-rate viscosity (η_∞^* is taken to be zero), λ^* is a time constant, $\dot{\gamma}^*$ is the rate-of -strain tensor, n is the "power law exponent" and a is a dimensionless parameter that describes the transitlon between the zero-shear-rate region and the power law region. This constitutive relationship can be rewritten as

$$\frac{\eta^* - \eta_\infty^*}{\eta_0^* - \eta_\infty^*} = [1 + (De \, \dot{\gamma}^*)^a]^{(n-1)/a} \tag{2.2}$$

where $De := \lambda^* \Omega^*$ (Deborah number).

2.3 Cylinderical Annulus

Consider a vertical annulus of inner radius R_i^*, outer radius R_o^* and height H^* (Fig. 1.1c) completely filled with a Boussinesq fluid. The inner and lower surfaces of the cavity (denoted as Γ_4 and Γ_3, respectively) are heated and rotating at a constant angular velocity Ω^*, while the other two surfaces (denoted as Γ_1 and Γ_2) are cooled and stationary. This case simulates the upper annulus in an electric motor [10, 14]. The scale factors are: ΔR^* for length, $\Delta R^* \Omega^*$ for velocity, and $\Delta T^* = T_i^* - T_o^*$ for the temperature difference. The nondimensional temperature is defined as $\Theta := (T^* - T_o^*)/(T_i^* - T_o^*)$. The geometry is determined by two characteristic parameters: the aspect ratio $\kappa := \frac{H^*}{\Delta R^*}$ and the radii ratio $\delta := \frac{R_o^*}{R_i^*}$. The other two physical parameters which define the flow are: $Ra := \frac{g^* \beta^* \Delta T^* \Delta R^{*3}}{\nu^* \alpha^*}$ (Rayleigh number) and $Re := \frac{\Omega_i^* \Delta R^{*2}}{\nu^*}$ (Reynolds number). The finite element code was checked by a comparison with previously published numerical data [10, 14] for this test case.

2.4 Sphere-Capsule System

Consider a sphere of inner radius R_i^*, rotating at a constant angular velocity Ω_i^* and enclosed in a capsule with outer radius R_o^*. The distance between the centers is denoted by e^* (Fig. 1.1d). The gap is filled with an incompressible fluid. The scale factors are: R_i^* for length, $R_i^* \Omega_i^*$ for velocity. The physical parameters are: $Re := \frac{\Omega_i^* R_i^{*2}}{\nu^*}$ (Reynolds number); $\delta := \frac{R_i^*}{R_o^*}$ (radii ratio) or, alternatively $s = 1 - \delta$ (dimensionless gap); $\epsilon := \frac{e^*}{\Delta R^*}$ ("eccentricity" ratio). This enclosure is designed for demonstrating possible interaction between Taylor vortices which appear at the the equatorial region and the vortex breakdown bubble which arises at the north and south poles.

3. Numerical Formulation

The penalty Galerkin finite element method, described in [1, 6], has been employed for the numerical solution of the rotationally symmetric continuity, momentum, and energy equations in their primitive form with proper boundary conditions. A biquadratic Lagrangian interpolation is employed for the velocity and temperature fields and a linear interpolation for the pressure field. The discrete nonlinear set of algebraic equations that results from the finite element formulation can be written as $\mathcal{F}(U, \chi) = 0$, where U is the N-dimensional vector that consists of the nodal values of the velocity and temperature fields, and χ is the n-dimensional parameter set. The solution of this set of nonlinear algebraic equations is an n-dimensional surface in $\Re^{N+n}$. Solutions by low order continuation algorithms are used to trace branches in the parameter space along which lie steady flow states. In the present study, a continuation in Re, and Ra, was used. At each continuation step, a nonlinear set of algebraic equations was solved by a Newton iteration, and the approximation to the solution at each iteration was found by a modified vectorized version of the frontal solver for in-core solution of large, sparse nonsymmetric systems of the linearized set of algebraic equations. Numerical solutions for the disk-cylinder case are calculated by three methods: finite element, finite volume, and Galerkin spectral methods [13].

4. Results and Discussion

4.1 Disk-Cylinder System

The details of the vortex breakdown structure are much more demanding on a numerical method than the relatively simple structure of the outer flow. Therefore, the three different methods (finite element, finite volume and Galerkin spectral methods) were checked by a comparison with previously published numerical and experimental data for a range of values of Reynolds number ($1000 \leq Re \leq 2494$) and aspect ratio ($1.58 \leq \kappa \leq 3.25$) [1, 13]. Streamline, ψ, and moment of azimuthal velocity, $\mathcal{M}_\phi = r v_\phi$ (solid lines), and the corresponding most unstable perturbation modes (dashed lines) are displayed in Fig. 4.1 (the solution is presented in a compact form: each frame is composed of a right and left section which depicts the ψ- and $\mathcal{M}_\phi$- isolines, respectively), for aspect ratio, $\kappa = 1.5, 2.5, 3.25$. The corresponding values of the critical Reynolds number are: Re_c= 2724, 2705, 3204. The numerical results were calculated by the Galerkin spectral method using 30 modes.

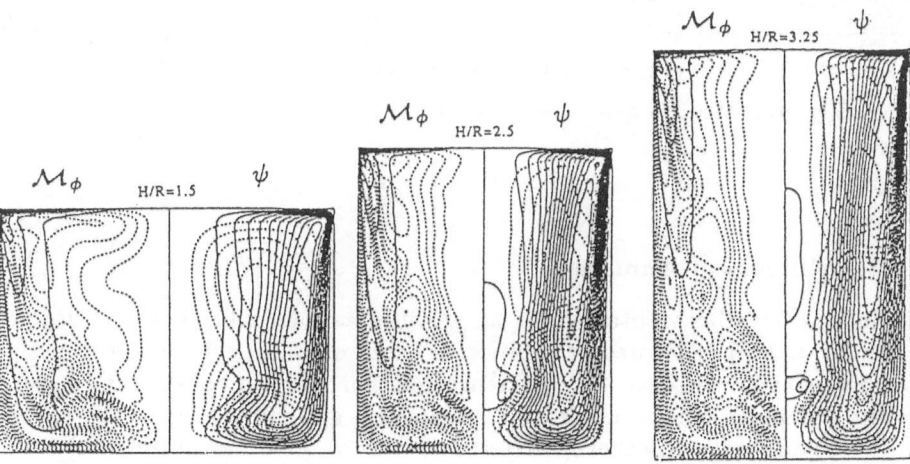

Fig. 4.1.

4.2 Spherical Annulus

The new $0 \rightarrow 1 \rightarrow 2$ *steady state* transition depicted in Fig. 4.2 (results are shown for $Re = 1000, 1100, 1200, 1300$; $Ra = 50000$; $\xi = 0$), indicates that in mixed-convection, unlike to incompressible flows, a gradual steady state transition from the basic flow to a flow with two pairs of vortices via a flow with one pair of vortices is possible. Also, the abrupt change from a flow with two pairs of vortices to a flow with three pairs of vortices which occurs at $Ra_c \approx 3.75 \times 10^5$ can be achieved by a gradual transition along the following computational path (Fig. 4.3): (i) $Re = 1100, Ra \times 10^{-5} = 1.0, 2.0, 3.0, 3.6, 4.0$; (ii) $Re = 1100, 1300, 1500$; $Ra = 4 \times 10^5$.

The effect of shear-thinning of fluid properties on the onset of the vortex breakdown bubble is demonstrated in Fig. 4.4 (streamlines are shown for $Re = 4500$, $De = 0.0, 0.6, 4.0, 6.0$; $\xi = 0$).

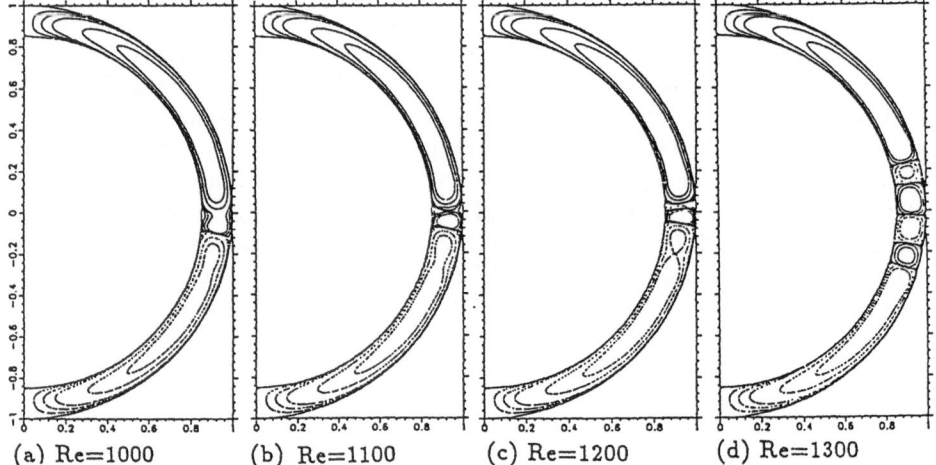

(a) Re=1000   (b) Re=1100   (c) Re=1200   (d) Re=1300

Fig. 4.2. Medium gap regime - the $0 \rightarrow 1 \rightarrow 2$ transition ($R_a = 50000$)

4.3 Cylinderical Annulus

Solutions have been obtained for an air-filled cavity (Table 1; $\kappa = 1, Re = 10^2, Ra = 10^4$). The present solution is based on a 15×15 uniform mesh of biquadratic elements. The reference solution is based on a second order finite difference scheme using 31×31 uniform grid [10, 14]. Comparison of the streamline and isotherm patterns also show good agreement with the reference solution [11].

Table 4.1. Cylindrical annulus – extremum stream function values.

δ	[10]		This work	
	Ψ_{max}	Ψ_{min}	Ψ_{max}	Ψ_{min}
1.2	0.999	-1.8×10^{-4}	0.992	-2.5×10^{-4}
1.5	0.212	-1.06×10^{-2}	0.21	-2×10^{-2}
2.0	5.6×10^{-2}	-5.9×10^{-2}	5.9×10^{-2}	-5.4×10^{-2}
4.0	1.4×10^{-3}	-6.2×10^{-2}	1×10^{-3}	-6×10^{-2}
8.0	2.55×10^{-4}	-4.9×10^{-2}	2.4×10^{-4}	-5×10^{-2}

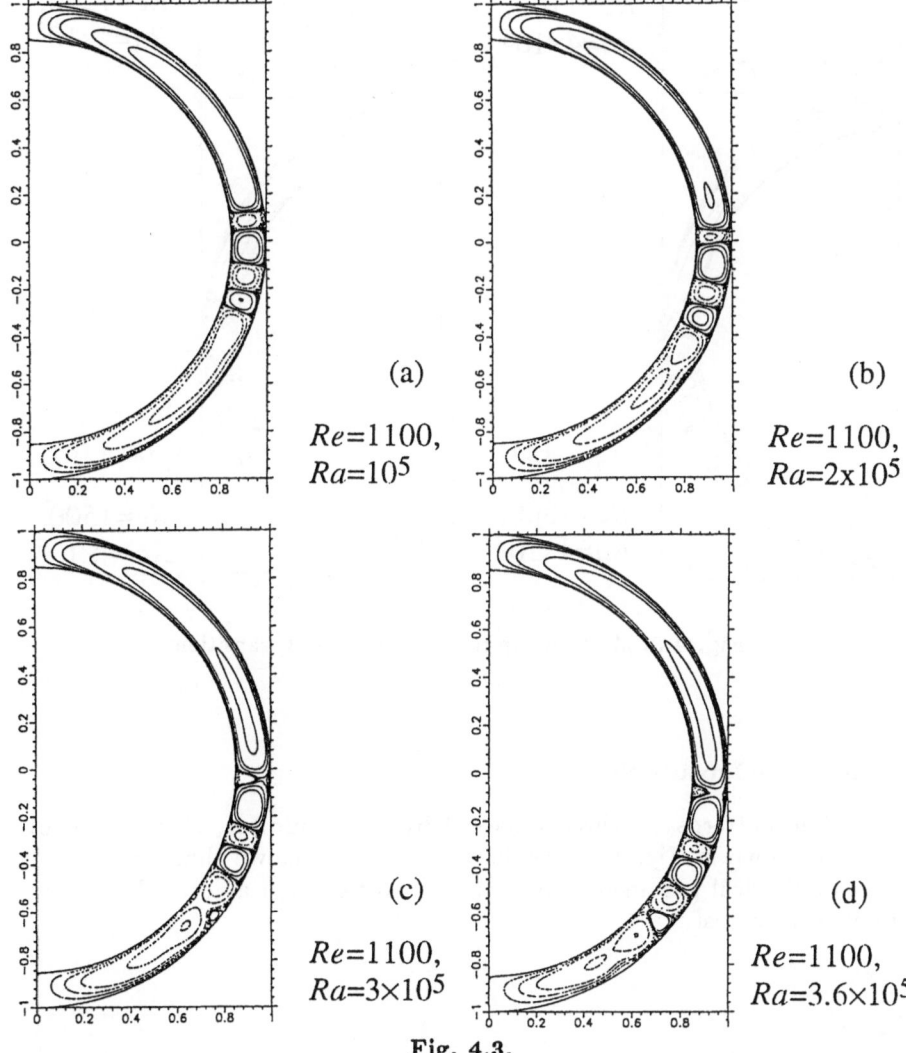

(a) $Re=1100$, $Ra=10^5$

(b) $Re=1100$, $Ra=2 \times 10^5$

(c) $Re=1100$, $Ra=3 \times 10^5$

(d) $Re=1100$, $Ra=3.6 \times 10^5$

Fig. 4.3.

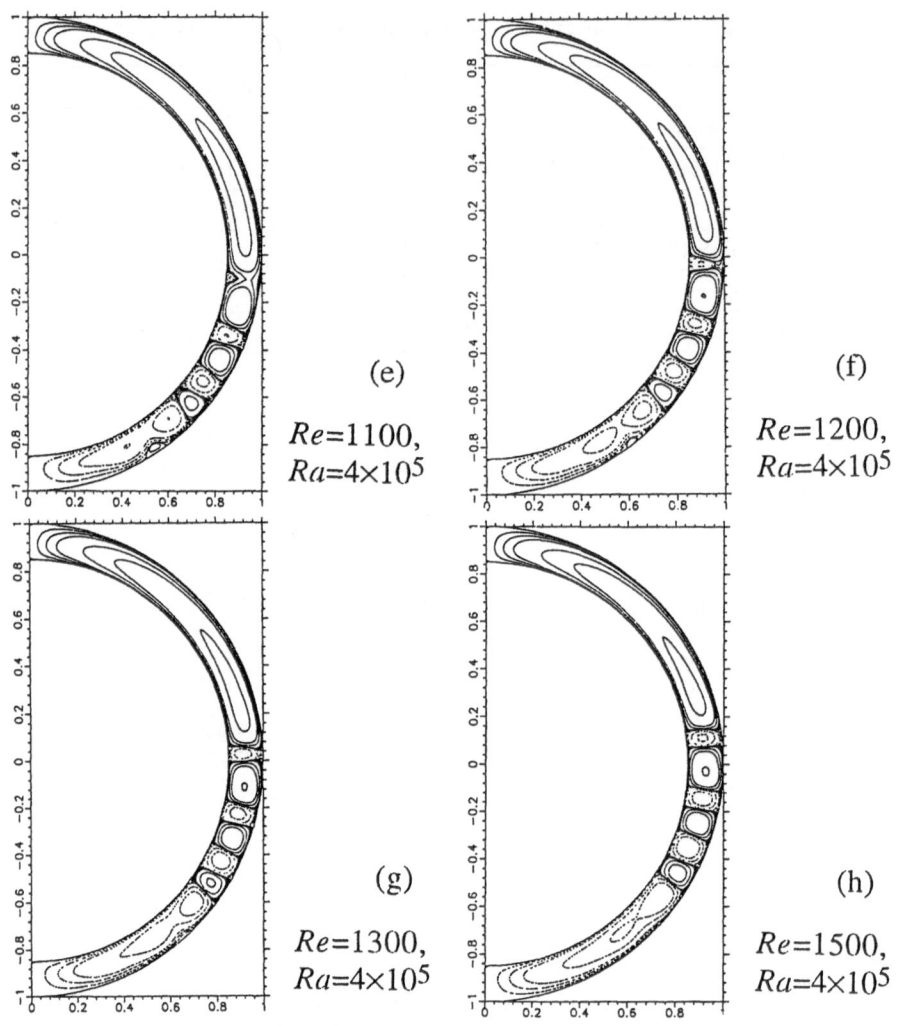

(e) $Re{=}1100,$ $Ra{=}4{\times}10^5$

(f) $Re{=}1200,$ $Ra{=}4{\times}10^5$

(g) $Re{=}1300,$ $Ra{=}4{\times}10^5$

(h) $Re{=}1500,$ $Ra{=}4{\times}10^5$

Fig. 4.3. Medium gap regime - the $2 \to 3$ transition

4.4 Sphere-Capsule System

Streamline patterns for three values of Reynolds number, $Re{=}1895$, 4061, 4964 are shown in Fig. 4.5 ($s{=}0.05$, $\epsilon{=}0.55$). It is shown that for a certain range of physical parameters, both Taylor vortices and a vortex breakdown bubble can coexist.

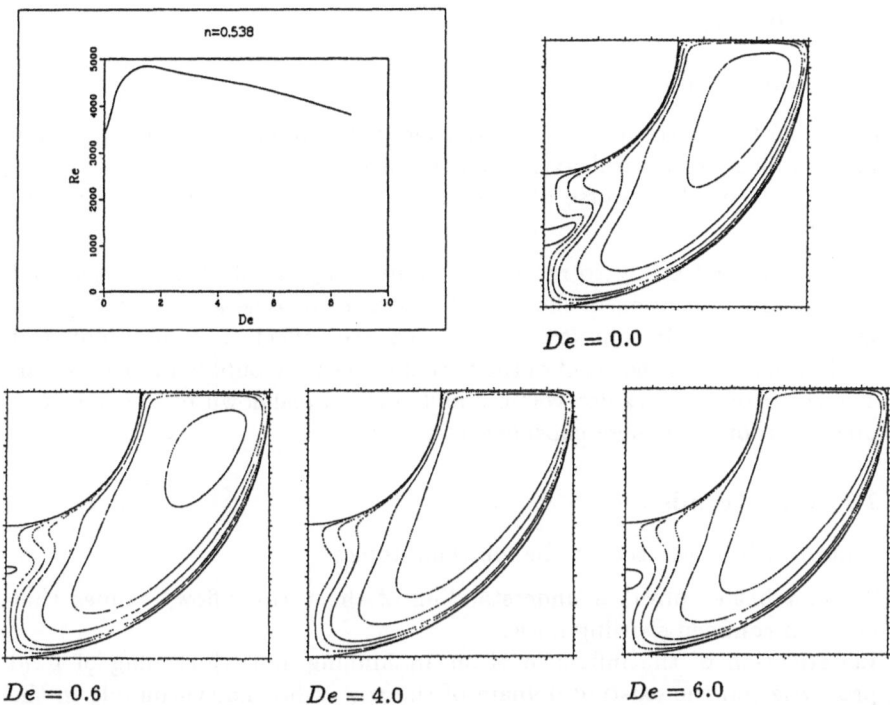

Fig. 4.4. Wide gap regime: Non-Newtonian effects on the onset of single vortex breakdown bubble

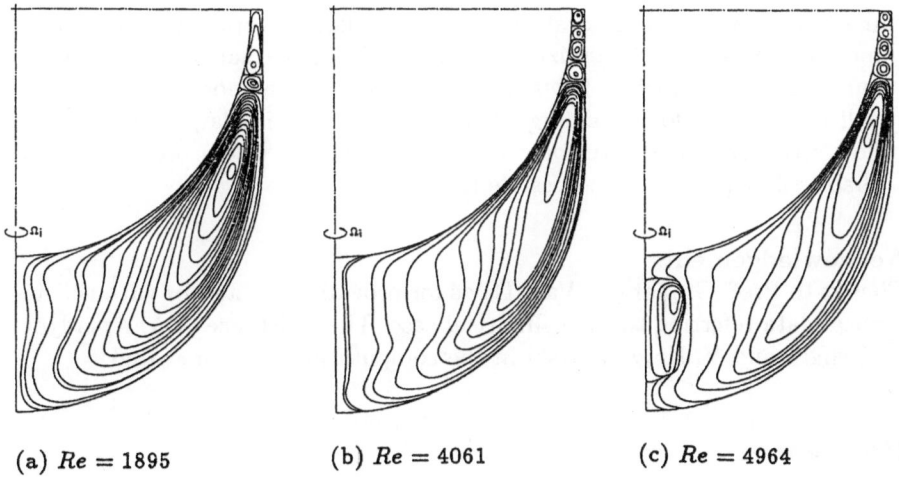

Fig. 4.5. Sphere-Capsule system: Coexistence of Taylor vortices and a vortex breakdown bubble

5. Conclusions

5.1 Present Study

The laminar flow patterns and heat transfer for Newtonian and non-Newtonian fluids contained in several axisymmetric enclosures have been studied numerically. During the course of investigation several new phenomena were explored:

- This is the first study to reproduce numerically the $0 \rightarrow 1 \rightarrow 2$ gradual steady state transition in the case of medium spherical gap($s = 0.15$).
- In the case of wide spherical gap($s = 0.5$), the effect of shear-thinning of fluid properties on the onset of the vortex breakdown bubble was examined.
- The coexistence of Taylor vortices and vortex breakdown phenomena was presented for the sphere-capsule system.

5.2 Future Trends

Attention will be focused on the following topics:

- To gain better physical understanding of the various flow regimes that occur in confined swirling flows.
- Investigation of the influence of shear-thinning and -thickening of fluid properties on the existing domain of the vortex breakdown bubble in the space of physical parameters.
- Developing high order continuation methods for revealing the solution behaviour for those cases where the present numerical approach failed to converge.
- Investigation of the temporal evolution of the flow structure in order to gain more insight into the stabilizing and destabilizing mechanisms that can occur in confined swirling flows. It seems that high temporal discretization will be required for obtaining accurate numerical simulation of unsteady flow structures in the regions of parameter space, where oscillatory and chaotic flow patterns are possible [15].

Acknowledgements
Thanks to Prof. Graham de Vahl Davis for providing the author with the numerical data referring to the cylindrical case. The assistance of A. Arkadyev, A. Gelfgat and Y. Kryzhanovski in the computations is appreciated.

References

1. Arkadyev, A., Bar-Yoseph, P., Solan, A. Roesner, K.G.: Thermal effects on axisymmetric vortex breakdown in a spherical gap. Phys. Fluids A **5** (1993) 1211–1223

2. Bar-Yoseph, P., Seelig, S., Solan, A., Roesner, K.G.: Vortex breakdown in spherical gap. Phys. Fluids **30** (1987) 1581–1583
3. Bar-Yoseph, P., Solan, A. Roesner, K.G.: Sekundärstromung in axialsymmetrischen hohlräumen. Z. Angew. Math. Mech. (ZAMM) **70** (1990) 442–444
4. Bar-Yoseph, P., Roesner, K.G., Solan, A.: Vortex breakdown in the polar region between rotating spheres. Phys. Fluids A **4** (1992) 1677–1686
5. Bar-Yoseph, P., Even-Sturlesi, G., Arkadyev,A., Solan, A., Roesner, K.G.: Mixed-convection of rotating fluids in spherical annuli. in *Proc. 13th International Conference on Numerical Methods in Fluid Dynamics*, ed. M. Napolitano, F. Sabetta (Lecture Notes in Physics ,**414**, Springer-Verlag, New York), (1992) 381–385
6. Bar-Yoseph, P., Solan, A., Hillen, H., Roesner, K.G.: Taylor vortex flow between eccentric coaxial rotating spheres. Phys. Fluids A **2** (1993) 1564–1573
7. Bar-Yoseph, P., Kryzhanovski,Y.: Non-Newtonian effects on axisymmetric vortex breakdown in spherical annuli. in *2nd European Fluid Dynamics Conference*, Warsaw, Poland, 20-24 September (1994)
8. Bar-Yoseph, P.: On multiple flow patterns and vortex breakdown phenomena in confined rotating flows. Computational Fluid Dynamics J. (to appear)
9. Bird,R.B., Armstrong, R.C., Hassager, O.: *Dynamics of Polymeric Liquids* **1** , (Wiley, New York) (1987)
10. de Vahl Davis,G., Leonardi, E., Reizes, J.A.: Convection in a rotating annular cavity. in *Heat and Mass Transfer in Rotating Machinery*, ed. D.E. Metzger, N.H. Afgan (Hemisphere, Washington) (1984) 131–142
11. Even-Sturelsi, G.: *Finite Element Solution of Compressible Navier-Stokes Equations in Rotating Flow* , (Mechanical Engineering, Technion, Haifa) (1991)
12. Gelfgat, A., Bar-Yoseph, P., Solan, A.: Stability of a confined swirling flow. in *2nd European Fluid Dynamics Conference*, Warsaw, Poland, 20-24 September (1994)
13. Gelfgat, G., Bar-Yoseph, P. and Solan, A.: Stability of confined swirling flow with and without vortex breakdown. (submitted for publication)
14. Hessami, M.A., de Vahl Davis, G., Leonardi, E., Reizes, J.A.: Mixed convection in vertical, cylindrical annuli. Int. J. Heat Mass Transfer **30** (1987) 151–164
15. Lopez, J.M.: Axisymmetric vortex breakdown Part 3. Onset of periodic flow and chaotic advection. J. Fluid Mech. **234** (1992) 449–471

Eshelbian Continuum Mechanics and Nonlinear Waves

G.A. Maugin

CNRS-Laboratoire de Modélisation en Mécanique, Université Pierre et Marie Curie, Paris

Summary. Problems in fracture mechanics and elastic-defect theory, whether quasi-static or dynamical, find their best expression on an abstract material manifold rather than in physical space. This we called Eshelbian mechanics. We have shown that the additional balance laws in material space can be used not only for checking the accurancy of numerical schemes but also as a means to capture the essential solitonic features of systems which are only nearly integrable. In such systems the solutions are viewed as quasi-particles which are governed by perturbed Newtonian or Einsteinian equations of motion on the material manifold. Perturbation forces in these equations (which necessarily reduce to true conservation laws for the related completely integrable systems) have the same nature as the pseudo-forces that act on defects. This is illustrated by examples originating essentially from crystal dynamics.

1. Introduction

In recent works ([1]-[4]) we have presented a formulation of nonlinear continuum mechanics which, in order to make explicit smooth material inhomogegeities, employs a *complete* projection of balance laws, including that of momentum, onto the *material manifold*, i.e., that abstact manifold whose elements are the "particles" of continuum mechanics. This somewhat abstract approach which gives a special status to *material* coordinates (not to be mistaken for the usual *Lagrangian* coordinates of fluid mechanics[1]) is particularly well suited to the formulation of elastic-defect theory and that of the macroscopic theory of fracture [3],[4]. We have further shown that the same formalism provided useful additional balance laws that help one in the study of characteristic properties of nonlinear waves, such as *solitary waves* and *solitons*, which are supported by some nonlinear elastic dispersive systems [6]. The purpose of the present contribution is to elaborate on this last aspect and, with two companion papers - one devoted to *fracture* [7], and the other to *defects* and *phase-transition layers* [8] - to complete and extend

---

[1] It seems that G. Piola to whom we owe the notions of "Piola" finite strain and "Piola-(Kirchhoff)" stress, was the first to make such a clear distinction between an abstract *reference* configuration (which he took of uniform unit density) and a Lagrangian configuration - the latter is a configuration actually occupied by the material at a specific instant of time often called t_0 or simply $t = 0$ (thus it belongs to the time sequence of configurations occupied by the body in physical space, what is *not necessarily* the case of a reference configuration [5]).

the contents of our resent book [4]. Before entering details of the approach
we must recall our immense debt to J.D. Eshelby [9]-[10] and D. Rogula [11]
to whom the basic notions of "force on a singularity" and "material force"
belong; hence the coinage of *Eshelbian mechanics* for that mechanics which
has for arena the *material manifold*.

2. Momentum and Pseudo-Momentum; Force and Pseudo-Force

We start with a specification of the notions of *momentum* and *pseudo-momentum* in a general material continuum subjected to deformation and
motion. *Linear momentum*, here called *physical momentum*, strictly is the
"quatity" of motion in Descartes' original sense, i.e., the product of the "quantity of matter", *density* for a continuum, and *velocity* (physical velocity: the
time rate of change of the *actual position* or *placement x* in *physical space*).
More precisely, the *physical momentum* per unit volume of the *reference* configuration $\mathcal{K}_R$ of a material body $\mathcal{B}$ is given by (e.g., [12])

$$\mathbf{p}(X,t) = \rho_0(\mathbf{X})\mathbf{v}(\mathbf{X},t), \qquad \mathbf{v} \equiv \left.\frac{\partial \chi}{\partial t}\right|_{\mathbf{x}}, \qquad (2.1)$$

where $\mathbf{x} = \chi(\mathbf{X},t)$ is the smooth motion of the material between its *reference*
configuration (of *material* coordinates $\mathbf{X} = \{\mathbf{X}^K, K = 1,2,3\}$) , and the
actual configuration in physical space at time $t, \mathcal{K}_t$, of *Eulerian* coordinates
$\mathbf{x} = \{\mathbf{x}^i, i = 1,2,3\}$). The *material configuration*, in the case of a solid-like
behavior, is ideal in the sense that it is both stress-free and strain-free; it is a
so-called *natural* configuration[2]. The possible dependence of ρ_0, the reference
matter density, on $\mathbf{X}$ (and not on $\mathbf{x}$!) indicates a possible (here supposed to
be smooth) *material inhomogeneity* in the *inertial* properties of our material.
the $\mathbf{X}$s and t are independent space-time variables, which is *not* the case of
$\mathbf{x}$ and t. The *finite-deformation* matrix between $\mathcal{K}_R$ and $\mathcal{K}_t$, $\mathbf{F}$, is defined
by $\mathbf{f} = \left.\frac{\partial \chi}{\partial \mathbf{X}}\right|_{t\,fixed}$ and has the components F_K^i. According to recent works
([1]-[4]), *pseudomomentum* $\mathcal{P}$, per unit volume of $\mathcal{K}_R$, of a material is defined
alternately by the two definitions

$$\mathcal{P}(\mathbf{X},t) = -\rho_0(\mathbf{X})\mathbf{F}^T \cdot \mathbf{v} = \rho_0(\mathbf{X})\mathbf{C} \cdot \mathbf{V}, \qquad (2.2)$$

where the superscript T indicates transposition, and we emphasize the presence of a *minus* sign in the first definition. Thus $\mathcal{P}$ is, in terms of the theory
of *geometrical objects on manifolds*, a **material covector** on the *material*

---

[2] This coinage is somewhat misleading; for instance, none of the configurations
that we may think of on earth is *natural* as gravitation is always operative and
it varies point-wise, so that no strictly stress-free state exists

manifold defined at $\mathcal{K}_R$ (and *not* in physical space). The second definition contained in (2.2) just means the following: The fields $\mathbf{V}$ and $\mathbf{C}$ are defined by

$$\mathbf{V} = \left. \frac{\partial \chi^{-1}}{\partial t} \right|_{\mathbf{x}}, \qquad \mathbf{C} = \mathbf{F}^T \cdot \mathbf{F}, \qquad (2.3)$$

so that $\mathbf{V}$ is a *material* **contravariant** vector, the so-called *material* velocity, obtained by taking the partial time derivative of the inverse $\chi^{-1} = \chi^{-1}(\mathbf{x}, t)$ of the motion χ (provided $J \equiv \det \mathbf{F} > 0$ always), at fixed actual position, and $\mathbf{C}$ may be considered alternately as a *material* measure of finite strains or a *deformed metric* which helps one to convert the contravariant object $\mathbf{V}$ into the covariant one $\mathcal{P}$ up to the factor ρ_0. This is the precise meaning of pseudomomentum. In the absence of dissipative processes, i.e. when all equations are derivable from Hamilton's formulation of Lagrange analytical mechanics: *pseudomomentum is the conjugate of the material coordinates* $\mathbf{X}$ (see below). [3]

In the same way as the *local balance of physical momentum* (referred to as the first equation of motion of Euler and Cauchy in continuum mechanics) is generated, via the notion of *virtual work*, by an infinitesimal change in *actual position* in physical space (keeping the "labelling" of the "particles", i.e. $\mathbf{X}$, fixed), a local balance of pseudomomentum will be generated by an infinitesimal change in the $\mathbf{X}$s, keeping the actual position $\mathbf{x}$ fixed. The two variations, obviously noted by the symbolisms $\delta_{\mathbf{X}}$ and $\delta_{\mathbf{x}}$, are consistent with one another, and are related by

$$\delta_{\mathbf{X}} \chi + \mathbf{F} \cdot \delta_{\mathbf{x}} \chi^{-1} = 0, \qquad (2.4)$$

in the same way as $\mathbf{v}$ and $\mathbf{V}$ can be shown to be related by

$$\mathbf{v} + \mathbf{F} \cdot \mathbf{V} = 0, \qquad \text{or} \qquad \mathbf{V} = -\mathbf{F}^{-1} \cdot \mathbf{v}, \qquad (2.5)$$

if $\mathbf{F}^{-1}$ is the inverse of $\mathbf{F}$, hence the equivalence between the two definitions (2.2).

In classical nonlinear continuum mechanics the time rate of change of physical momentum is balanced by the material divergence of the so-called Piola-Kirchhoff stress tensor $\mathbf{T}$ and external, also called *body* or *physical* forces $\mathbf{f}^{\text{ext}}$ (such as gravity; they depend on $\mathbf{x}$). The time rate of change of *pseudomomentum* is balanced by the material (with respect to $\mathbf{X}$) divergence of the so-called *Eshelby* (energy-momentum) stress tensor $\mathbf{b}$ - of which we shall give an expression later on) and possibly "forces" that we may call *pseudo-forces*

---

[3] In analytical *point mechanics*, $\mathbf{p} = m\mathbf{v}$, while $\mathcal{P}$ is the *canonical momentum*, i.e. the derivative $\partial \mathcal{L}/\partial \mathbf{v}$ of the Langrangian with respect to the velocity. It is easy to imagine cases where $\mathcal{P} \neq \mathbf{p}$, e.g. when the point particle is electrically charged and is acted upon by a magnetic field. In *analytical continuum mechanics*, the situation is worst because $\mathcal{P}$ is the derivative of the Lagrangian density with respect to $\mathbf{V}$, the *material* velocity, and not $\mathbf{v}$, so that $\mathcal{P} \neq \mathbf{p}$ even in the purely mechanical case, unless there is no deformation at all (rigid-body case).

and of which *material inhomogeneity forces* $\mathbf{f}^{inh}$ provide an example (they depend on $\mathbf{X}$; hence their name). That is, we have, respectively,

$$\left.\frac{\partial \mathbf{p}}{\partial t}\right|_{\mathbf{X}} - \text{div}_R \mathbf{T} = \mathbf{f}^{ext}, \qquad (2.6)$$

for physical momentum and

$$\left.\frac{\partial \mathcal{P}}{\partial t}\right|_{\mathbf{X}} - \text{div}_R \mathbf{b} = \mathbf{f}^{inh}, \qquad (2.7)$$

for pseudomomentum, in which both $\mathcal{P}$ and $\mathbf{f}^{inh}$ are material covectors. Physical forces are originally acting *per unit mass of material* in *physical space*, whereas *pseudo-forces* act on inhomogeneities (e.g. defects or field singularities) in *material space*. The latter are *not* Newtonian in character. We qualify them of "Eshelbian".

The *displacement filed* $\mathbf{u}$ can be introduced in continuum mechanics by writing

$$\mathbf{x} = \chi(\mathbf{X}, t) = \mathbf{1}_s \cdot \mathbf{X} + \mathbf{u}(\mathbf{X}, t), \qquad (2.8)$$

where $\mathbf{1}_s$ is a so-called shifter. Whence ($\mathbf{1}$ is the identity transformation)

$$\mathbf{F} = \mathbf{1} + \nabla_{\mathbf{X}} \mathbf{u}. \qquad (2.9)$$

Then from (2.2) we have

$$\mathcal{P} = -\mathbf{1}_s \cdot \mathbf{p} + \mathcal{P}^f, \qquad (2.10)$$

where

$$\mathcal{P}^f = -\rho_0 (\nabla_{\mathbf{X}} \mathbf{u}) \cdot \frac{\partial \mathbf{u}}{\partial t}, \qquad (2.11)$$

is referred to as the *field-momentum density*. Consequently,

$$\mathcal{P} + \mathbf{1}_s \cdot \mathbf{p} = \mathcal{P}^f, \qquad (2.12)$$

which is a general decomposition law when (2.8) applies. For *small strains*, as it is the case is *crystals*, we need not to distinguish between lower case and upper case Latin indices, i.e. we use only one basis to refer to both actual and reference configuration. Then (2.11) takes on the following (Carthesian) component form:

$$\mathcal{P}_i^f = -\rho_0 \dot{u}_j u_{j,i}, \qquad (2.13)$$

where a superposed dot now denotes the partial time derivative. The notion of momentum (2.13) is that one considered in *crystal physics* under the name of *crystal momentum* or *wave momentum* [13], when wavelike disturbances of small amplitude are considered. For *harmonic motions*, it is this momentum which is connected via *de Broglie's relation* to the wave vector (see, e.g. [2]) and thus participates in the quantization of low-energy lattice vibrations in crystals. Eqs. (2.2) and (2.11) or (2.13) in fact show that both pseudomo-mentum and field- or wave-momentum are quantities which are *quadratic* in the basic fields, so that they a akin to energy up to a physical constant. This "quadratic" feature is essential both in the derivation of *de Broglie's relation* and in the justification of the use of the pseudomomentum balance in global form for various purposes (see below).

3. Field Formulation

We consider the case of *anisotropic, inhomogeneous, first-grade hyperelasticity* (i.e. energy-based nonlinear elasticity involving only the first gradient of the motion). Let $W(\mathbf{F}; \mathbf{X})$ be the strain-energy function per unit volume in $\mathcal{K}_R$. As indicated, this may depend *explicitly* on $\mathbf{X}$. Thus *elastic* inhomogeneities and *inertial* inhomogeneities, via $\rho_0(\mathbf{X})$, are present. In this case we have [14]

$$\mathbf{T} = \left(\frac{\partial W}{\partial \mathbf{F}}\right)^T, \qquad \mathbf{T}\mathbf{F}^T = \mathbf{F}\mathbf{T}^T, \qquad (3.1)$$

the second of which expresses the rotational invariance of W with respect to the actual frame $\{x_i\}$. For the sake of example left $\mathbf{f}^{\text{ext}} \equiv \mathbf{0}$ (neglect of phy-sical forces). Then on multiplying (2.6) to the left by $\mathbf{v}$ and $\mathbf{F}^T$, integrating by parts and accounting for kinematic identities such as

$$\left(\frac{\partial \mathbf{F}}{\partial t}\right)\bigg|_{\mathbf{X}} = (\nabla_{\mathbf{X}}\mathbf{v})^T, \qquad (3.2)$$

we are led to the two equations (of which the second has three independent components in general)

$$\frac{\partial \mathcal{H}}{\partial t}\bigg|_{\mathbf{X}} - \text{div}_R \mathbf{Q} = 0 \qquad (3.3)$$

and

$$\frac{\partial \mathcal{P}}{\partial t}\bigg|_{\mathbf{X}} - \text{div}_R \mathbf{b} = \mathbf{f}^{\text{inh}}, \qquad (3.4)$$

wherein we have defined the following quatities:

$$\mathcal{H} := \frac{1}{2}\rho_0(\mathbf{X})\mathbf{v}^2 + W(\mathbf{F}; \mathbf{X}), \qquad \mathbf{Q} := \mathbf{T} \cdot \mathbf{v} \qquad (3.5)$$

and

$$\mathcal{P} := -\rho_0(\mathbf{X})\mathbf{F}^T \cdot \mathbf{v}, \qquad \mathbf{b} := -(\mathcal{L}\mathbf{1}_R + \mathbf{F}^T\mathbf{T}) \qquad (3.6)$$

together with

$$\mathcal{L} := \frac{1}{2}\rho_0(\mathbf{X})\mathbf{v}^2 - W(\mathbf{F}; \mathbf{X}) \qquad (3.7)$$

$$\mathbf{f}^{\text{inh}} := \left(\frac{\partial \mathcal{L}}{\partial \mathbf{X}}\right)_{expl} = \left(\frac{1}{2}\mathbf{v}^2\right)\frac{\partial \rho_0}{\partial \mathbf{X}} - \left(\frac{\partial W}{\partial \mathbf{X}}\right)_{expl}. \qquad (3.8)$$

Although no elements of analytical continuum mechanics were so far introduced, one easily recognizes in $\mathcal{L}$ and $\mathcal{H}$ densities of Lagrangian and Hamiltonian (energy), respectively, while $\mathbf{Q}$ is the material energy flux due to elastic stresses, and $\mathbf{b}$ is the fully *elasto-dynamical Eshelby stress*, which was already formally introduced in Eq. (2.7). The definitions (3.7) to (3.8) provide the expressions of material *inhomogeneity force*: it is the *explicit material gradient* of $\mathcal{L}$. This, obviously, hints at the fact that Eqs. (2.6), (3.3) and (3.4) are indeed directly derivable from variational formulations of the Hamilton-Lagrange type (for details we refer the reader to [3] and [4], chapter 5). In brief: (*i*) considering a direct-motion variation $\delta_{\mathbf{X}}\chi$ and starting from the Langrangian density (3.7), one derives in a direct conventional variational way Eq. (2.6), - here with $\mathbf{f}^{\text{ext}} = 0$ - while the application of *Noether's* celebrated theorem for t and $\mathbf{X}$ considered as parameters yield Eqs. (3.3) and (3.4), respectively, and (*ii*) considering and *inverse-motion variation* $\delta_{\mathbf{x}}\chi^{-1}$ and starting from the Lagrangian density

$$\hat{\mathcal{L}} = \frac{1}{2}\rho_0(\mathbf{X})\mathbf{V}\cdot\mathbf{C}\cdot\mathbf{V} - \hat{W}(\mathbf{F}^{-1};\mathbf{X}), \qquad (3.9)$$

one directly derives Eq. (3.4) while the application of Noether's theorem for t and $\mathbf{x}$ considered as parameters yields Eqs. (3.3) and (2.6) respectively.

The above-stated result shows that the dynamical elasticity considered is a true *field theory*, and that Eqs. (2.6) and (3.4) are but two different *representations* of the same physical statement, the balance of momentum, whether *physical* or *"pseudo"*, but one emphasizes the presence of *physical forces* – when they exist – and the other that of inhomogeneity forces. The uniqueness of the *energy balance* (3.3) is obvious as it indeed is the same in both formulations. As a matter of fact, multiplying (2.6) by $\mathbf{v}$ or (3.4) by $\mathbf{V}$, one indeed obtains the *same* Eq. (3.3), as shown by using the following identity:

$$\mathbf{F}^T \cdot \left(\frac{\partial \mathbf{p}}{\partial t} - \text{div}_R \mathbf{T}\right) + \left(\frac{\partial \mathcal{P}}{\partial t} - \text{div}_R \mathbf{b} - \mathbf{f}^{\text{inh}}\right) = 0, \qquad (3.10)$$

which generalizes the identity obtained by Ericksen [14] for finite homogeneous elastostatics. Multiplying (3.10) by $\mathbf{v}$ and using (2.5) proves our statement. The latter in fact shows that *local* pseudo-forces $\mathbf{f}^{\text{inh}}$ do not produce any dissipation as Eq. (3.3) is a strict conservation law. Altogether, the above described approach also emphasizes that there are situations in which the external *physical force* is zero while the pseudo-force is *not* zero. The converse situation may also happen. This was fully realized in the penetrating analysis given by Sir Rudolph Peierls [16] in the framework of optics and acoustics.

The above recalled variational treatment and field-theoretical framework extends automatically to *higher-grade* elasticity such as the *second-grade* one for which W depends not only on $\mathbf{F}$ but also on its material gradient $\nabla_{\mathbf{X}}F$ in addition to $\mathbf{X}$ itself, i.e., the second material gradient of the direct motion χ, hence its name. We then have (cf. [3],[4]):

$$\mathbf{T} = \left(\frac{\delta W}{\delta \mathbf{F}}\right)^{T} = \bar{\mathbf{T}} - \operatorname{div}_{R}\bar{\mathbf{M}}, \qquad (3.11)$$

and

$$\mathbf{b} = -\left\{\mathcal{L}\mathbf{1}_{R} + \mathbf{F}^{T} \cdot \bar{\mathbf{T}} + 2\left(\nabla_{\mathbf{X}}\mathbf{F}\right)^{T} : \bar{\mathbf{M}}\right\} - \operatorname{div}_{R}(\mathbf{F}^{T} \cdot \bar{\mathbf{M}}), \qquad (3.12)$$

wherein

$$\bar{\mathbf{T}} = \left(\frac{\partial W}{\partial \mathbf{F}}\right)^{T}, \qquad \bar{\mathbf{M}} = \frac{\partial W}{\partial (\nabla_{\mathbf{X}}\mathbf{F}^{T})}, \qquad W = W(\mathbf{F}, \nabla_{\mathbf{X}}\mathbf{F}; \mathbf{X}). \quad (3.13)$$

The reader can check for himself that the celebrated *Boussinesq equation* of crystal elasticity in one-space dimension is a special case of Eqs. (2.6) and (3.11) (in an obvious notation);

$$\frac{\partial^{2} u}{\partial t^{2}} - c^{2}\frac{\partial^{2} u}{\partial x^{2}}\left(1 + \beta\frac{\partial u}{\partial x}\right) - c^{2}\delta^{2}\frac{\partial^{4} u}{\partial x^{4}} = 0. \qquad (3.14)$$

Here c is a characteristic speed (corresponding to a linear mode), β is a non-dimensional parameter accounting for *nonlinearity*, and δ is a characteristic length accounting for *microstructure* or *dispersion* (second-gradient-theory!). It is well known that the combination of appropriate amounts of nonlinearity and dispersion in Eq. (3.14) results in this equation presenting *solitary-wave* and *soliton* solutions. This is better known for the "fluid" *Boussinesq equation* which for small δ (i.e. weak dispersion) is deduced from (3.14) via a naive perturbation scheme as (this is also called the *improved* Boussinesq equation in soliton theory)

$$\frac{\partial^{2} u}{\partial t^{2}} - c^{2}\frac{\partial^{2} u}{\partial x^{2}}\left(1 + \beta\frac{\partial u}{\partial x}\right) - \delta^{2}\frac{\partial^{4} u}{\partial x^{2}\partial t^{2}} = 0. \qquad (3.15)$$

It obviously is of interest to look for the implifications of the corresponding equation of balance of pseudomomentum based on the Eshelby stress (3.12). Before doing so we need to discuss the relative significance and practicality of the *global* forms of Eqs. (2.6) and (3.4).

4. Non-Equivalence Between Global Formulations

Eq. (3.10) clearly shows the equivalence between the *local* balances of physical and pseudo-momenta, for sufficiently smooth notions. But the situation is at variance with this if we consider the corresponding *global* balance laws obtained by summing Eqs. (2.6) and (3.4) over a finite region - or the whole (if this makes physical sense at all - of the material manifold $\mathcal{M}$. Let $\mathcal{B}$ be this region of $\mathcal{M}$ in $\mathcal{K}_R$ with regular boundary $\partial\mathcal{B}$ equipped with outward unit normal $\mathbf{N}$. Consider the case where $\mathbf{f}^{\text{ext}} \equiv 0$. With (2.6) we ca associate the following *global* balance law

$$\frac{d}{dt} \int_{\mathcal{B}} \mathbf{p}\, dV = \int_{\partial\mathcal{B}} \mathbf{N} \cdot \mathbf{T}\, dA, \qquad (4.1)$$

where it must be understood that this represents the set of three *components* on the vectorial basis $\{e_i\}, i = 1, 2, 3$ on $\chi(\mathcal{B}, t)$. Similyrly, in the absence of material inhomogeneities, $\mathbf{f}^{inh} \equiv 0$, the same manipulation performed on (3.4) yields

$$\frac{d}{dt} \int_{\mathcal{B}} \mathcal{P}\, dV = \int_{\partial\mathcal{B}} \mathbf{b} \cdot \mathbf{N}\, dA \qquad (4.2)$$

This, also, if to be viewed as a *tensorial relation*, has to be considered by *component* on a (material) basis $\{E_I\}, I = 1, 2, 3$. But $\{E_I\}$ and $\{e_i\}$, obviously, are not placed in correspondence by *convection*. In other words, "*convection is lost in passing from local to global*" [17]. it is this property, together with the "quadratic" nature of both $\mathcal{P}$ and $\mathbf{b}$ in the basic fields, which grants a very special role to the *global balance of pseudomomentum* and to *global material forces* in both *fracture* and *nonlinear wave propagation*. In the first instance that is nor our primary interest here, the situation is easily grasped [7] with the following exapmle. If it happens, in a two-dimensional problem (say, plane stresses) that the fields $\mathbf{T}$ and $\mathbf{v}$ present a singularity in $\sqrt{r}$ about a point $\mathbf{X}_0$ in $\mathcal{B}$, then this singularity is not "felt" in the global law (4.1) where each integrand remains integrable so that all components go to zero with $\mathcal{B}$ shrinking to point $\mathbf{X}_0$. But this is not the case in (4.2) so that the quantity (defined before use of any Green-Gauss and transport theorem)

$$\mathcal{F} := \int_{\mathcal{B}_0} \left(\frac{\partial \mathcal{P}}{\partial t} - \operatorname{div}_R \mathbf{b} \right) dV, \qquad (4.3)$$

$\mathbf{X}_0 \in \mathcal{B}_0 \subset \mathcal{B}$, $\mathcal{B}_0$ shrinking to $\mathbf{X}_0$, which may be viewed as a global *inhomogeneity force* (compare to the volume integral of Eq. (3.4) with $\mathbf{f}^{inh} \neq 0$; singularity points are material inhomogeneities in their own right), will have *nonzero* (diverging) components. We may say that such a quantity, $\lim \mathcal{F}$, *captures in the singularity* in the elastic field. This finds direct application in the macroscopic theory of fracture [4], [7], [17]. In particular, although the material does not present any *local* irreversible behavior at any point in the bulk and *local* inhomogeneity forces - as defined by Eq. (3.9) - do not dissipate, the irreversible progress "en bloc" of a notch or a crack (with stress-free faces) causes a global dissipation (e.g. the fractured body cannot solder back, hence the global irreversibility), what can be expressed by means of a global *dissipation-rate inequality* of the form

$$\Phi(\mathcal{B}) = \mathcal{F} \cdot \bar{\mathbf{V}} \geq 0, \tag{4.4}$$

where $\bar{\mathbf{V}}$ is the spatially uniform (material) vectorial velocity of the progressing defect, and $\mathcal{F}$ is the associated global material force. Eq. (4.4) allows one, using basic principles of irreversible thermodynamics (as exposed, e.g., in [18]), to study the non-equilibrium thermodynamics of *global* material forces and to fromulate thermodynamically admissible (i.e., respecting the "Arrow of time") laws of advancement of defects such as notches, cracks, dislocations, cavities, and phas-transition layers (see [8]). For comparison purposes with the developments to come, we shall retain from this section the evaluation of global quantities such as (4.3) from the knowledge of local solutions established on the basis of boundary-initial value problems associated with the more classical field Eq. (2.6). The same procedure holds true for nonlinear waves.

5. The Role of Pseudomomentum and Energy in Nonlinear-Wave Propagation

5.1 General Features

It is well known in the study of numerical schemes for dynamical systems of partial differential equations that an often imposed, or wished for, requirement is *energy conservation*. This is essential in treating hyperbolic systems by finite-difference schemes where, indeed, the fulfillment of an energy-conservation criterion (with a certain accuracy) is conclusive [19]. The general idea of the following developments borrows this idea by considering that if Eq. (2.6) is satisfied by any means, i.e., either we know exactly an analytical solution or we have approximately constructed a numerical one, then not only the volume integral form of the energy balance (3.3), i.e., for a sufficiently regular solution,

$$\frac{d}{dt} \int_B \mathcal{H} \, dV = \int_{\partial B} \mathbf{Q} \cdot \mathbf{N} \, dA, \tag{5.1}$$

must be enforced, but this must also be the case of the volume integral of the pseudomomentum balance, i.e. Eq. (4.2), because both equations have an equal ontological status. The following view is adopted. Either these two equations are automatically satisfied by the known analytical solution or they can be viewed as *criteria* qualifying the validity of the numerical scheme, or else (see below) they can be exploited in a perturbation procedure of known exact solutions. This applies particularly well to systems exhibiting solitary-wave solutions and solitons (e.g., Eq. (3.14). Indeed, if such solutions, i.e., extremely localized nonuniform dynamical solutions (pulses, bumps or kinks) which propagate at constant speed and "en bloc" (this is the case of solitary-like constant-profile solutions), then these solutions usually have field derivatives which are essentially zero at infinity or, for practical purpose, far enough from the distinctly localized signal. If this is the case with solutions presenting no divergence within the domain of integration, then Eq. (5.1) and (4.2) will yield, with homogenous conditions at infinity, the two *global conservation laws*:

$$\frac{dE}{dt} = 0, \qquad \frac{d\mathbf{P}}{dt} = 0 \qquad (5.2)$$

with (in practice the integration volume is finite but "large enough")

$$E = \int_{\mathcal{R}^3} \mathcal{H} dV, \qquad \mathbf{P} = \int_{\mathcal{R}^3} \mathcal{P} dV. \qquad (5.3)$$

Eqs. (5.2) provide *four* equations which are the *canonical* equations of *energy* and *momentum*. It is readily checked that Eq. $(5.2)_1$ also holds true for *field momentum*, i.e. *wave momentum*

$$\frac{d\mathbf{P}^f}{dt} = 0, \qquad \mathbf{P}^f = \int_{\mathcal{R}^3} \mathcal{P}^f dV. \qquad (5.4)$$

If it is possible in some way to grant a total *mass*, $\mathbf{M}$, to the solution of (2.6) over $\mathcal{R}^3$, - e.g. in well known one-dimensional cases, this mass is the total number of *phonons*, the integral of the solution field over $\mathcal{R}$, etc - , so that this scalar quantity $\mathbf{M}$ is also conserved, i.e.

$$\frac{d\mathbf{M}}{dt} = 0, \qquad (5.5)$$

and $\mathbf{M}, \mathbf{P}^f$ and E relate to one another as they should in Galilean or relativistic dynamics, then Eqs. (5.6), $(5.2)_1$ and $(5.4)_1$ are but the three *conservation equations* that describe a uniformly moving *point particle* in Galilean or relativistic dynamics, depending on the case. Any additional contributions in the right-hand sides of Eqs. $(5.2)_1$ and $(5.4)_1$ due to *dissipation* and *forcing* in the first equation, and to dissipation, forcing and material inhomogeneities (defect) for the second equation, if *small enough* (in a sense to be precisely specified), may be considered as perturbations on the strict conservation laws. In other words, in such circumstances, $(5.2)_1$ and $(5.4)_1$ will be replaced by the Newtonian-like equations

$$\frac{dE}{dt} = \varepsilon Q^d, \qquad (5.6)$$

$$\frac{d\mathbf{P}^f}{dt} = \varepsilon \mathcal{F}, \qquad (5.7)$$

where the ordering parameter ε emphasizes the smallness of the right-hand contributions, Q^d is a global *heat source* and $\mathcal{F}$ is a *global material force* in the language of previous sections. Pure inhomogeneity forces, however, will alter only Eq. (5.7), as they are *not* dissipative! Thus, Eqs. (5.7) are in principle directly amenable by means of analysis through a perturbation method. This can be illustrated on various cases which, for the time being, remain one-dimensional in space as the computation involved in treating Eqs. (5.7) may be far from easy except in very peculiar situations.

5.2 First Example: The "Good" Boussinesq (GoB) Equation

This is Eq. (3.14) in normalized form and with a change in sign in front of the fourth-order derivative so that it agrees with Eq. (3.15) from the point of view of dispersion (when one studies harmonic propagation in the absence of nonlinearities), i.e., in an obvious notation:

$$u_{tt} - u_{xx} - (u^2 - u_{xx})_{xx} = 0. \tag{5.8}$$

Introducing auxiliary variables q and w by

$$u_t = q_x, \qquad\qquad w = u_{xx}, \tag{5.9}$$

Eq. (5.8) is shown to be equivalent to the *evolution system* composed of Eqs. (5.10) and

$$q_t = w + (u^2)_x - w_{xx}. \tag{5.10}$$

The total "mass", field-wave momentum, and energy of the solution given by the triplet $(u, q, w) \in \mathcal{R}^3$ of (5.9-(5.10) read

$$M\{u\} = \int_{\mathcal{R}} u dx, \qquad P^f\{u, q\} = -\int_{\mathcal{R}} uq dx,$$

$$\tag{5.11}$$

$$E\{u, q, w\} = \int_{\mathcal{R}} \left(\frac{1}{2}q^2 + w^2 + u^2 + \frac{2}{3}u^3 \right) dx.$$

With

$$u = \bar{u}_x, \qquad\qquad \bar{u}_t(x = -\infty) = 0 \tag{5.12}$$

it is checked that

$$\bar{u}_t = q, \qquad P^f = -\bar{u}_x \bar{u}_t, \qquad \frac{1}{2}q^2 = \frac{1}{2}\bar{u}_t^2, \tag{5.13}$$

and

$$M\{u\} = [\bar{u}]_{-\infty}^{+\infty} := \bar{u}(+\infty) - \bar{u}(-\infty). \tag{5.14}$$

Thus the mass here is the difference between limit values, the "jump" of the potential $\bar{u}$ of the solution. P^f and $q^2/2$ are indeed the local field momentum and kinetic energy for the $\bar{u}$ solution. The above are satisfied by solitary-wave solutions of the "good" Boussinesq Eq. (5.8) as is easily checked. They agree with the work of Manoranjan *et al* [20].

5.3 Second Example: The "Generalized" Boussinesq Equation (GB)

This is an equation which finds its origin in the dynamical study of phase-transition patterns in the ferroelastic ans shape-memory alloys [21],[22]:

$$s_{tt} - c_0^2 s_{xx} - [F(s) - \beta s_{xx} + s_{xxxx}] = 0, \qquad \beta > 0. \tag{5.15}$$

Here s is a shear strain, i.e. $s = v_x$, where v is a displacement in a direction orthogonal to x. Eq. (5.15) represents a very *stiff* system, with a sixth-order space derivative whose purpose is to correct the bad dispersion feature of the fourth-order derivative (cf. Eq. (5.8)]; $F(s)$ is a polynomial in s starting with second degree, which indicates a possible *nonconvexity* of the strain-energy function with respect to s. On setting

$$F(s) = -\frac{d\mathcal{U}(s)}{ds}, \tag{5.16}$$

We show that (5.15) is eqivalent to the following evolution system [22], [23]

$$\begin{aligned} s_t &= q_{xx}, & w &= s_{xx}, \\ q_t &= c_0^2 s + F(s) - \beta w + w_{xx}, \end{aligned} \tag{5.17}$$

for the triplet $\{s, q, w\} \in \mathcal{R}^3$; and the total mass, field-wave momentum, and energy of this triplet are given by

$$\begin{aligned} M\{s\} &= \int_{\mathcal{R}} s\, dx, & P^f\{s, q\} = -\int_{\mathcal{R}} s q_x\, dx, \\ E\{s, q, w\} &= \int_{\mathcal{R}} \frac{1}{2}\left(c_0^2 s^2 + q_x^2 - 2\mathcal{U}(s) + \beta s_x^2 + w^2\right) dx. \end{aligned} \tag{5.18}$$

Again, introducing the potential $\bar{u}$ which is nothing but the *true* displacement v, so that $s = v_x$, for solutions sytisfying $v_t(x = -\infty) = 0$, we have $P^f = -v_x v_t$ the local wave momentum, $q_x^2/2 = v_t^2/2$, the local kinetic energy, M takes the form (5.14), and we have the global "Newtonian" conservation equations:

$$\frac{dM}{dt} = 0, \qquad \frac{dE}{dt} = 0, \tag{5.19}$$

but

$$\frac{dP^f}{dt} = \mathcal{F} := -\left[s_{xx}^2\right]_{-\infty}^{+\infty}, \tag{5.20}$$

so that total field-wave momentum will be conserved only if the "driving" pseudo-force $\mathcal{F}$ is zero, i.e. for solutions which satisfy limit conditions such as $s_{xx}^2(+\infty) = s_{xx}^2(-\infty)$ or, a fortiori, $s_{xx}(\mp\infty) = 0$, which is the case for solitary-wave solutions exhibited in analytical form by Christov and Maugin [22]. This example shows what are the limit conditions (or boundary conditions on a finite but large interval) which entail the true conservation of global wave momentum. From Maugin and Cadet [21], we have the following more complicated, but also more realistic, model for which s couples to the *elongation* $e = u_x$, where u is the longitudinal displacement:

$$\begin{aligned} s_{tt} - c_T^2 s_{xx} + \left(s^3 - s^5 + 2\gamma se + \alpha s_{xx}\right)_{xx} &= 0, \\ e_{tt} - c_L^2 s_{xx} + \gamma(s^2)_{xx} = 0, \end{aligned} \tag{5.21}$$

where γ is a coupling coefficient. The sixth-order space derivative has been discarded for the sake of simplicity. Now we introduce auxiliary variables q and r by

$$s = v_x, \quad e = u_x, \quad s_t = q_{xx}, \quad e_t = c_L r_x, \tag{5.22}$$

and this, together with

$$
\begin{aligned}
q_t &= c_T^2 s - s^3 + s^5 - 2\gamma se - \alpha s_{xx}, \\
r_t &= c_L e_x - \left(\frac{\gamma}{c_L}\right)(s^2)_x,
\end{aligned}
\tag{5.23}
$$

provides an *evolution system* that is equivalent to the original equations (5.21). Then

$$M\{s\} = \int_{\mathcal{R}} s\, dx = [v]_{-\infty}^{+\infty}, \quad P^f\{s, q, e, r\} = \int_{\mathcal{R}} (sq_x + c_L er)\, dx,$$

$$
\begin{aligned}
E\{s, q, e, r\} = \int_{\mathcal{R}} \Bigg\{ &\frac{1}{2}(q_x^2 + c_L^2 r^2) + \frac{1}{2}(c_T^2 s^2 + c_L^2 e^2) \\
&-\gamma es^2 - \frac{1}{4}s^4 + \frac{1}{6}s^6 + \frac{\alpha}{2}(s_x)^2 \Bigg\}\, dx,
\end{aligned}
\tag{5.24}
$$

and we have conservation of total "mass" and energy while

$$\mathcal{P}^f = -(u_x u_t + v_x v_t), \tag{5.25}$$

and

$$\frac{dP^f}{dt} = \mathcal{F} \equiv -\frac{1}{2}\left[\alpha s_x^2 + c_L^2 e^2\right]_{-\infty}^{+\infty}. \tag{5.26}$$

Here we have an example of field momentum for two degrees of freedom, whereas Eq. (5.26) dictates the *good* additional limit comditions to guatantee the conservation of total field-wave momentum, e.g $v_{xx}(x = \mp\infty) = 0, u_x(x = \mp\infty) = 0$. This is verified by analytical solitary-wave solutions obtained by Maugin and Cadet [21].

Numerically, the systems (5.8), (5.15) and (5.21) above are simulated on the basis of their evolution-system variants. The quality of the numerical scheme used can be checked by verifying the global conservation of energy and field momentum. As a matter of fact, the good "*solitonic*" nature of the above-given systems, although some are *not* exactly (other authors say "completely") integrable in the sense of soliton theory, is checked by simulating interactions of two initially sufficiently set apart individual waves and verifying that energy and pseudomomentum are practically conserved in these interactions. That is, at least three of the first conservation laws of the infinite hierarchy exhibited by exactly (or completely) integrable systems are practically checked, so that the systems are suspected to be *close to exact integrability*. We may say that such systems, e.g. (5.21) are *nearly integrable* (compare to Kivshar and Malomed [24]) for all practical purposes (see below for other such systems).

5.4 Material Global Forces as Perturbations

Consider the following apparently simple PDE:

$$u_{tt} - u_{xx} - \sin u = \varepsilon f(u, u_t, \ldots; x). \tag{5.27}$$

This is a perturbed *sine-Gordon equation*. the term $\sin u$ contains both non-linearity *and* dispersion but it cannot be considered in the framework of elasticity as coming from the action of internal forces (direct dependence on displacement u is precluded in the strain-energy function in reason of translational invariance). Thus it results here from an external body force (a sinusoidal substrate potential within the Frenkel-Kontorova model of dislocations). For $f = 0$, the resulting sine-Gordon equation is an excellent "solitonic" system for which one can write (see e.g. [24] or chapter 5 in [25]):

$$
\begin{aligned}
u(x,t) &= u(\xi) = 4\tan^{-1}\left(e^{\mp\gamma(\xi-\xi_0)}\right), \\
\xi &= x - ct, \qquad \gamma = (1-c^2)^{-\frac{1}{2}}, \qquad |c| < 1, \tag{5.28}
\end{aligned}
$$

and (these are *kink* solitons)

$$E = 8\gamma, \qquad P^f = 8\gamma c = Mc, \qquad M = \gamma M_0, \qquad M_0 = 8, \tag{5.29}$$

$$\frac{dM}{dt} = 0, \qquad \frac{dE}{dt} = 0, \qquad \frac{dP^f}{dt} = 0, \qquad P^f = -u_x u_t. \tag{5.30}$$

With $f \neq 0$ in (5.27) we replace $(5.30)_{2-3}$ by the perturbed equations

$$\frac{dE}{dt} = \varepsilon \int_{\mathcal{R}} u_t f \, dx, \qquad \frac{dP^f}{dt} = -\varepsilon \int_{\mathcal{R}} u_x f \, dx, \tag{5.31}$$

and this in theory provides two perturbation equations to study the variation of two characteristic parameters of the soliton solution (5.28), but only one parameter, c, here is left free is so far as $|c| < 1$ (subsonic soliton); but there is no contradiction between Eqs. (5.31) because there holds the relativistic energy equation $E^2 = M_0^2 + P_f^2$, but M_0 here is fixed and $u_t = -cu_x$ for progressive-wave solutions. Thus both Eqs. (5.31) yield one *and only one* equation to determine the variation of c. This follows from fact that solitons governed by (5.27) are *topological solitons* (the amplitude, or equivalently M_0, is fixed once and for all). For $f =$ const. Eqs. (5.31) show that the perturbed soliton behaves like a *uniformly accelerated relativistic point particle in a constant force field*. The relativistic features observed follow from the obvious *Lorentz invariance* of Eq. (2.6) - see [26].

The more general system now known as the *sine-Gordon-d'Alembert system* for the doublet (ϕ, u) and given by

$$
\begin{aligned}
\phi_{tt} - \phi_{xx} - \sin\phi &= \eta u_x \cos\phi, \\
u_{tt} - c_T^2 u_{xx} &= -\eta(\sin\phi)_x, \tag{5.32}
\end{aligned}
$$

was first obtained while studying domain-wall motions in elastic ferromagnets [27] and elastic ferroelectrics of the molecular-group type [28]. It is a two-degree of freedom system where η is a coupling coefficient. This is no longer exactly integrable, but this system *does* present certain properties, such as energy and field-momentum conservation [6], which are shared by exactly integrable systems: it is only *nearly integrable* [29]. Its local field momentum obviously reads [compare to the last of Eqs. (5.30)]

$$\mathcal{P}^f = -(\phi_x \phi_t + u_x u_t). \tag{5.33}$$

Perturbations can be studied according to the above sketched out procedure for we know exact solitary-wave solutions of (5.32), [28], [24].

Another system of interest which does appear while studying the propagation of surface solitary-waves on layered nonlinear elastic structures [30] is the *nearly integrable "generalized Zakharov"* (GZ) system (λ and μ are coefficients) given by

$$
\begin{aligned}
ia_t + a_{xx} + \lambda |a|^2 a + 2an &= 0, \\
n_{tt} - c_0^2 n_{xx} + \mu \left(|a|^2\right)_{xx} &= 0, \\
n &= u_x,
\end{aligned}
\tag{5.34}
$$

which couples the slowly varying complex-valued envelope a of a carrier wave in a transverse elastic displacement (so called SH wave) - evaluated at the top of the structure- to a real-valued longitudinal mode u. This apparently complicated system is *not* exactly integrable, but we know one-soliton solutions of (5.34) (see [31]). For system (5.34) such solutions conserves mass, energy, and field-wave momentum, these quantities being given by the following expressions:

$$
\begin{aligned}
M\{a\} &= \int_{\mathcal{R}} |a|^2 \, dx, \\
P^f\{a, u\} &= \int_{\mathcal{R}} [i\,(aa_t^* - a^* a_t) - u_x u_t] \, dx, \\
E\{a, u\} &= \int_{\mathcal{R}} \frac{1}{2} \left(|a|^2 - \lambda |a|^4 - 2u_x |a|^2 + \mu^{-1} \left(u_t^2 + c_0^2 u_x^2\right)\right) dx,
\end{aligned}
\tag{5.35}
$$

where $*$ denotes complex conjugacy. Note that here M is the total *wave action* (a typical feature of the nonlinear Schrödinger (NLS) equation, $n = 0$, and of the pure Zakharov (Z) system obtained for $\lambda = 0$ in plasma physics). But in the specific application treated by the author, it may be considered as the total *number of surface phonons* in the a-subsystem. P^f is formally the same for the GZ and pure Z systems. Both pure NLS and Z systems are exactly integrable. The study of the influence of dissipation in the a-subsytem is more or less trivial. However, the study of the influence of a dissipation (viscosity) in the n-subsystem provides a good example of application of the above-recalled perturbation method. Although tedious, the perturbation in total field-momentum, i.e. Eq. (5.7), can be exploited, resulting in the demonstration that due to the nonlinearity in the system, such a small perturbation may have drastic effects. In fact, it yields a critical behavior referred to as "perestroika" by Hadouaj *et al.* [31], in reason of an abrupt rearrangement of the soliton solution when a critical value of field momentum is reached, an effect predicted through the analysis and confirmed by a direct numerical simulation.

Present developments consider systems which are stiffer than those above but still one-dimensional in space, and systems which are *two-dimensional* in space, and the numerical implementation of the conservation of energy and pseudomomentum for such systems.

6. A Newtonian Mechanics for Global Material Forces?

From the developments given in section 5 we can deduce the flow chart in Fig. 6.1 in which, usually starting from **lattice** *discrete* **models** we pass to a continuum vision by the working hypothesis of long wavelengths (top right). In modern teaching practice, however, **continuum equations** of balance are often directly formulated in global form (e.g. Eq. (4.1)) and it is a *localization argument* relying on a sufficient smoothness of the fields that yields the classical **local balance equations** (e.g. Eq. (2.6)). But these equations, in the absence of dissipation, can as well be deduced from Lagrangian-Hamiltonian approach in which due exploitation of Noether's theorem yields the **local canonical balance laws** (energy and pseudomomentum). But these can also be obtained by manipulating the local continuum balance laws. From the latter and associated boundary and initial conditions, with some ingenuity we may possibly deduce nonlinear wave solutions of the "solitonic" type which *do* behave as **quasi-particles** (hence the name of solitons) *on the material manifold*. These quasi-particles have a Newtonian (or relativistic) motion which is in fact governed by the **global canonical balance laws**. The latter, in turn, can be used in a perturbation scheme to find out the **change in motion** due to a variety of causes (dissipation, forcing, inhomogeneities) all treated as *heat sources* and *material forces*, and thus manifesting themselves as contributions in the right-hand sides of Eqs. (5.6)-(5.7). The number of such meaningful (in a mechanical frame of mind) equations is obviously limited to four, one for energy, and three for momentum, and these are the only attributes of a *classical quasi-particle* apart form "mass" and, perhaps, electric charge. This number is limited to two in the one-dimensional case in space. Thus only the bare essential, the **change in motion** (i.e. phase and momentum or speed) of the particle-like soliton, can be reached. In more than one dimension *moment of pseudomomentum* would also be of interest: we have at hand the basic equations for that (see chapter R in [4]). In order to reach the change in the morphology (inner structure) of solitons, one

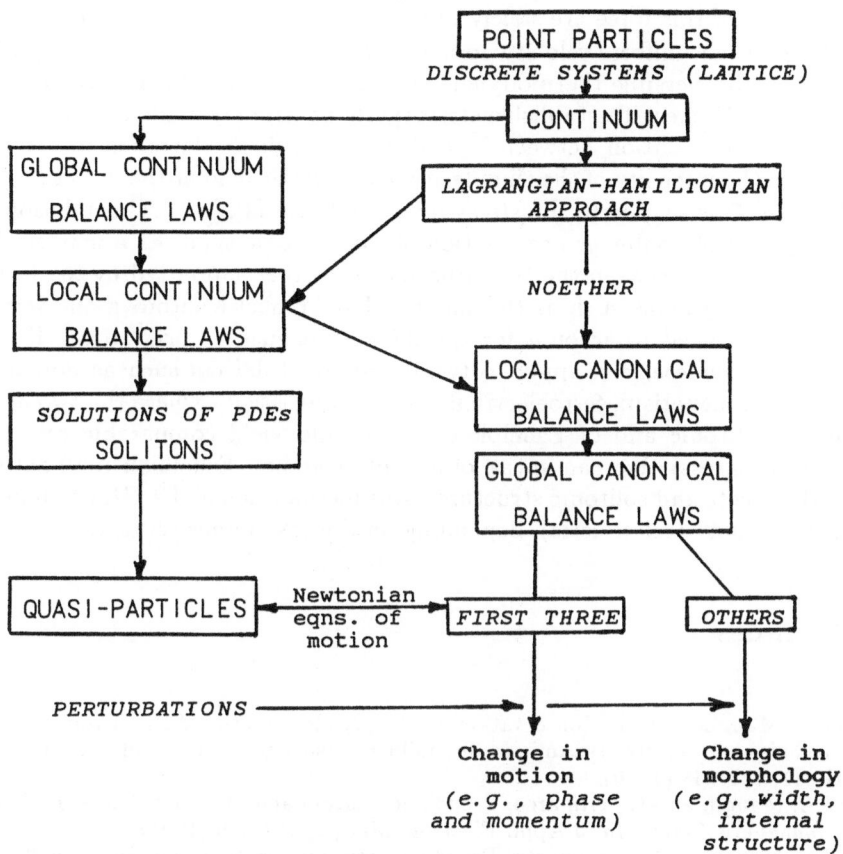

Fig. 6.1. Flow chart: From particles in physical space to quasi-particles on the material maifold

should use higher-order conservation equations in the infinite hierachy known in soliton theory. But these extra equations have no clear physical meaning (for the time being!) and, to our knowledge, no one has ever taken this drastic step, being quite satisfied with the gross changes, i.e. those in particle motion only. But the loop is now closed as we started with the "particles" of the lattice (top right) to finally arrive at larger structures reffered to as (quasi)-particles.

Now a fundamental question is: are the developments of section 5 trully disconnected from those recalled in section 4 when dealing with defects such as in fracture? The answer is *no*. For on the one hand, the *global field* (or pseudo-, for that matter) momentum $(5.2)_2$ for *solitary-wave* solution corresponds to a uniform motion of the non-distorted signal viewed as a quasi-particle at *constant velocity*, and perturbations to this are indeed *global material forces*, or *pseudo forces*, in the sense of section 2. Furthermore, velocity and intensity of this force are stricly related. On the other hand, defects moving "en bloc" at uniform velocity do this under the action of a global material force. When irreversible thermodynamics is applied to the global dissipation inequality (4.4) one will establish a strict (perhaps strongly nonlinear, or even not one-to-one) relation between $\bar{V}$ and $\mathcal{F}$, of which the latter in fact is a measure of the intensity of the singularity of the elastic field at the defect (e.g. , in fracture $\mathcal{F}$ is related to the *stress-intensity factor* [4], [18]). This relationship is typical of nonlinear propagation of the solitonic type. As a matter of fact, the analogy between the two situations is further supported by the fact that solitonic systems such as the sine-Gordon/Frenkel-Kontorova one were initially introduced to account for specific defects such as dislocations [32]. The same considerations apply to two-dimensional defects such as domain walls in feromagnetism, ferroelectricity, and ferroelasticity where the two approaches, solitonic and of Eshelbian-type, should yield comparable results (work in progress). We can speak of a trully common Eshelbian mechanics for both defects and solitonic structures; this mechanics may be "Newtonian" or "Einsteinian" (relativistic), depending on the case under scrutiny.

References

1. G.A. Maugin, Sur la conservation de la pseudo-quantité de mouvement en mécanique et électrodynamique des milieux continus, C.R. Acad. Sci. Paris, II-311, 763-768 (1990).
2. G.A. Maugin and C. Trimarco, Pseudomomentum and Material Forces in Electromagnetic Solids, Int.J.Appl. Electrom.Mat., 2, 207-216 (1991).
3. G.A. Maugin and C. Trimarco, Pseudomomentum and Material Forces in Nonlinear Elasticity: Variational Formulations and Application to Brittle Fracture, Acta Mechanica, 94, 1-28 (1992).
4. G.A. Maugin, Material Inhomogeneities in Elasticity, Chapman and Hall, London (1993).
5. G. Piola, Intorno alle equazioni fondementali del monvimento di corpi qualsivogliono, considerate secondo la naturale loro forma e costituzione, Mem. Mat.Soc. Ital. Modena, 24, (1), 1-166 (1848).
6. G.A. Maugin, Application of an Energy-Momentum Tensor in Nonlinear Elastodynamics (Pseudomomentum and Eshelby Stress in Solitonic Elastic Systems), J. Mech. Phys. Solids., 40, 1543-1558 (1992).
7. G.A. Maugin, Variations on a Theme of A.A. Griffith, Griffith's Centenary Volume, ed. G.P. Cherepanov, USA (1994).
8. G.A. Maugin and C. Trimarco, The Global Dissipation of Configurational Forces in Defective Elastic Bodies, Zeit. angew. Math. Phys., (P.M. Naghdi's Anniversary Volume), 1994.
9. J.D. Eshelby, The Force on an Elastic Singularity, Phil. Trans. Roy. Soc. Lond., A244, 87-112 (1951).
10. J.D. Eshelby, The Elastic Energy-Momentum Tensor, J.Elasticity, 5, 321-335 (1975).
11. D. Rogula, Forces in Material Space, Arch. Mechanics, 29, 705-715 (1977).
12. A.C. Eringen and G.A. Maugin, Electrodynamics of Continua, Vol. I, Springer-Verlag, Berlin (1990).

13. W. Brenig, Besitzen Schallwellen einen Impuls, Zeit.Phys. 143, 168-172 (1955).
14. C.A. Truesdell and R.A. Toupin, The Classical Field Theories, in: Handbuch der Physik, Bd.III/1, ed. S. Flügge, Springer-Verlag, Berlin (1960).
15. J.D. Ericksen, Special Topics in Elastostatics, in: Advances in Applied Mechanics, ed. C.-S.Yih, Vol.17, pp. 189-244, Academic Press, New York (1977).
16. R. Peierls, Momentum and Pseudomomentum of Light and Sound, in: Highlights of Condensed-matter Physics, Vol LXXXIX, Corso, ed. M. Tosi, pp. 237-255, Soc. Ital. Fisica, Bologna (1985).
17. C. Dascalu and G.A. Maugin, Material Forces and Energy-Release Rates in Homogeneous Elastic Bodies with Defects, C.R.Acad.Sci. Paris, II-317 (1993).
18. G.A. Maugin, The Thermomechanics of Plasticity and Fracture, Cambridge University Press, U.K. (1992).
19. R.D. Richtmayer and K.W. Morton Difference Methods for Initial Value Problems, 2nd Edition, Interscience, New York (1967).
20. V.S. Manoranjan, T. Ortega and J.M. Sanz-Serna, Soliton and Anitsoliton Interactions in the "Good" Boussinesq Equation, J. Math. Phys., 29, 1964-1968.
21. G.A. Maugin and S. Cadet, Existence of Solitary Waves in Martensitic Alloys, Int. J. Engng. Sci., 29, 243-255 (1991).
22. C.I. Christov and G.A. Maugin, Long-Time Evolution of Acoustic Signals in Nonlinear Crystals, in: Advances in Nonlinear Acoustics, ed. H. Hobaek, pp. 467-472, World Scientific, Singapore (1993).
23. G.A. Maugin, Pseudomomentum and Nonlinear Waves in Solids, Proc. IUTAM Symp. on Nonlinear Waves in Solids, Victoria, June 1993, in: Appl. Mech. Rev., (1994).
24. Yu.S. Kivshar and B.A. Malomed, Dynamics of Solitons in Nearly Integrable Systems, Rev. Mod. Phys., 61, 763-915 (1989).
25. G.A. Maugin, J. Pouget, R. Drouot and B. Collet, Nonlinear Electromechanical Couplings, J. Wiley, New York (1992).
26. J. Pouget and G.A. Maugin, Influence of an External Electric Field on the Motion of a Ferroelectric Domain Wall, Physics Lett., A109, 389-392 (1985).
27. G.A. Maugin and A. Miled, Solitary Waves in Elastic Ferromagnets, Phys. Rev., B33, 1477-1499 (1986).
28. J. Pouget and G.A. Maugin, Solitons and Electroacoustic Interactions in Ferroelectric Crystals-I, Phys. Rev., B30, 5306-5325 (1984)
29. G.A. Maugin, Analytical and Numerical Problems for Nonlinear Wave Propagation in "Nearly" Integrable Systems, in: Mathematical and Numerical Aspects of Wave Propagation (2), eds. R.E. Kleinman et al, pp.338-351, S.I.A.M., Philadelphia (1993).
30. G.A. Maugin, H. Hadouaj and B.A. Malomed, Nonlinear Coupling Between SH Surface Solitons and Rayleigh Modes on Elastic Structures, Phys.Rev., B45, 9688-9694 (1992).
31. H. Hadouaj, B.A. Malomed, and G.A. Maugin, Dynamics of a Soliton in the Generalized Zakharov System, Phys.Rev., A44, 3925-3931 (1992).
32. J. Frenkel and T. Kontorova, On the Theory of Plastic Deformation and Twinning, Phys. Sowjet Union, 13, 1 (1938).